MSM 362
Kleman, Spring 83
12:45 → 2:00 Tues + Thurs.

Final: Monday May 16, 1983 @ 3:30 - 6:30.

Undergraduate Texts in Mathematics

John A. Thorpe

Elementary Topics in Differential Geometry

Springer-Verlag
New York Heidelberg Berlin

J. A. Thorpe
Department of Mathematics
State University of New York
Stony Brook, New York 11794
USA

AMS Subject Classification: 53-01

With 126 Figures

Library of Congress Cataloging in Publication Data

Thorpe, John A
Elementary topics in differential geometry.

(Undergraduate texts in mathematics)
Bibliography: p.
Includes index.
1. Geometry, Differential. I. Title.
QA641.T36 516'.36 78-23308

Printed in the United States of America.

9 8 7 6 5 4 3 2 1

ISBN 0-387-90357-7 Springer-Verlag New York
ISBN 3-540-90357-7 Springer-Verlag Berlin Heidelberg

To my parents

whose love, support, and encouragement
over the years have to a large extent
made the writing of this book possible.

Preface

In the past decade there has been a significant change in the freshman/sophomore mathematics curriculum as taught at many, if not most, of our colleges. This has been brought about by the introduction of linear algebra into the curriculum at the sophomore level. The advantages of using linear algebra both in the teaching of differential equations and in the teaching of multivariate calculus are by now widely recognized. Several textbooks adopting this point of view are now available and have been widely adopted. Students completing the sophomore year now have a fair preliminary understanding of spaces of many dimensions.

It should be apparent that courses on the junior level should draw upon and reinforce the concepts and skills learned during the previous year. Unfortunately, in differential geometry at least, this is usually not the case. Textbooks directed to students at this level generally restrict attention to 2-dimensional surfaces in 3-space rather than to surfaces of arbitrary dimension. Although most of the recent books do use linear algebra, it is only the algebra of $\mathbb{R}^3$. The student's preliminary understanding of higher dimensions is not cultivated.

This book develops the geometry of n-dimensional surfaces in $(n + 1)$-space. It is designed for a 1-semester differential geometry course at the junior-senior level. It draws significantly on the contemporary student's knowledge of linear algebra, multivariate calculus, and differential equations, thereby solidifying the student's understanding of these subjects. Indeed, one of the reasons that a course in differential geometry is so valuable at this level is that it does turn out students with a thorough understanding of several variable calculus.

Another reason that differential geometry regularly attracts students is that it contains ideas which are not only beautiful in themselves but are basic for both advanced mathematics and theoretical physics. It has been the author's experience that students taking his course have been more or less evenly divided between mathematics and physics majors. The approach adopted in this book, describing surfaces as solution sets of equations, seems to be especially attractive to physicists.

The book considers from the outset the geometry of orientable hypersurfaces in $\mathbb{R}^{n+1}$, exhibited as inverse images of regular values of smooth functions. By considering only such hypersurfaces for the first half of the book, it is possible to move rapidly into interesting global geometry without getting hung up on the development of sophisticated machinery. Thus, for example, charts (coordinate patches) are not introduced until after the initial discussions of geodesics, parallelism, curvature, and convexity. When charts are introduced, it is as a tool for computation. However, they then lead the development naturally into the study of focal points and surfaces of arbitrary codimension.

One of the advantages of treating the geometry of n-dimensions from the outset is that one can then illustrate each concept simultaneously in each of the low dimensions. Thus, for example, the student's understanding of the Gauss map and its (spherical) image is aided by the possibility of studying 1-dimensional examples, where the spherical image is a subset of the unit circle.

The main tool used in developing the theory is that of the calculus of vector fields. This seems to be the most natural tool for studying differential geometry as well as the one most familiar to undergraduate students of mathematics and physics. Differential forms are not introduced until fairly late in the book, and then only as needed for use in integration.

Students who have completed a good 2-year calculus sequence including linear algebra and differential equations should be adequately prepared to study this book. There are occasional places (e.g., in Chapter 13 on convexity) where some exposure to the ideas of mathematical analysis would be helpful, but not essential.

There is probably more material here than can be covered comfortably in one semester except by students with unusually strong backgrounds. Chapters 1–12, 14, 15, 22, and 23 contain the core of basic material which should be covered in every course. Most instructors will probably also want to cover at least parts of Chapters 17, 19, and 24.

The interdependence of the chapters is as follows:

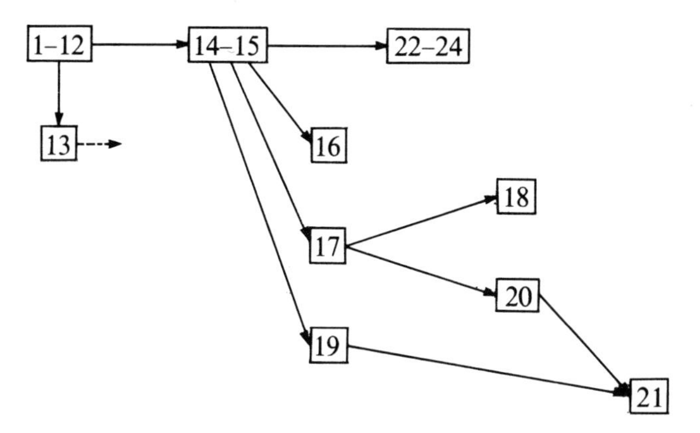

A few concepts in the early part of Chapter 13 are used in later chapters but these may be studied, by those skipping Chapter 13, as needed.

Like the author of any textbook, I owe a considerable debt to researchers and textbook writers who have preceded me and to teachers, colleagues, and students who have influenced me. While I cannot explicitly acknowledge all these, I must at least credit M. do Carmo and E. Lima whose paper, Isometric immersions with semi-definite second quadratic forms, *Arch. Math. 20* (1969) 173–175, inspired the treatment of convex surfaces in Chapter 13, and S. S. Chern whose paper, A simple intrinsic proof of the Gauss–Bonnet formula for closed Riemannian manifolds, *Ann. of Math.* (2) *45* (1944) 747–752, inspired the treatment of the Gauss–Bonnet theorem in Chapter 21. In addition, special thanks are due to Wolfgang Meyer whose comments on the manuscript have been extremely helpful.

Stony Brook, New York
November, 1978

JOHN A. THORPE

Contents

Graphs and Level Sets 1

Associated with each real valued function of several real variables is a collection of sets, called level sets, which are useful in studying qualitative properties of the function. Given a function $f: U \to \mathbb{R}$, where $U \subset \mathbb{R}^{n+1}$, its *level sets* are the sets $f^{-1}(c)$ defined, for each real number c, by

$$f^{-1}(c) = \{(x_1, \ldots, x_{n+1}) \in U : f(x_1, \ldots, x_{n+1}) = c\}.$$

The number c is called the *height* of the level set, and $f^{-1}(c)$ is called the level set *at height* c. Since $f^{-1}(c)$ is the solution set of the equation $f(x_1, \ldots, x_{n+1}) = c$, the level set $f^{-1}(c)$ is often described as "the set $f(x_1, \ldots, x_{n+1}) = c$."

The "level set" and "height" terminologies arise from the relation between the level sets of a function and its graph. The *graph* of a function $f: U \to \mathbb{R}$ is the subset of $\mathbb{R}^{n+2}$ defined by

$$\text{graph}(f) = \{(x_1, \ldots, x_{n+2}) \in \mathbb{R}^{n+2} : (x_1, \ldots, x_{n+1}) \in U$$
$$\text{and } x_{n+2} = f(x_1, \ldots, x_{n+1})\}.$$

For $c \geq 0$, the level set of f at height c is just the set of all points in the domain of f over which the graph is at distance c (see Figure 1.1). For $c < 0$, the level set of f at height c is just the set of all points in the domain of f under which the graph lies at distance $-c$.

For example, the level sets $f^{-1}(c)$ of the function $f(x_1, \ldots, x_{n+1}) = x_1^2 + \cdots + x_{n+1}^2$ are empty for $c < 0$, consist of a single point (the origin) if $c = 0$, and for $c > 0$ consist of two points if $n = 0$, circles centered at the origin with radius $\sqrt{c}$ if $n = 1$, spheres centered at the origin with radius $\sqrt{c}$ if $n = 2$, etc (see Figures 1.1 and 1.2).

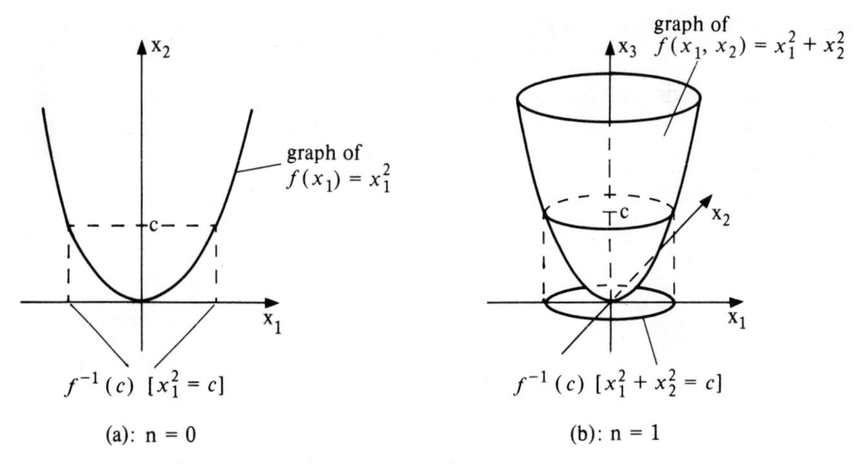

Figure 1.1 The level sets $f^{-1}(c)$ $(c > 0)$ for the function $f(x_1, \ldots, x_{n+1}) = x_1^2 + \cdots + x_{n+1}^2$.

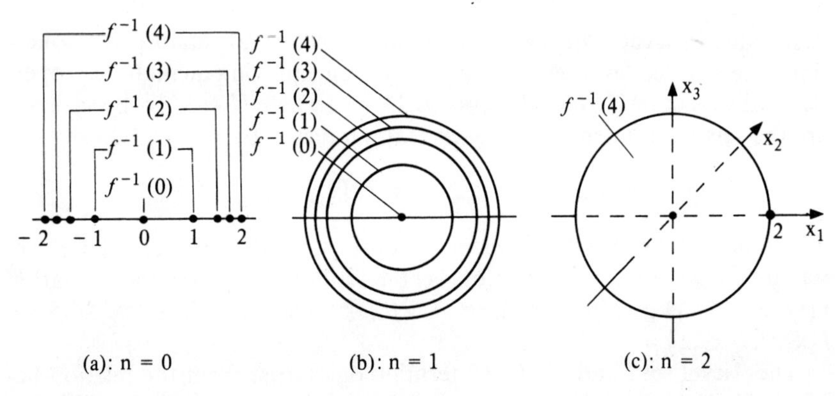

Figure 1.2 Level sets for the function $f(x_1, \ldots, x_{n+1}) = x_1^2 + \cdots + x_{n+1}^2$.

For $n = 1$, level sets are (at least for non-constant differentiable functions) generally curves in $\mathbb{R}^2$. These curves play the same roles as contour lines on a topographic map. If we think of the graph of f as a land, with local maxima representing mountain peaks and local minima representing valley bottoms, then we can construct a topographic map of this land by projecting orthogonally onto $\mathbb{R}^2$. Then all points on any given level curve $f^{-1}(c)$ correspond to points on the land which are at exactly height c above "sea level" $(x_3 = 0)$.

Just as contour maps provide an accurate picture of the topography of a land, so does a knowledge of the level sets and their heights accurately portray the graph of a function. For functions $f\colon \mathbb{R}^2 \to \mathbb{R}$, study of the level curves can facilitate the sketching of the graph of f. For functions $f\colon \mathbb{R}^3 \to \mathbb{R}$, the graph lies in $\mathbb{R}^4$, prohibiting sketches and leaving the level sets as the best tools for studying the behavior of the function.

One way of visualizing the graph of a function $f: U \to \mathbb{R}$, $U \subset \mathbb{R}^2$, given its level sets, is as follows. Think of a plane, parallel to the (x_1, x_2)-plane, moving vertically. When it reaches height c this plane, $x_3 = c$, cuts the graph of f in the translate to this plane of the level set $f^{-1}(c)$. As the plane moves, these sets generate the graph of f (see Figure 1.3).

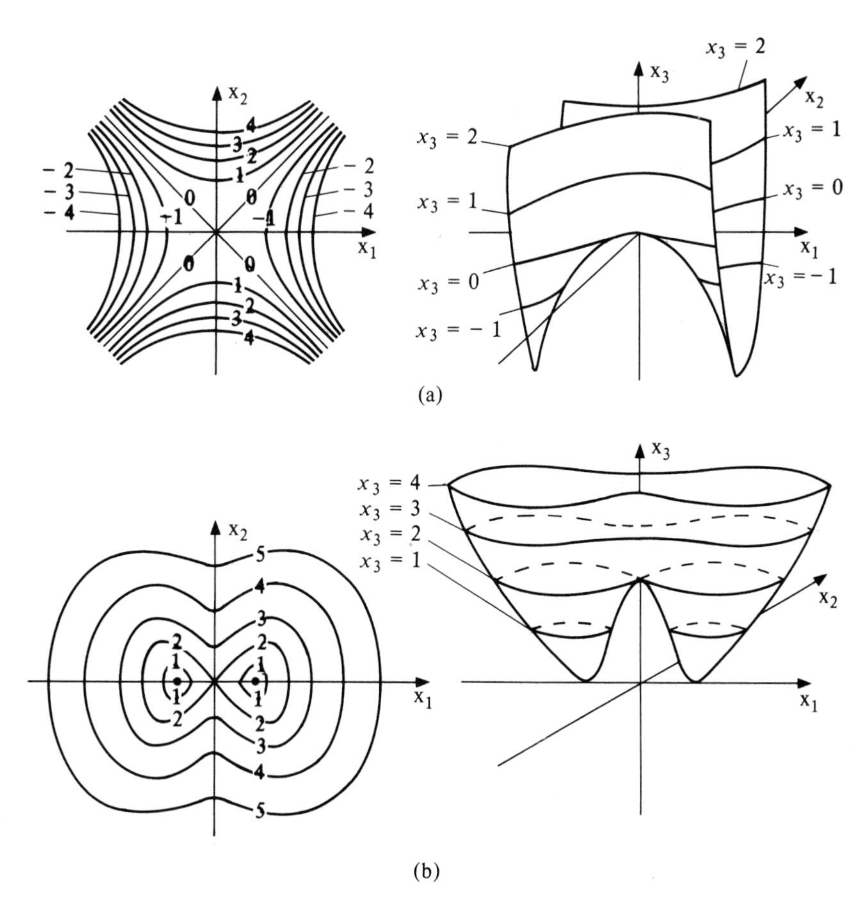

Figure 1.3 Level sets and graphs of functions $f: \mathbb{R}^2 \to \mathbb{R}$. The label on each level set indicates its height. (a) $f(x_1, x_2) = -x_1^2 + x_2^2$. (b) A function with two local minima.

The same principle can be used to help visualize level sets of functions $f: U \to \mathbb{R}$, where $U \subset \mathbb{R}^3$. Each plane $x_i =$ constant will cut the level set $f^{-1}(c)$ (c fixed) in some subset, usually a curve. Letting the plane move, by changing the selected value of the x_i-coordinate, these subsets will generate the level set $f^{-1}(c)$ (see Figure 1.4).

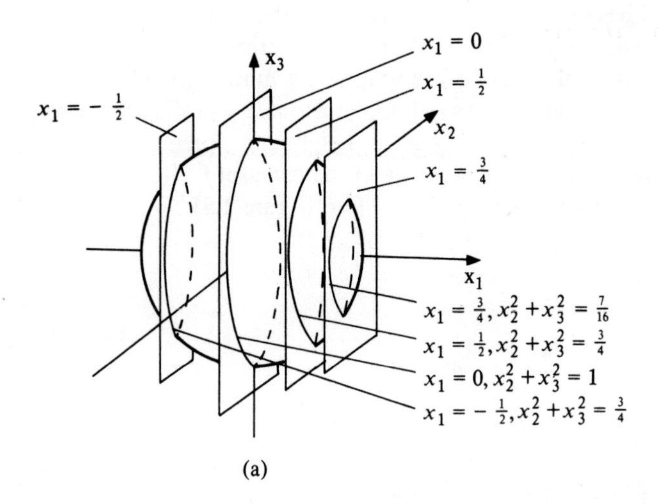

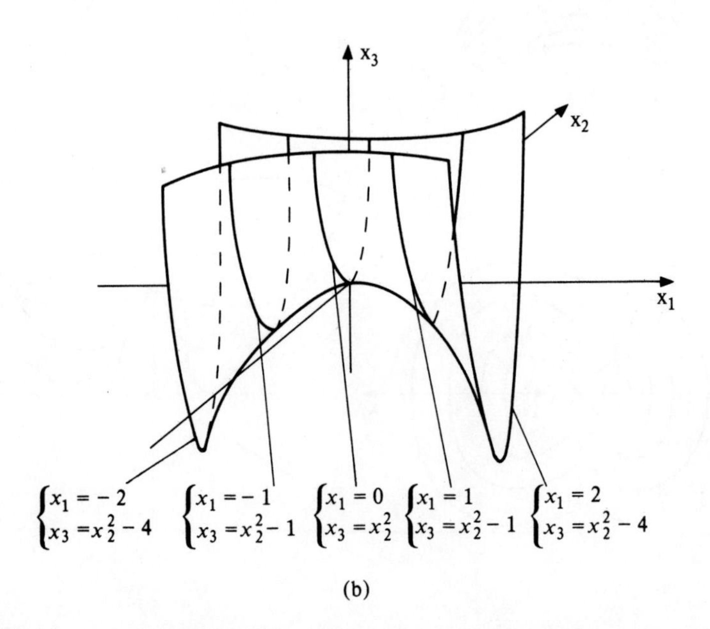

Figure 1.4 Level sets in $\mathbb{R}^3$, as generated by intersections with the planes $x_1 = \text{constant}$. (a) $x_1^2 + x_2^2 + x_3^2 = 1$. (b) $x_1^2 - x_2^2 + x_3 = 0$.

Exercises

In Exercises 1.1–1.4 sketch typical level curves and the graph of each function.

1.1. $f(x_1, x_2) = x_1$.

1.2. $f(x_1, x_2) = x_1 - x_2$.

1.3. $f(x_1, x_2) = x_1^2 - x_2^2$.

1.4. $f(x_1, x_2) = 3r^8 - 8r^6 + 6r^4$ where $r^2 = x_1^2 + x_2^2$. [*Hint*: Find and identify the critical points of f as a function of r.]

In exercises 1.5–1.9 sketch the level sets $f^{-1}(c)$, for $n = 0$, 1, and 2, of each function at the heights indicated.

1.5. $f(x_1, x_2, \ldots, x_{n+1}) = x_{n+1}$; $c = -1, 0, 1, 2$.

1.6. $f(x_1, x_2, \ldots, x_{n+1}) = 0x_1^2 + x_2^2 + \cdots + x_{n+1}^2$; $c = 0, 1, 4$.

1.7. $f(x_1, x_2, \ldots, x_{n+1}) = x_1 - x_2^2 - \cdots - x_{n+1}^2$; $c = -1, 0, 1, 2$.

1.8. $f(x_1, x_2, \ldots, x_{n+1}) = x_1^2 - x_2^2 - \cdots - x_{n+1}^2$; $c = -1, 0, 1$.

1.9. $f(x_1, x_2, \ldots, x_{n+1}) = x_1^2 + x_2^2/4 + \cdots + x_{n+1}^2/(n+1)^2$; $c = 1$.

1.10. Show that the graph of any function $f: \mathbb{R}^n \to \mathbb{R}$ is a level set for some function $F: \mathbb{R}^{n+1} \to \mathbb{R}$.

2 Vector Fields

The tool which will allow us to study the geometry of level sets is the calculus of vector fields. In this chapter we develop some of the basic ideas.

A *vector at a point* $p \in \mathbb{R}^{n+1}$ is a pair $\mathbf{v} = (p, v)$ where $v \in \mathbb{R}^{n+1}$. Geometrically, think of $\mathbf{v}$ as the vector v translated so that its tail is at p rather than at the origin (Figure 2.1). The vectors at p form a vector space $\mathbb{R}_p^{n+1}$ of

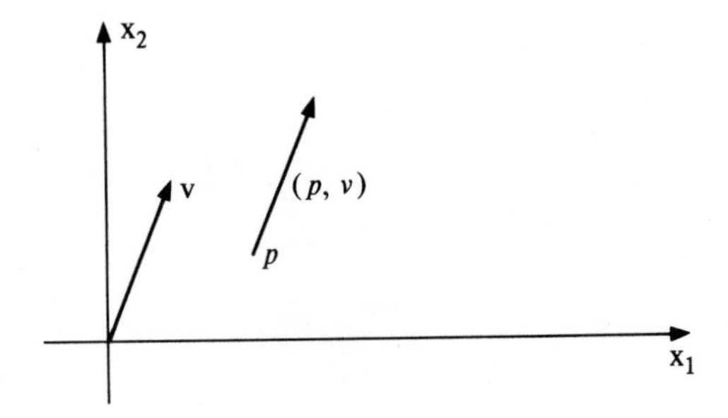

Figure 2.1 A vector at p.

dimension $n + 1$, with addition defined by $(p, v) + (p, w) = (p, v + w)$ (Figure 2.2) and scalar multiplication by $c(p, v) = (p, cv)$. The set $\{(p, v_1), \ldots, (p, v_{n+1})\}$ is a basis for $\mathbb{R}_p^{n+1}$ where $\{v_1, \ldots, v_{n+1}\}$ is any basis for $\mathbb{R}^{n+1}$. The set of all vectors at all points of $\mathbb{R}^{n+1}$ can be identified (as a set) with the Cartesian product $\mathbb{R}^{n+1} \times \mathbb{R}^{n+1} = \mathbb{R}^{2n+2}$. However note that our rule of addition does not permit the addition of vectors at different points of $\mathbb{R}^{n+1}$.

Given two vectors (p, v) and (p, w) at p, their *dot product* is defined, using the standard dot product on $\mathbb{R}^{n+1}$, by $(p, v) \cdot (p, w) = v \cdot w$. When (p, v) and

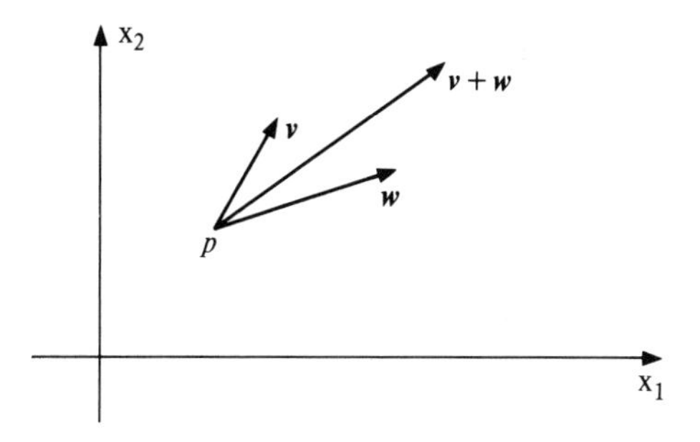

Figure 2.2 Addition of vectors at p.

$(p, w) \in \mathbb{R}_p^3$, $p \in \mathbb{R}^3$, the *cross product* is also defined, using the standard cross product on $\mathbb{R}^3$, by $(p, v) \times (p, w) = (p, v \times w)$.

Using the dot product, the *length* $\|\mathbf{v}\|$ of a vector $\mathbf{v} = (p, v)$ at p and the *angle* θ between two vectors $\mathbf{v} = (p, v)$ and $\mathbf{w} = (p, w)$ are defined by

$$\|\mathbf{v}\| = (\mathbf{v} \cdot \mathbf{v})^{1/2}$$

$$\cos \theta = \mathbf{v} \cdot \mathbf{w}/\|\mathbf{v}\|\ \|\mathbf{w}\| \qquad 0 \leq \theta < \pi$$

A *vector field* $\mathbf{X}$ on $U \subset \mathbb{R}^{n+1}$ is a function which assigns to each point of U a vector at that point. Thus

$$\mathbf{X}(p) = (p, X(p))$$

for some function $X: U \to \mathbb{R}^{n+1}$. Vector fields on $\mathbb{R}^{n+1}$ are often most easily described by specifying this associated function X. Three typical vector fields on $\mathbb{R}^2$ are shown in Figure 2.3.

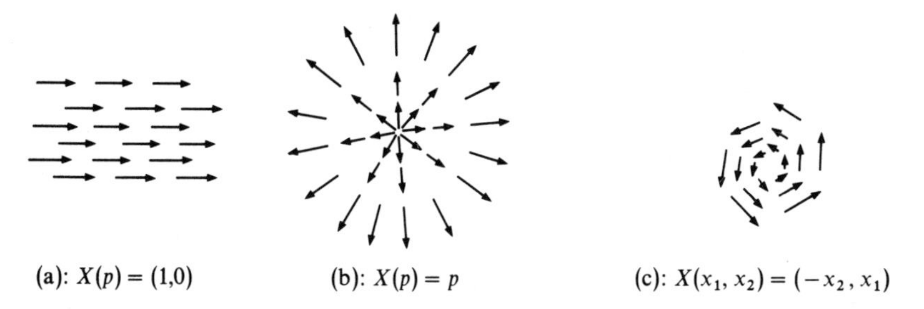

Figure 2.3 Vector fields on $\mathbb{R}^2$: $\mathbf{X}(p) = (p, X(p))$.

We shall deal in this text mostly with functions and vector fields that are *smooth*. A function $f: U \to \mathbb{R}$ (U an open† set in $\mathbb{R}^{n+1}$) is smooth if all its partial derivatives of all orders exist and are continuous. A function $f: U \to \mathbb{R}^k$ is smooth if each component function $f_i: U \to \mathbb{R}$ ($f(p) = (f_1(p), \ldots, f_k(p))$ for $p \in U$) is smooth. A vector field $\mathbf{X}$ on U is smooth if the associated function $X: U \to \mathbb{R}^{n+1}$ is smooth.

† Recall that $U \subset \mathbb{R}^{n+1}$ is *open* if for each $p \in U$ there is an $\varepsilon > 0$ such that $q \in U$ whenever $\|q - p\| < \varepsilon$.

Associated with each smooth function $f\colon U \to \mathbb{R}$ (U open in $\mathbb{R}^{n+1}$) is a smooth vector field on U called the *gradient* ∇f of f, defined by

$$(\nabla f)(p) = \left(p, \frac{\partial f}{\partial x_1}(p), \ldots, \frac{\partial f}{\partial x_{n+1}}(p)\right).$$

We shall see that this vector field plays an important role in the study of the level sets of f.

Vector fields often arise in physics as velocity fields of fluid flows. Associated with such a flow is a family of parametrized curves called flow lines. These "flow lines" are in fact associated with any smooth vector field and are important in geometry as well as in physics. In geometry these flow lines are called "integral curves".

A *parametrized curve* in $\mathbb{R}^{n+1}$ is a smooth function $\alpha\colon I \to \mathbb{R}^{n+1}$, where I is some open interval in $\mathbb{R}$. By smoothness of such a function is meant that α is of the form $\alpha(t) = (x_1(t), \ldots, x_{n+1}(t))$ where each x_i is a smooth real valued function on I.

The *velocity vector* at time t ($t \in I$) of the parametrized curve $\alpha\colon I \to \mathbb{R}^{n+1}$ is the vector at $\alpha(t)$ defined by

$$\dot{\alpha}(t) = \left(\alpha(t), \frac{d\alpha}{dt}(t)\right) = \left(\alpha(t), \frac{dx_1}{dt}(t), \ldots, \frac{dx_{n+1}}{dt}(t)\right).$$

This vector is tangent to the curve α at $\alpha(t)$ (see Figure 2.4.). If $\alpha(t)$ represents

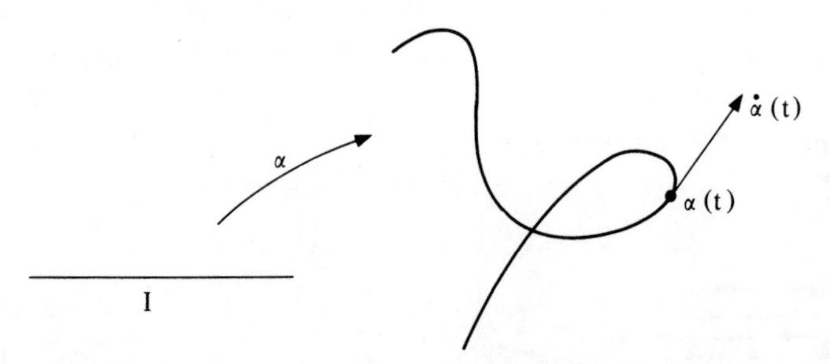

Figure 2.4 Velocity vector of a parametrized curve in $\mathbb{R}^2$.

for each t the position at time t of a particle moving in $\mathbb{R}^{n+1}$ then $\dot{\alpha}(t)$ represents the velocity of this particle at time t.

A parametrized curve $\alpha\colon I \to \mathbb{R}^{n+1}$ is said to be an *integral curve* of the vector field $\mathbf{X}$ on the open set U in $\mathbb{R}^{n+1}$ if $\alpha(t) \in U$ and $\dot{\alpha}(t) = \mathbf{X}(\alpha(t))$ for all $t \in I$. Thus α has the property that its velocity vector at each point of the curve coincides with the value of the vector field at that point (see Figure 2.5).

Theorem. *Let* $\mathbf{X}$ *be a smooth vector field on an open set* $U \subset \mathbb{R}^{n+1}$ *and let* $p \in U$. *Then there exists an open interval* I *containing* 0 *and an integral curve* $\alpha\colon I \to U$ *of* $\mathbf{X}$ *such that*

(i) $\alpha(0) = p$.

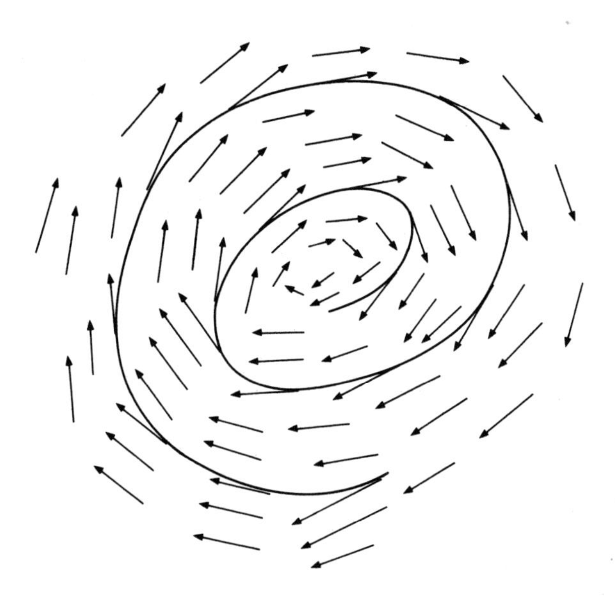

Figure 2.5 An integral curve of a vector field.

(ii) *If* $\beta\colon \tilde{I} \to U$ *is any other integral curve of* $\mathbf{X}$ *with* $\beta(0) = p$, *then* $\tilde{I} \subset I$ *and* $\beta(t) = \alpha(t)$ *for all* $t \in \tilde{I}$.

Remark. The integral curve α is called the *maximal integral curve of* $\mathbf{X}$ *through* p, or simply *the* integral curve of $\mathbf{X}$ through p.

PROOF. This theorem is a reformulation of the fundamental existence and uniqueness theorem for solutions of systems of first order differential equations. $\mathbf{X}$, being a smooth vector field on U, has the form

$$\mathbf{X}(p) = (p, X_1(p), \ldots, X_{n+1}(p))$$

where the $X_i\colon U \to \mathbb{R}$ are smooth functions on U. A parametrized curve $\alpha\colon I \to \mathbb{R}^{n+1}$ has the form

$$\alpha(t) = (x_1(t), \ldots, x_{n+1}(t))$$

where the $x_i\colon I \to \mathbb{R}$ are smooth functions on I. The velocity of α is

$$\dot{\alpha}(t) = \left(\alpha(t), \frac{dx_1}{dt}(t), \ldots, \frac{dx_{n+1}}{dt}(t)\right).$$

The requirement that α be an integral curve of $\mathbf{X}$ says that $\dot{\alpha}(t) = \mathbf{X}(\alpha(t))$, or

$$\text{(E)} \quad \begin{cases} \dfrac{dx_1}{dt}(t) = X_1(x_1(t), \ldots, x_{n+1}(t)) \\ \vdots \\ \dfrac{dx_{n+1}}{dt}(t) = X_{n+1}(x_1(t), \ldots, x_{n+1}(t)). \end{cases}$$

This is a system of $n+1$ first order ordinary differential equations in $n+1$ unknowns. By the existence theorem for the solutions of such equations,† there exists an open interval I_1 about 0 and a set $x_i: I_1 \to \mathbb{R}$ of smooth functions satisfying this system subject to the initial conditions $x_i(0) = p_i$ for $i \in \{1, \ldots, n+1\}$, where $p = (p_1, \ldots, p_{n+1})$. Setting $\beta_1(t) = (x_1(t), \ldots, x_{n+1}(t))$ for this choice of functions yields an integral curve $\beta_1: I_1 \to U$ of $\mathbf{X}$ with $\beta_1(0) = p$.

By the uniqueness theorem for the solutions of first order ordinary differential equations,† if $\tilde{x}_i: I_2 \to \mathbb{R}$ is another set of functions satisfying the system (E) together with the initial conditions $\tilde{x}_i(0) = p_i$, then $\tilde{x}_i(t) = x_i(t)$ for all $t \in I_1 \cap I_2$. In other words, if $\beta_2: I_2 \to U$ is another integral curve of $\mathbf{X}$ with $\beta_2(0) = p$ then $\beta_1(t) = \beta_2(t)$ for all $t \in I_1 \cap I_2$. It follows from this that there is a unique maximal integral curve α of $\mathbf{X}$ with $\alpha(0) = p$ (its domain is the union of the domains of all integral curves of $\mathbf{X}$ which map 0 to p) and that if $\beta: \tilde{I} \to U$ is any other integral curve of $\mathbf{X}$ with $\beta(0) = p$ then β is simply the restriction of α to the smaller interval $\tilde{I}$. □

Example. Let $\mathbf{X}$ be the vector field $\mathbf{X}(p) = (p, X(p))$ where $X(x_1, x_2) = (-x_2, x_1)$ (Figure 2.3.(c)). A parametrized curve $\alpha(t) = (x_1(t), x_2(t))$ is an integral curve of $\mathbf{X}$ if and only if the functions $x_1(t)$ and $x_2(t)$ satisfy the differential equations

$$\begin{cases} \dfrac{dx_1}{dt} = -x_2 \\[2ex] \dfrac{dx_2}{dt} = x_1. \end{cases}$$

The general solution of this pair of equations is

$$x_1(t) = C_1 \cos t + C_2 \sin t$$
$$x_2(t) = C_1 \sin t - C_2 \cos t.$$

Thus the integral curve of $\mathbf{X}$ through the point (1, 0) (with $x_1(0) = 1$ and $x_2(0) = 0$) is

$$\alpha(t) = (\cos t, \sin t),$$

whereas the integral curve through an arbitrary point (a, b) (with $x_1(0) = a$ and $x_2(0) = b$) is

$$\beta(t) = (a \cos t - b \sin t, a \sin t + b \cos t)$$

(see Figure 2.6).

† See e.g. W. Hurewicz, *Lectures on Ordinary Differential Equations*, Cambridge, Mass.: M.I.T. Press (1958), p. 28.

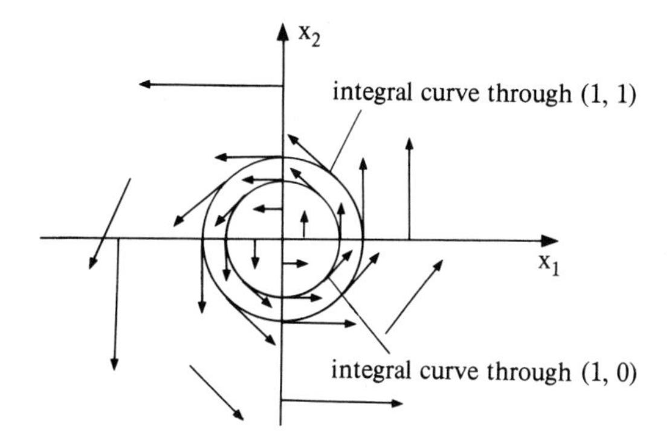

Figure 2.6 Integral curves of the vector field $\mathbf{X}(x_1, x_2) = (x_1, x_2, -x_2, x_1)$.

Exercises

2.1. Sketch the following vector fields on $\mathbb{R}^2$: $\mathbf{X}(p) = (p, X(p))$ where

(a) $X(p) = (0, 1)$
(b) $X(p) = -p$
(c) $X(x_1, x_2) = (x_2, -x_1)$
(d) $X(x_1, x_2) = (x_2, x_1)$
(e) $X(x_1, x_2) = (-2x_2, \frac{1}{2}x_1)$.

2.2. Find and sketch the gradient field of each of the following functions:

(a) $f(x_1, x_2) = x_1 + x_2$
(b) $f(x_1, x_2) = x_1^2 + x_2^2$
(c) $f(x_1, x_2) = x_1 - x_2^2$
(d) $f(x_1, x_2) = (x_1^2 - x_2^2)/4$.

2.3. The *divergence* of a smooth vector field $\mathbf{X}$ on $U \subset \mathbb{R}^{n+1}$,

$$\mathbf{X}(p) = (p, X_1(p), \ldots, X_{n+1}(p)) \quad \text{for } p \in U,$$

is the function div $\mathbf{X}$: $U \to \mathbb{R}$ defined by div $\mathbf{X} = \sum_{i=1}^{n+1} (\partial X_i/\partial x_i)$. Find the divergence of each of the vector fields in Exercises 2.1 and 2.2.

2.4. Explain why an integral curve of a vector field cannot cross itself as does the parametrized curve in Figure 2.4.

2.5. Find the integral curve through $p = (1, 1)$ of each of the vector fields in Exercise 2.1.

2.6. Find the integral curve through $p = (a, b)$ of each of the vector fields in Exercise 2.1.

2.7. A smooth vector field $\mathbf{X}$ on an open set U of $\mathbb{R}^{n+1}$ is said to be *complete* if for each $p \in U$ the maximal integral curve of $\mathbf{X}$ through p has domain equal to $\mathbb{R}$. Determine which of the following vector fields are complete:

(a) $\mathbf{X}(x_1, x_2) = (x_1, x_2, 1, 0)$, $U = \mathbb{R}^2$.
(b) $\mathbf{X}(x_1, x_2) = (x_1, x_2, 1, 0)$, $U = \mathbb{R}^2 - \{(0, 0)\}$.
(c) $\mathbf{X}(x_1, x_2) = (x_1, x_2, -x_2, x_1)$, $U = \mathbb{R}^2 - \{(0, 0)\}$.
(d) $\mathbf{X}(x_1, x_2) = (x_1, x_2, 1 + x_1^2, 0)$, $U = \mathbb{R}^2$.

2.8. Let U be an open set in $\mathbb{R}^{n+1}$, let $p \in U$, and let $\mathbf{X}$ be a smooth vector field on U. Let $\alpha: I \to U$ be the maximal integral curve of $\mathbf{X}$ through p. Show that if $\beta: \tilde{I} \to U$ is any integral curve of $\mathbf{X}$, with $\beta(t_0) = p$ for some $t_0 \in \tilde{I}$, then $\beta(t) = \alpha(t - t_0)$ for all $t \in \tilde{I}$. [*Hint*: Verify that if $\tilde{\beta}$ is defined by $\tilde{\beta}(t) = \beta(t + t_0)$ then $\tilde{\beta}$ is an integral curve of $\mathbf{X}$ with $\tilde{\beta}(0) = p$.]

2.9. Let U be an open set in $\mathbb{R}^{n+1}$ and let $\mathbf{X}$ be a smooth vector field on U. Suppose $\alpha: I \to U$ is an integral curve of $\mathbf{X}$ with $\alpha(0) = \alpha(t_0)$ for some $t_0 \in I$, $t_0 \neq 0$. Show that α is periodic; i.e., show that $\alpha(t + t_0) = \alpha(t)$ for all t such that both t and $t + t_0 \in I$. [*Hint*: See Exercise 2.8.]

2.10. Consider the vector field $\mathbf{X}(x_1, x_2) = (x_1, x_2, 1, 0)$ on $\mathbb{R}^2$. For $t \in \mathbb{R}$ and $p \in \mathbb{R}^2$, let $\varphi_t(p) = \alpha_p(t)$ where α_p is the maximal integral curve of $\mathbf{X}$ through p.

(a) Show that, for each t, φ_t is a one to one transformation from $\mathbb{R}^2$ onto itself. Geometrically, what does this transformation do?
(b) Show that

$$\varphi_0 = \text{identity}$$

$$\varphi_{t_1+t_2} = \varphi_{t_1} \circ \varphi_{t_2} \quad \text{for all } t_1, t_2 \in \mathbb{R}$$

$$\varphi_{-t} = \varphi_t^{-1} \quad \text{for all } t \in \mathbb{R}.$$

[Thus $t \mapsto \varphi_t$ is a homomorphism from the additive group of real numbers into the group of one to one transformations of the plane.]

2.11. Repeat Exercise 2.10 for the vector fields

(a) $\mathbf{X}(x_1, x_2) = (x_1, x_2, -x_2, x_1)$
(b) $\mathbf{X}(x_1, x_2) = (x_1, x_2, x_1, x_2)$
(c) $\mathbf{X}(x_1, x_2) = (x_1, x_2, x_2, x_1)$.

2.12. Let $\mathbf{X}$ be any smooth vector field on U, U open in $\mathbb{R}^{n+1}$. Let $\varphi_t(p) = \alpha_p(t)$ where α_p is the maximal integral curve of $\mathbf{X}$ through p. Use the uniqueness of integral curves to show that $\varphi_{t_1}(\varphi_{t_2}(p)) = \varphi_{t_1+t_2}(p)$ and $\varphi_{-t}(p) = \varphi_t^{-1}(p)$ for all t, t_1, and t_2 for which all terms are defined. [φ_t is called the *local* 1-*parameter group* associated to $\mathbf{X}$.]

3 The Tangent Space

Let $f\colon U \to \mathbb{R}$ be a smooth function, where $U \subset \mathbb{R}^{n+1}$ is an open set, let $c \in \mathbb{R}$ be such that $f^{-1}(c)$ is non-empty, and let $p \in f^{-1}(c)$. A vector at p is said to be *tangent to the level set* $f^{-1}(c)$ if it is a velocity vector of a parametrized curve in $\mathbb{R}^{n+1}$ whose image is contained in $f^{-1}(c)$ (see Figure 3.1).

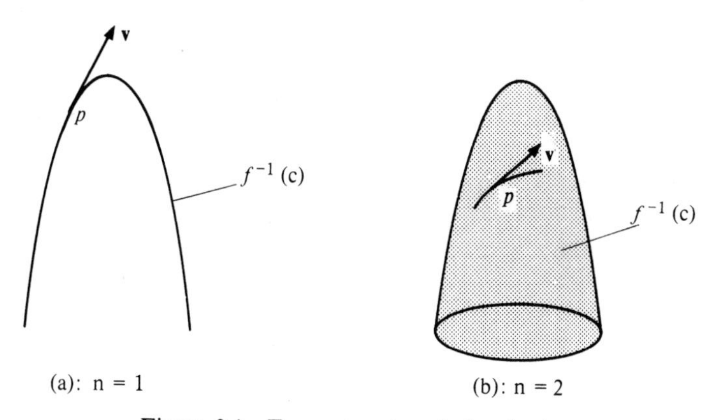

Figure 3.1 Tangent vectors to level sets.

Lemma. *The gradient of f at $p \in f^{-1}(c)$ is orthogonal to all vectors tangent to $f^{-1}(c)$ at p.*

PROOF. Each vector tangent to $f^{-1}(c)$ at p is of the form $\dot{\alpha}(t_0)$ for some parametrized curve $\alpha\colon I \to \mathbb{R}^{n+1}$ with $\alpha(t_0) = p$ and Image $\alpha \subset f^{-1}(c)$. But Image $\alpha \subset f^{-1}(c)$ implies $f(\alpha(t)) = c$ for all $t \in I$ so, by the chain rule,

$$0 = \frac{d}{dt}(f \circ \alpha)(t_0) = \nabla f(\alpha(t_0)) \cdot \dot{\alpha}(t_0) = \nabla f(p) \cdot \dot{\alpha}(t_0). \qquad \square$$

If $\nabla f(p) = \mathbf{0}$, this lemma says nothing. But if $\nabla f(p) \neq \mathbf{0}$, it says that the set of all vectors tangent to $f^{-1}(c)$ at p is contained in the n-dimensional vector subspace $[\nabla f(p)]^\perp$ of $\mathbb{R}^{n+1}_p$ consisting of all vectors orthogonal to $\nabla f(p)$. A point $p \in \mathbb{R}^{n+1}$ such that $\nabla f(p) \neq \mathbf{0}$ is called a *regular point* of f.

Theorem. *Let U be an open set in $\mathbb{R}^{n+1}$ and let $f\colon U \to \mathbb{R}$ be smooth. Let $p \in U$ be a regular point of f, and let $c = f(p)$. Then the set of all vectors tangent to $f^{-1}(c)$ at p is equal to $[\nabla f(p)]^\perp$.*

PROOF. That every vector tangent to $f^{-1}(c)$ at p is contained in $[\nabla f(p)]^\perp$ was proven as the lemma above. Thus it suffices to show that, if $\mathbf{v} = (p, v) \in [\nabla f(p)]^\perp$, then $\mathbf{v} = \dot{\alpha}(0)$ for some parametrized curve α with Image $\alpha \subset f^{-1}(c)$. To construct α, consider the constant vector field $\mathbf{X}$ on U defined by $\mathbf{X}(q) = (q, v)$. From $\mathbf{X}$ we can construct another vector field $\mathbf{Y}$ by subtracting from $\mathbf{X}$ the component of $\mathbf{X}$ along ∇f:

$$\mathbf{Y}(q) = \mathbf{X}(q) - \frac{\mathbf{X}(q) \cdot \nabla f(q)}{\|\nabla f(q)\|^2} \nabla f(q).$$

The vector field $\mathbf{Y}$ has domain the open subset of U where $\nabla f \neq 0$. Since p is a regular point of f, p is in the domain of $\mathbf{Y}$. Moreover, since $\mathbf{X}(p) = \mathbf{v} \in [\nabla f(p)]^\perp$, $\mathbf{Y}(p) = \mathbf{X}(p)$. Thus we have obtained a smooth vector field $\mathbf{Y}$ such that $\mathbf{Y}(q) \perp \nabla f(q)$ for all $q \in$ domain $(\mathbf{Y})$, and $\mathbf{Y}(p) = \mathbf{v}$.

Now let α be an integral curve of $\mathbf{Y}$ through p. Then $\alpha(0) = p$, $\dot{\alpha}(0) = \mathbf{Y}(\alpha(0)) = \mathbf{Y}(p) = \mathbf{X}(p) = \mathbf{v}$ and

$$\frac{d}{dt} f(\alpha(t)) = \nabla f(\alpha(t)) \cdot \dot{\alpha}(t) = \nabla f(\alpha(t)) \cdot \mathbf{Y}(\alpha(t)) = 0$$

(first equality: chain rule; second: since α is an integral curve of $\mathbf{Y}$; third: since $\mathbf{Y} \perp \nabla f$)

for all $t \in$ domain α, so that $f(\alpha(t)) =$ constant. Since $f(\alpha(0)) = f(p) = c$, this means that Image $\alpha \subset f^{-1}(c)$, as required. □

Thus we see that at each regular point p on a level set $f^{-1}(c)$ of a smooth function there is a well defined *tangent space* consisting of all velocity vectors at p of all parametrized curves in $f^{-1}(c)$ passing through p, and this tangent space is precisely $[\nabla f(p)]^\perp$ (see Figure 3.2).

EXERCISES

3.1. Sketch the level sets $f^{-1}(-1)$, $f^{-1}(0)$, and $f^{-1}(1)$ for $f(x_1, \ldots, x_{n+1}) = x_1^2 + \cdots + x_n^2 - x_{n+1}^2$; $n = 1, 2$. Which points p of these level sets fail to have tangent spaces equal to $[\nabla f(p)]^\perp$?

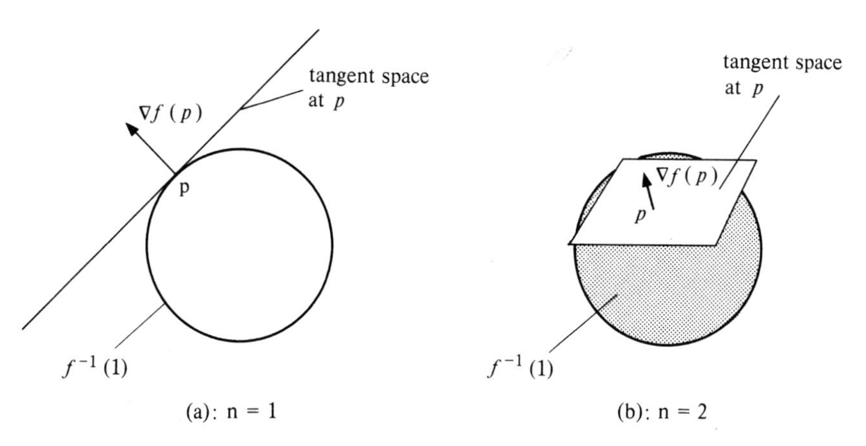

Figure 3.2 Tangent space at a typical point of the level set $f^{-1}(1)$, where $f(x_1, \ldots, x_{n+1}) = x_1^2 + \cdots + x_{n+1}^2$.

3.2. Show by example that

(a) The set of vectors tangent at a point p of a level set need not in general be a vector subspace of $\mathbb{R}_p^{n+1}$.
(b) The set of vectors tangent at a point p of a level set might be all of $\mathbb{R}_p^{n+1}$.

3.3. Sketch the level set $f^{-1}(0)$ and typical values $\nabla f(p)$ of the vector field ∇f for $p \in f^{-1}(0)$, when

(a) $f(x_1, x_2) = x_1^2 + x_2^2 - 1$
(b) $f(x_1, x_2) = x_1^2 - x_2^2 - 1$
(c) $f(x_1, x_2) = x_1^2 - x_2^2$
(d) $f(x_1, x_2) = x_1 - x_2^2$.

3.4. Let $f\colon U \to \mathbb{R}$ be a smooth function, where $U \subset \mathbb{R}^{n+1}$ is an open set, and let $\alpha\colon I \to U$ be a parametrized curve. Show that $f \circ \alpha$ is constant (i.e., the image of α is contained in a level set of f) if and only if α is everywhere orthogonal to the gradient of f (i.e., if and only if $\dot{\alpha}(t) \perp \nabla f(\alpha(t))$ for all $t \in I$).

3.5. Let $f\colon U \to \mathbb{R}$ be a smooth function and let $\alpha\colon I \to U$ be an integral curve of ∇f.

(a) Show that $(d/dt)(f \circ \alpha)(t) = \|\nabla f(\alpha(t))\|^2$ for all $t \in I$.
(b) Show that for each $t_0 \in I$, the function f is increasing faster along α at $\alpha(t_0)$ then along any other curve passing through $\alpha(t_0)$ with the same speed (i.e., show that if $\beta\colon \tilde{I} \to U$ is such that $\beta(s_0) = \alpha(t_0)$ for some $s_0 \in \tilde{I}$ and $\|\dot{\beta}(s_0)\| = \|\dot{\alpha}(t_0)\|$ then $(d/dt)(f \circ \alpha)(t_0) \geq (d/dt)(f \circ \beta)(t_0)$).

4 Surfaces

A *surface of dimension n*, or *n-surface*, in $\mathbb{R}^{n+1}$ is a non-empty subset S of $\mathbb{R}^{n+1}$ of the form $S = f^{-1}(c)$ where $f\colon U \to \mathbb{R}$, U open in $\mathbb{R}^{n+1}$, is a smooth function with the property that $\nabla f(p) \neq \mathbf{0}$ for all $p \in S$. A 1-surface in $\mathbb{R}^2$ is also called a *plane curve*. A 2-surface in $\mathbb{R}^3$ is usually called simply a *surface*. An n-surface in $\mathbb{R}^{n+1}$ is often called a *hypersurface*, especially when $n > 2$.

By the theorem of the previous chapter, each n-surface S has at each point $p \in S$ a *tangent space* which is an n-dimensional vector subspace of the space $\mathbb{R}_p^{n+1}$ of all vectors at p. This tangent space will be denoted by S_p. It is important to notice that this tangent space S_p depends only on the set S and is independent of the function f which is used to define S. Indeed, S_p is characterized as the set of all vectors at p which can be obtained as velocity vectors of parametrized curves in $\mathbb{R}^{n+1}$ with images lying completely in S. If f is any smooth function such that $S = f^{-1}(c)$ for some $c \in \mathbb{R}$ and $\nabla f(p) \neq 0$ for all $p \in S$ (by definition of n-surface, there must exist one such a function; in fact there are many such functions for each n-surface S) then S_p may also be described as $[\nabla f(p)]^\perp$.

EXAMPLE 1. The unit *n-sphere* $x_1^2 + \cdots + x_{n+1}^2 = 1$ is the level set $f^{-1}(1)$ where $f(x_1, \ldots, x_{n+1}) = x_1^2 + \cdots + x_{n+1}^2$ (Figure 3.2). It is an n-surface because $\nabla f(x_1, \ldots, x_{n+1}) = (x_1, \ldots, x_{n+1}, 2x_1, \ldots, 2x_{n+1})$ is not zero unless $(x_1, \ldots, x_{n+1}) = (0, \ldots, 0)$ so in particular $\nabla f(p) \neq \mathbf{0}$ for $p \in f^{-1}(1)$. [Warning! Beware that for a vector $(p, v) \in \mathbb{R}_p^{n+1}$ to be zero it is only necessary that $v = 0$; thus $(x_1, \ldots, x_{n+1}, 2x_1, \ldots, 2x_{n+1}) = 0$ implies $2x_1 = \cdots = 2x_{n+1} = 0$ so $(x_1, \ldots, x_{n+1}) = 0$.] When $n = 1$, the unit n-sphere is the *unit circle*.

EXAMPLE 2. For $0 \neq (a_1, \ldots, a_{n+1}) \in \mathbb{R}^{n+1}$ and $b \in \mathbb{R}$, the *n-plane* $a_1 x_1 + \cdots + a_{n+1} x_{n+1} = b$ is the level set $f^{-1}(b)$ where $f(x_1, \ldots, x_{n+1}) =$

$a_1 x_1 + \cdots + a_{n+1} x_{n+1}$. It is an n-surface for each $b \in \mathbb{R}$ since

$$\nabla f(x_1, \ldots, x_{n+1}) = (x_1, \ldots, x_{n+1}, a_1, \ldots, a_{n+1})$$

is never zero. A 1-plane is usually called a *line* in $\mathbb{R}^2$, a 2-plane is usually called simply a *plane* in $\mathbb{R}^3$, and an n-plane for $n > 2$ is sometimes called a *hyperplane* in $\mathbb{R}^{n+1}$. Two different values of b with the same value of $(a_1, \ldots, a_{n+1})$ define *parallel* n-planes (see Figure 4.1).

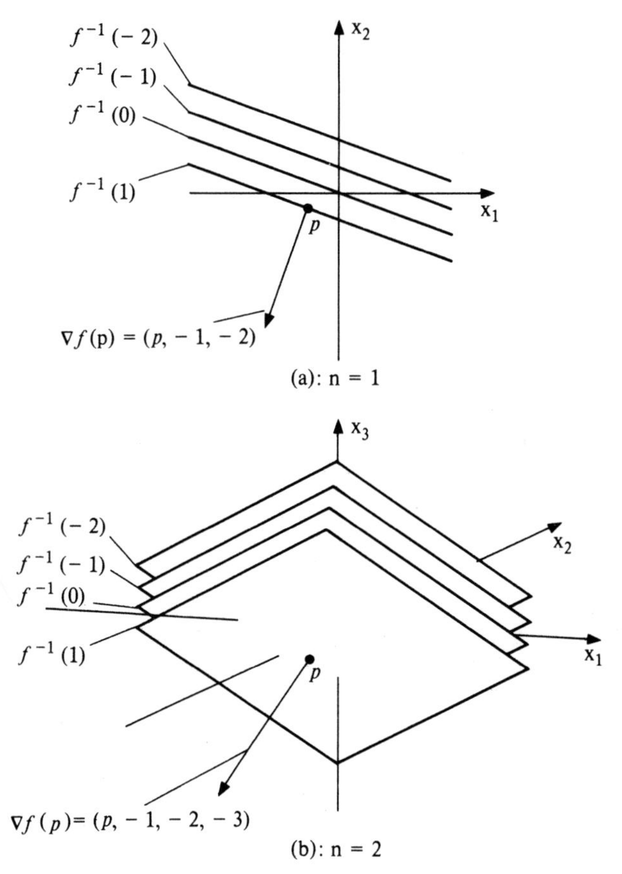

Figure 4.1 Parallel n-planes $f^{-1}(b)$, $b = -2, -1, 0, 1$, where $f(x_1, \ldots, x_{n+1}) = -x_1 - 2x_2 - 3x_3 - \cdots - (n+1)x_{n+1}$.

Example 3. Let $f\colon U \to \mathbb{R}$ be a smooth function on U, U open in $\mathbb{R}^n$. The graph of f,

$$\text{graph}(f) = \{(x_1, \ldots, x_{n+1}) \in \mathbb{R}^{n+1} : x_{n+1} = f(x_1, \ldots, x_n)\}$$

is an n-surface in $\mathbb{R}^{n+1}$ since graph $(f) = g^{-1}(0)$ where

$$g(x_1, \ldots, x_{n+1}) = x_{n+1} - f(x_1, \ldots, x_n)$$

and $\nabla g(x_1, \ldots, x_{n+1}) = (x_1, \ldots, x_{n+1}, \; -\partial f/\partial x_1, \ldots, \; -\partial f/\partial x_n, 1)$ is never zero.

EXAMPLE 4. Let S be an $(n-1)$-surface in $\mathbb{R}^n$, given by $S = f^{-1}(c)$, where $f\colon U \to \mathbb{R}$ (U open in $\mathbb{R}^n$) is such that $\nabla f(p) \neq \mathbf{0}$ for all $p \in f^{-1}(c)$. Let $g\colon U_1 \to \mathbb{R}$, where $U_1 = U \times \mathbb{R} = \{(x_1, \ldots, x_{n+1}) \in \mathbb{R}^{n+1}\colon (x_1, \ldots, x_n) \in U\}$, be defined by

$$g(x_1, \ldots, x_{n+1}) = f(x_1, \ldots, x_n).$$

Then $g^{-1}(c)$ is an n-surface in $\mathbb{R}^{n+1}$ because

$$\nabla g(x_1, \ldots, x_{n+1}) = \left(x_1, \ldots, x_{n+1}, \frac{\partial f}{\partial x_1}, \ldots, \frac{\partial f}{\partial x_n}, 0\right)$$

and $\partial f/\partial x_1, \ldots, \partial f/\partial x_n$ cannot be simultaneously zero when $g(x_1, \ldots, x_{n+1}) = f(x_1, \ldots, x_n) = c$ because $\nabla f(x_1, \ldots, x_n) \neq \mathbf{0}$ whenever $(x_1, \ldots, x_n) \in f^{-1}(c)$. The n-surface $g^{-1}(c)$ is called the *cylinder* over S (see Figure 4.2).

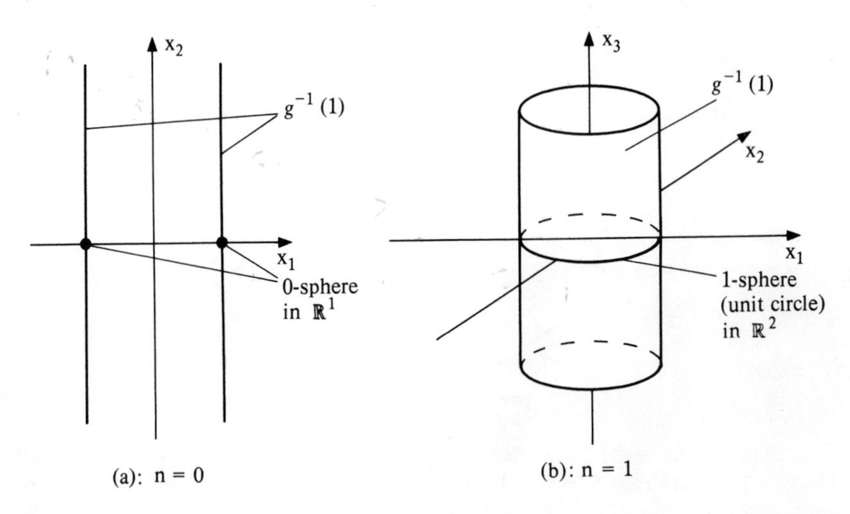

Figure 4.2 The cylinder $g^{-1}(1)$ over the n-sphere: $g(x_1, \ldots, x_{n+1}) = x_1^2 + \cdots + x_n^2$.

EXAMPLE 5. Let C be a curve in $\mathbb{R}^2$ which lies above the x_1-axis. Thus $C = f^{-1}(c)$ for some $f\colon U \to \mathbb{R}$ with $\nabla f(p) \neq 0$ for all $p \in C$, where U is contained in the upper half plane $x_2 > 0$. Define $S = g^{-1}(c)$ where $g\colon U \times \mathbb{R} \to \mathbb{R}$ by $g(x_1, x_2, x_3) = f(x_1, (x_2^2 + x_3^2)^{1/2})$. Then S is a 2-surface (Exercise 4.7). Each point $p = (a, b) \in C$ generates a circle of points of S, namely the circle in the plane $x_1 = a$ consisting of those points $(x_1, x_2, x_3) \in \mathbb{R}^3$ such that $x_1 = a$, $x_2^2 + x_3^2 = b^2$. S is called the *surface of revolution* obtained by rotating the curve C about the x_1-axis (see Figure 4.3).

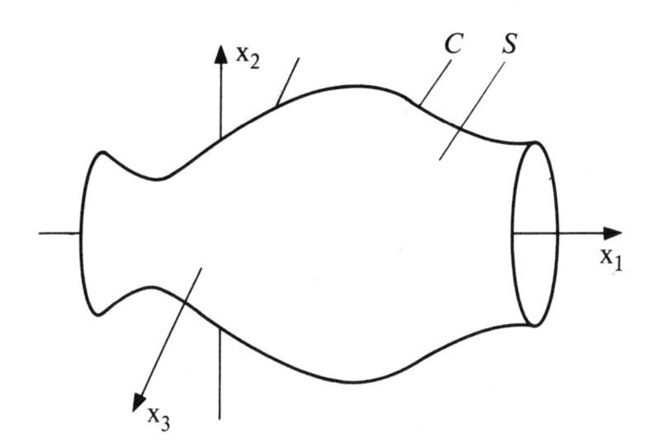

Figure 4.3 The surface of revolution S obtained by rotating the curve C about the x_1-axis.

Theorem. *Let S be an n-surface in $\mathbb{R}^{n+1}$, $S = f^{-1}(c)$ where $f\colon U \to \mathbb{R}$ is such that $\nabla f(q) \neq \mathbf{0}$ for all $q \in S$. Suppose $g\colon U \to \mathbb{R}$ is a smooth function and $p \in S$ is an extreme point of g on S; i.e., either $g(q) \le g(p)$ for all $q \in S$ or $g(q) \ge g(p)$ for all $q \in S$. Then there exists a real number λ such that $\nabla g(p) = \lambda \nabla f(p)$.* (The number λ is called a *Lagrange multiplier*.)

PROOF. The tangent space to S at p is $S_p = [\nabla f(p)]^\perp$. Hence $S_p^\perp$ is the 1-dimensional subspace of $\mathbb{R}_p^{n+1}$ spanned by $\nabla f(p)$. It follows, then, that $\nabla g(p) = \lambda \nabla f(p)$ for some $\lambda \in \mathbb{R}$ if (and only if) $\nabla g(p) \in S_p^\perp$; i.e., if (and only if) $\nabla g(p) \cdot \mathbf{v} = 0$ for all $\mathbf{v} \in S_p$. But each $\mathbf{v} \in S_p$ is of the form $\mathbf{v} = \dot{\alpha}(t_0)$ for some parametrized curve $\alpha\colon I \to S$ and $t_0 \in I$ with $\alpha(t_0) = p$. Since $p = \alpha(t_0)$ is an extreme point of g on S, t_0 is an extreme point of $g \circ \alpha$ on I. Hence

$$0 = (f \circ \alpha)'(t_0) = \nabla g(\alpha(t_0)) \cdot \dot{\alpha}(t_0) = \nabla g(p) \cdot \mathbf{v}$$

for all $\mathbf{v} \in S_p$ and so $\nabla g(p) = \lambda \nabla f(p)$ for some λ, as required. □

Remark. If S is *compact* (closed and bounded†) then every smooth function $g\colon U \to \mathbb{R}$ attains a maximum on S and a minimum on S. The above theorem can then be used to locate candidates for these extreme points. If S is not compact, there may be no extrema.

EXAMPLE. Let S be the unit circle $x_1^2 + x_2^2 = 1$ and define $g\colon \mathbb{R}^2 \to \mathbb{R}$ by $g(x_1, x_2) = ax_1^2 + 2bx_1x_2 + cx_2^2$ where $a, b, c \in \mathbb{R}$ (see Figure 4.4). Then $S = f^{-1}(1)$ where $f(x_1, x_2) = x_1^2 + x_2^2$,

$$\nabla f(x_1, x_2) = (x_1, x_2, 2x_1, 2x_2),$$

and

$$\nabla g(x_1, x_2) = (x_1, x_2, 2ax_1 + 2bx_2, 2bx_1 + 2cx_2),$$

† S is *closed* if $\mathbb{R}^{n+1} - S$ is open; S is *bounded* if there exists $M \in \mathbb{R}$ such that $\|p\| < M$ for all $p \in S$.

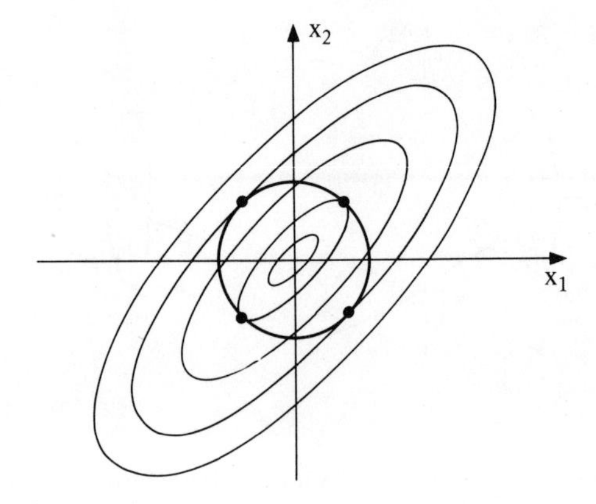

Figure 4.4 Level curves of the function $g(x_1, x_2) = ax_1^2 + 2bx_1x_2 + cx_2^2$ $(ac - b^2 > 0)$. The four points where these curves are tangent to the unit circle S are the extreme points of g on S.

so $\nabla g(p) = \lambda \nabla f(p)$ for $p = (x_1, x_2) \in S$ if and only if

$$\begin{cases} 2ax_1 + 2bx_2 = 2\lambda x_1 \\ 2bx_1 + 2cx_2 = 2\lambda x_2 \end{cases}$$

or

$$\begin{pmatrix} a & b \\ b & c \end{pmatrix}\begin{pmatrix} x_1 \\ x_2 \end{pmatrix} = \lambda \begin{pmatrix} x_1 \\ x_2 \end{pmatrix}.$$

Thus the extreme points of g on S are eigenvectors of the symmetric matrix $\left(\begin{smallmatrix} a & b \\ b & c \end{smallmatrix}\right)$. Note that if

$$\begin{pmatrix} x_1 \\ x_2 \end{pmatrix}$$

is an eigenvector of $\left(\begin{smallmatrix} a & b \\ b & c \end{smallmatrix}\right)$ then

$$\begin{aligned} ax_1^2 + 2bx_1x_2 + cx_2^2 &= (x_1 x_2)\begin{pmatrix} a & b \\ b & c \end{pmatrix}\begin{pmatrix} x_1 \\ x_2 \end{pmatrix} \\ &= (x_1 x_2)\lambda \begin{pmatrix} x_1 \\ x_2 \end{pmatrix} = \lambda(x_1^2 + x_2^2) = \lambda \end{aligned}$$

so the eigenvalue λ is just $g(p)$, where $p = (x_1, x_2)$. Since a 2×2 matrix has only two eigenvalues, these eigenvalues are the maximum and minimum values of g on the compact set S.

Exercises

4.1. For what values of c is the level set $f^{-1}(c)$ an n-surface, where

(a) $f(x_1, \ldots, x_{n+1}) = x_1^2 + \cdots + x_{n+1}^2$
(b) $f(x_1, \ldots, x_{n+1}) = x_1^2 + \cdots + x_n^2 - x_{n+1}^2$
(c) $f(x_1, \ldots, x_{n+1}) = x_1 x_2 \cdots x_{n+1} + 1$

4.2. Show that the cylinder $x_1^2 + x_2^2 = 1$ in $\mathbb{R}^3$ can be represented as a level set of each of the following functions:

(a) $f(x_1, x_2, x_3) = x_1^2 + x_2^2$
(b) $f(x_1, x_2, x_3) = -x_1^2 - x_2^2$
(c) $f(x_1, x_2, x_3) = 2x_1^2 + 2x_2^2 + \sin(x_1^2 + x_2^2)$.

4.3. Show that if an n-surface S is represented both as $f^{-1}(c)$ and as $g^{-1}(d)$ where $\nabla f(p) \neq \mathbf{0}$ and $\nabla g(p) \neq \mathbf{0}$ for all $p \in S$, then for each $p \in S$, $\nabla f(p) = \lambda \nabla g(p)$ for some real number $\lambda \neq 0$.

4.4. Sketch the graph of the function $f\colon \mathbb{R}^2 \to \mathbb{R}$ given by $f(x_1, x_2) = x_2^3 - 3x_1^2 x_2$. [*Hint*: First find the level set $f^{-1}(0)$. In what region of the plane is $f > 0$? Where is $f < 0$?] The 2-surface graph (f) is called a *monkey saddle*. (Why?)

4.5. Sketch the cylinders $f^{-1}(0)$ where

(a) $f(x_1, x_2) = x_1$
(b) $f(x_1, x_2, x_3) = x_1 - x_2^2$
(c) $f(x_1, x_2, x_3) = (x_1^2/4) + (x_2^2/9) - 1$

4.6. Sketch the cylinder over the graph of $f(x) = \sin x$.

4.7. Verify that a surface of revolution (Example 5) is a 2-surface.

4.8. Sketch the surface of revolution obtained by rotating C about the x_1-axis, where C is the curve

(a) $x_2 = 1$ (cylinder)
(b) $-x_1^2 + x_2^2 = 1$, $x_2 > 0$ (1-sheeted hyperboloid)
(c) $x_1^2 + (x_2 - 2)^2 = 1$ (torus)

4.9. Show that the set S of all unit vectors at all points of $\mathbb{R}^2$ forms a 3-surface in $\mathbb{R}^4$. [*Hint*: $(x_1, x_2, x_3, x_4) \in S$ if and only if $x_3^2 + x_4^2 = 1$.]

4.10. Let $S = f^{-1}(c)$ be a 2-surface in $\mathbb{R}^3$ which lies in the half space $x_3 > 0$. Find a function $g\colon U \to \mathbb{R}$ (U open in $\mathbb{R}^4$) such that $g^{-1}(c)$ is the 3-surface obtained by rotating the 2-surface S about the (x_1, x_2)-plane.

4.11. Let $a, b, c \in \mathbb{R}$ be such that $ac - b^2 > 0$. Show that the maximum and minimum values of the function $g(x_1, x_2) = x_1^2 + x_2^2$ on the ellipse $ax_1^2 + 2bx_1x_2 + cx_2^2 = 1$ are of the form $1/\lambda_1$ and $1/\lambda_2$ where λ_1 and λ_2 are the eigenvalues of the matrix $\begin{pmatrix} a & b \\ b & c \end{pmatrix}$.

4.12. Show that the maximum and minimum values of the function $g(x_1, \ldots, x_{n+1}) = \sum_{i,j=1}^{n+1} a_{ij} x_i x_j$ on the unit n-sphere $x_1^2 + \cdots + x_{n+1}^2 = 1$, where (a_{ij}) is a symmetric $n \times n$ matrix of real numbers, are eigenvalues of the matrix (a_{ij}).

4.13. Show that if S is an n-surface in $\mathbb{R}^{n+1}$, $g\colon \mathbb{R}^{n+1} \to \mathbb{R}$ is a smooth function, and $p \in S$ is an extreme point of g on S, then the tangent space to the level set of g through p is equal to S_p, the tangent space to S at p, provided $\nabla g(p) \neq \mathbf{0}$ (see Figure 4.4).

4.14. Let S be an n-surface in $\mathbb{R}^{n+1}$ and let $p_0 \in \mathbb{R}^{n+1}$, $p_0 \notin S$. Show that the shortest line segment from p_0 to S (if one exists) is perpendicular to S; i.e., show that if $p \in S$ is such that $\|p_0 - p\|^2 \leq \|p_0 - q\|^2$ for all $q \in S$ then $(p, p_0 - p) \perp S_p$. [*Hint*: Use the Lagrange multiplier theorem.] Show also that the same conclusion holds for the longest line segment from p_0 to S (if one exists).

4.15. $\mathbb{R}^4$ may be viewed as the set of all 2×2 matrices with real entries by identifying the 4-tuple (x_1, x_2, x_3, x_4) with the matrix

$$\begin{pmatrix} x_1 & x_2 \\ x_3 & x_4 \end{pmatrix}.$$

The subset consisting of those matrices with determinant equal to 1 forms a group under matrix multiplication, called the special linear group $\mathbf{SL}(2)$. Show that $\mathbf{SL}(2)$ is a 3-surface in $\mathbb{R}^4$.

4.16. (a) Show that the tangent space $\mathbf{SL}(2)_p$ to $\mathbf{SL}(2)$ (Exercise 4.15) at $p = \left(\begin{smallmatrix} 1 & 0 \\ 0 & 1 \end{smallmatrix}\right)$ can be identified with the set of all 2×2 matrices of trace zero by showing that

$$\mathbf{SL}(2)_p = \left\{\left(p, \begin{pmatrix} a & b \\ c & d \end{pmatrix}\right) : a + d = 0\right\}.$$

[*Hint*: Show first that if

$$\alpha(t) = \begin{pmatrix} x_1(t) & x_2(t) \\ x_3(t) & x_4(t) \end{pmatrix}$$

is a parametrized curve in $\mathbf{SL}(2)$ with $\alpha(t_0) = \left(\begin{smallmatrix} 1 & 0 \\ 0 & 1 \end{smallmatrix}\right)$ then $(dx_1/dt)(t_0) + (dx_4/dt)(t_0) = 0$. Then use a dimension argument.]

(b) What is the tangent space to $\mathbf{SL}(2)$ at $q = \left(\begin{smallmatrix} 2 & 1 \\ 1 & 1 \end{smallmatrix}\right)$?

4.17. (a) Show that the set $\mathbf{SL}(3)$ of all 3×3 real matrices with determinant equal to 1 is an 8-surface in $\mathbb{R}^9$.

(b) What is the tangent space to $\mathbf{SL}(3)$ at

$$p = \begin{pmatrix} 1 & 0 & 0 \\ 0 & 1 & 0 \\ 0 & 0 & 1 \end{pmatrix}?$$

Vector Fields on Surfaces; Orientation

5

A *vector field* $\mathbf{X}$ *on an n-surface* $S \subset \mathbb{R}^{n+1}$ is a function which assigns to each point p in S a vector $\mathbf{X}(p) \in \mathbb{R}^{n+1}_p$ at p. If $\mathbf{X}(p)$ is tangent to S (i.e., $\mathbf{X}(p) \in S_p$) for each $p \in S$, $\mathbf{X}$ is said to be a *tangent vector field* on S. If $\mathbf{X}(p)$ is orthogonal to S (i.e., $\mathbf{X}(p) \in S_p^{\perp}$) for each $p \in S$, $\mathbf{X}$ is said to be a *normal vector field* on S (see Figure 5.1).

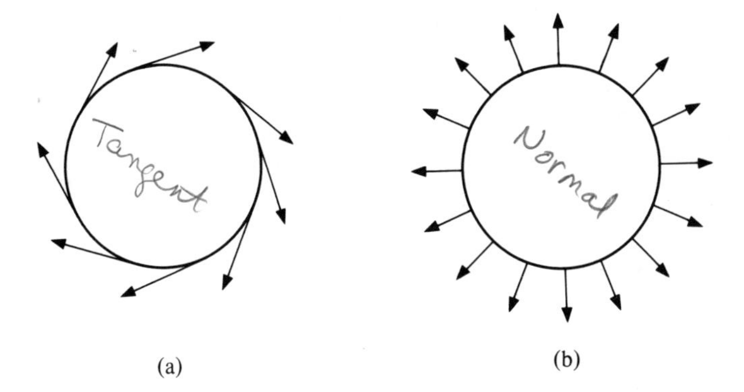

(a) (b)

Figure 5.1 Vector fields on the 1-sphere: (a) a tangent vector field, (b) a normal vector field.

As usual, we shall work almost exclusively with functions and vector fields which are *smooth*. A function $g\colon S \to \mathbb{R}^k$, where S is an n-surface in $\mathbb{R}^{n+1}$, is smooth if it is the restriction to S of a smooth function $\tilde{g}\colon V \to \mathbb{R}^k$ defined on some open set V in $\mathbb{R}^{n+1}$ containing S. Similarly, a vector field $\mathbf{X}$ on S is smooth if it is the restriction to S of a smooth vector field defined on some open set containing S. Thus, $\mathbf{X}$ is smooth if and only if $X\colon S \to \mathbb{R}^{n+1}$ is smooth, where $\mathbf{X}(p) = (p, X(p))$ for all $p \in S$.

The following theorem extends to n-surfaces the theorem of Chapter 2 on the existence and uniqueness of integral curves.

Theorem 1. *Let S be an n-surface in $\mathbb{R}^{n+1}$, let $\mathbf{X}$ be a smooth tangent vector field on S, and let $p \in S$. Then there exists an open interval I containing* 0 *and a parametrized curve $\alpha: I \to S$ such that*

(i) $\alpha(0) = p$
(ii) $\dot{\alpha}(t) = \mathbf{X}(\alpha(t))$ *for all $t \in I$*
(iii) *If $\beta: \tilde{I} \to S$ is any other parametrized curve in S satisfying* (i) *and* (ii), *then $\tilde{I} \subset I$ and $\beta(t) = \alpha(t)$ for all $t \in \tilde{I}$.*

A parametrized curve $\alpha: I \to S$ satisfying condition (ii) is called an *integral curve* of the tangent vector field $\mathbf{X}$. The unique α satisfying conditions (i)–(iii) is the *maximal integral curve* of $\mathbf{X}$ through $p \in S$.

PROOF. Since $\mathbf{X}$ is smooth, there exists an open set V containing S and a smooth vector field $\tilde{\mathbf{X}}$ on V such that $\tilde{\mathbf{X}}(q) = \mathbf{X}(q)$ for all $q \in S$. Let $f: U \to \mathbb{R}$ and $c \in \mathbb{R}$ be such that $S = f^{-1}(c)$ and $\nabla f(q) \neq \mathbf{0}$ for all $q \in S$. Let

$$W = \{q \in U \cap V: \nabla f(q) \neq \mathbf{0}\}.$$

Then W is an open set containing S, and both $\tilde{\mathbf{X}}$ and f are defined on W. Let $\mathbf{Y}$ be the vector field on W, everywhere tangent to the level sets of f, defined by

$$\mathbf{Y}(q) = \tilde{\mathbf{X}}(q) - (\tilde{\mathbf{X}}(q) \cdot \nabla f(q)/\|\nabla f(q)\|^2)\nabla f(q).$$

Note that $\mathbf{Y}(q) = \mathbf{X}(q)$ for all $q \in S$. Let $\alpha: I \to W$ be the maximal integral curve of $\mathbf{Y}$ through p. Then α actually maps I into S because

$$(f \circ \alpha)'(t) = \nabla f(\alpha(t)) \cdot \dot{\alpha}(t) = \nabla f(\alpha(t)) \cdot \mathbf{Y}(\alpha(t)) = 0,$$

and $f \circ \alpha(0) = f(p) = c$, so $f \circ \alpha(t) = c$ for all $t \in I$. Conditions (i) and (ii) are clearly satisfied, and condition (iii) is satisfied because any $\beta: \tilde{I} \to S$ satisfying (i) and (ii) is also an integral curve of the vector field $\mathbf{Y}$ on W so the theorem of Chapter 2 applies. □

Corollary. *Let $S = f^{-1}(c)$ be an n-surface in $\mathbb{R}^{n+1}$, where $f: U \to \mathbb{R}$ is such that $\nabla f(q) \neq \mathbf{0}$ for all $q \in S$, and let $\mathbf{X}$ be a smooth vector field on U whose restriction to S is a tangent vector field on S. If $\alpha: I \to U$ is any integral curve of $\mathbf{X}$ such that $\alpha(t_0) \in S$ for some $t_0 \in I$, then $\alpha(t) \in S$ for all $t \in I$.*

PROOF. Suppose $\alpha(t) \notin S$ for some $t \in I$, $t > t_0$. Let t_1 denote the greatest lower bound of the set

$$\{t \in I: t > t_0 \text{ and } \alpha(t) \notin S\}$$

Then $f(\alpha(t)) = c$ for $t_0 \leq t < t_1$ so, by continuity, $f(\alpha(t_1)) = c$; that is, $\alpha(t_1) \in S$. Let $\beta: \tilde{I} \to S$ be an integral curve through $\alpha(t_1)$ of the restriction of

$\mathbf{X}$ to S. Then β is also an integral curve of $\mathbf{X}$, sending 0 to $\alpha(t_1)$, as is the curve $\tilde{\alpha}$ defined by $\tilde{\alpha}(t) = \alpha(t + t_1)$. By uniqueness of integral curves, $\alpha(t) = \tilde{\alpha}(t - t_1) = \beta(t - t_1) \in S$ for all t such that $t - t_1$ is in the common domain of $\tilde{\alpha}$ and β. But this contradicts the fact that $\alpha(t) \notin S$ for values of t arbitrarily close to t_1. Hence $\alpha(t) \in S$ for all $t \in I$ with $t > t_0$. The proof for $t < t_0$ is similar. □

A subset S of $\mathbb{R}^{n+1}$ is said to be *connected* if for each pair p, q of points in S there is a continuous map α: $[a, b] \to S$, from some closed interval $[a, b]$ into S, such that $\alpha(a) = p$ and $\alpha(b) = q$. Thus S is connected if each pair of points in S can be joined by a continuous, but not necessarily smooth, curve which lies completely in S. Note, for example, that the n-sphere (Figure 5.2) is connected if and only if $n \geq 1$ (Exercise 5.1).

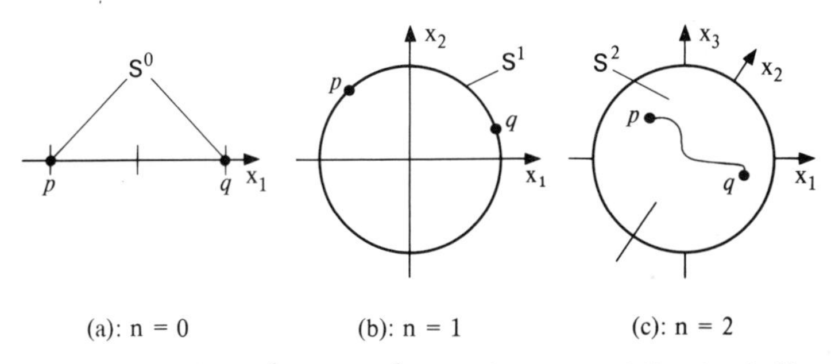

Figure 5.2 The n-sphere $x_1^2 + \cdots + x_{n+1}^2 = 1$ is connected if and only if $n \geq 1$.

In this book, we shall deal almost exclusively with connected n-surfaces. As we shall see in Chapter 15 (see Exercise 15.13), given any n-surface S and any $p \in S$, the subset of S consisting of all points of S which can be joined to p by a continuous curve in S is itself an n-surface, and it is connected. Hence we can study S by studying separately each of these "connected components" of S.

Theorem 2. *Let $S \subset \mathbb{R}^{n+1}$ be a connected n-surface in $\mathbb{R}^{n+1}$. Then there exist on S exactly two smooth unit normal vector fields $\mathbf{N}_1$ and $\mathbf{N}_2$, and $\mathbf{N}_2(p) = -\mathbf{N}_1(p)$ for all $p \in S$.*

PROOF. Let f: $U \to \mathbb{R}$ and $c \in \mathbb{R}$ be such that $S = f^{-1}(c)$ and $\nabla f(p) \neq \mathbf{0}$ for all $p \in S$. Then the vector field $\mathbf{N}_1$ on S defined by

$$\mathbf{N}_1(p) = \frac{\nabla f(p)}{\|\nabla f(p)\|}, \qquad p \in S$$

clearly has the required properties, as does the vector field $\mathbf{N}_2$ defined by $\mathbf{N}_2(p) = -\mathbf{N}_1(p)$ for all $p \in S$.

To show that these are the only two such vector fields, suppose $\mathbf{N}_3$ were another. Then, for each $p \in S$, $\mathbf{N}_3(p)$ must be a multiple of $\mathbf{N}_1(p)$ since both lie in the 1-dimensional subspace $S_p^\perp \subset \mathbb{R}_p^{n+1}$. Thus

$$\mathbf{N}_3(p) = g(p)\mathbf{N}_1(p)$$

where $g\colon S \to \mathbb{R}$ is a smooth function on S ($g(p) = \mathbf{N}_3(p) \cdot \mathbf{N}_1(p)$ for $p \in S$). Since $\mathbf{N}_1(p)$ and $\mathbf{N}_3(p)$ are both unit vectors, $g(p) = \pm 1$ for each $p \in S$. Finally, since g is smooth and S is connected, g must be constant on S (see Exercise 5.2). Thus either $\mathbf{N}_3 = \mathbf{N}_1$ or $\mathbf{N}_3 = \mathbf{N}_2$. □

A smooth unit normal vector field on an n-surface S in $\mathbb{R}^{n+1}$ is called an *orientation* on S. According to the theorem just proved, each connected n-surface in $\mathbb{R}^{n+1}$ has exactly two orientations. An n-surface together with a choice of orientation is called an *oriented* n-surface.

Remark. There are subsets of $\mathbb{R}^{n+1}$ which most people would agree should be called n-surfaces but on which there exist no orientations. An example is the Möbius band B, the surface in $\mathbb{R}^3$ obtained by taking a rectangular strip of paper, twisting one end through 180°, and taping the ends together (see Figure 5.3). That there is no smooth unit normal vector field on B can be

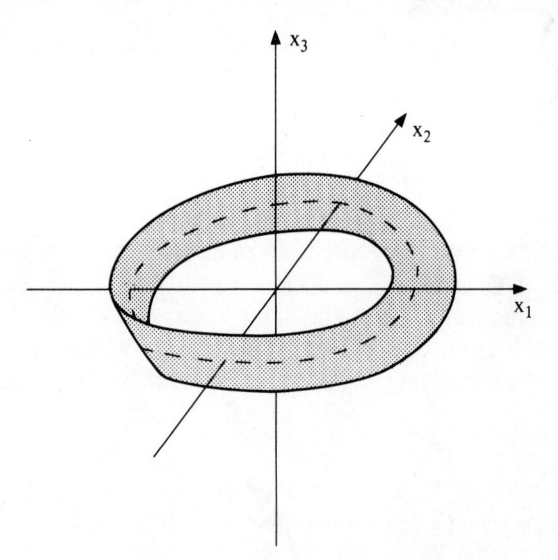

Figure 5.3 The Möbius band.

seen by picking a unit normal vector at some point on the central circle and trying to extend it continuously to a unit normal vector field along this circle. After going around the circle once, the normal vector is pointing in the opposite direction! Since there is no smooth unit normal vector field on B, B cannot be expressed as a level set $f^{-1}(c)$ of some smooth function $f\colon U \to \mathbb{R}$ with $\nabla f(p) \neq \mathbf{0}$ for all $p \in S$, and hence B is not a 2-surface according to our definition. B is an example of an "unorientable 2-surface". Until Chapter 14, we shall consider only "orientable" n-surfaces in $\mathbb{R}^{n+1}$.

A unit vector in $\mathbb{R}_p^{n+1}$ ($p \in \mathbb{R}^{n+1}$) is called a *direction* at p. Thus an orientation on an n-surface S in $\mathbb{R}^{n+1}$ is, by definition, a smooth choice of normal direction at each point of S.

On a plane curve, an orientation can be used to define a tangent direction at each point of the curve. The *positive tangent direction* at the point p of the oriented plane curve C is the direction obtained by rotating the orientation normal direction at p through an angle of $-\pi/2$, where the direction of positive rotation is counterclockwise (see Figure 5.4).

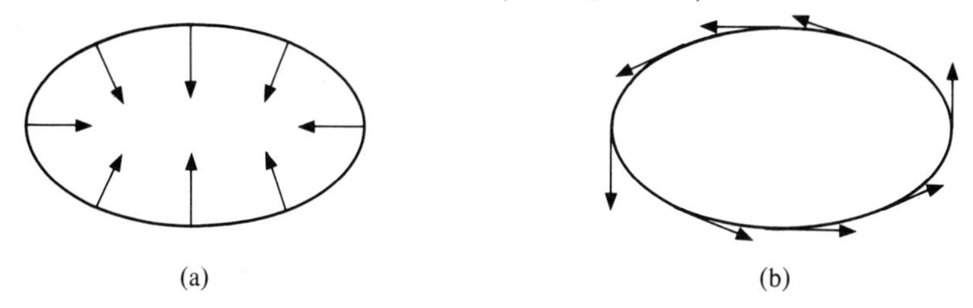

Figure 5.4 Orientation on a plane curve: (a) the chosen normal direction at each point determines (b) a choice of tangent direction at each point.

On a 2-surface in $\mathbb{R}^3$, an orientation can be used to define a direction of rotation in the tangent space at each point of the surface. Given $\theta \in \mathbb{R}$, the *positive θ-rotation* at the point p of the oriented 2-surface S is the linear transformation R_θ: $S_p \to S_p$ defined by $R_\theta(\mathbf{v}) = (\cos\theta)\mathbf{v} + (\sin\theta)\mathbf{N}(p) \times \mathbf{v}$ where $\mathbf{N}(p)$ is the orientation normal direction at p. R_θ is usually described as the "right-handed rotation about $\mathbf{N}(p)$ through the angle θ" (see Figure 5.5).

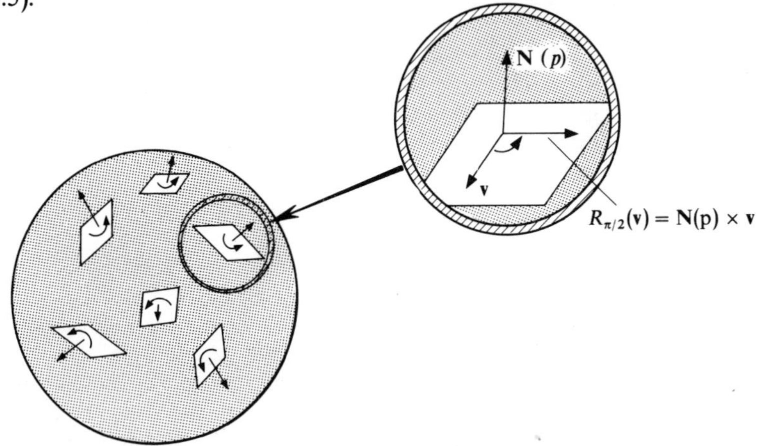

Figure 5.5 Orientation on the 2-sphere: at each point the chosen normal direction determines a sense of positive rotation in the tangent space. The satellite figure is an enlarged view of one tangent space.

On a 3-surface in $\mathbb{R}^4$, an orientation can be used to define a sense of "handedness" in the tangent space at each point of the surface. Given an oriented 3-surface S and a point $p \in S$, an ordered orthonormal basis $\{\mathbf{e}_1, \mathbf{e}_2, \mathbf{e}_3\}$ for the tangent space S_p to S at p is said to be *right-handed* if the determinant

$$\det\begin{pmatrix} e_1 \\ e_2 \\ e_3 \\ N(p) \end{pmatrix}$$

is positive, where $\mathbf{N}(p) = (p, N(p))$ is the orientation normal direction at p and $\mathbf{e}_i = (p, e_i)$ for $i \in \{1, 2, 3\}$; the basis is *left-handed* if the determinant is negative.

On an n-surface in $\mathbb{R}^{n+1}$ (n arbitrary), an orientation can be used to partition the collection of all ordered bases for each tangent space into two subsets, those consistent with the orientation and those inconsistent with the orientation. An ordered basis $\{\mathbf{v}_1, \ldots, \mathbf{v}_n\}$ (not necessarily orthonormal) for the tangent space S_p at the point p of the oriented n-surface S is said to be *consistent* with the orientation $\mathbf{N}$ on S if the determinant

$$\det\begin{pmatrix} v_1 \\ \cdots \\ v_n \\ N(p) \end{pmatrix}$$

is positive; the basis is *inconsistent* with $\mathbf{N}$ if the determinant is negative. Here, as usual, $\mathbf{v}_i = (p, v_i)$ and $\mathbf{N}(p) = (p, N(p))$.

Exercises

5.1. Show that the unit n-sphere $x_1^2 + \cdots + x_{n+1}^2 = 1$ is connected if $n > 1$.

5.2. Show that if S is a connected n-surface in $\mathbb{R}^{n+1}$ and $g: S \to \mathbb{R}$ is smooth and takes on only the values $+1$ and -1, then g is constant. [*Hint*: Let $p \in S$. For $q \in S$, let $\alpha: [a, b] \to S$ be continuous and such that $\alpha(a) = p$, $\alpha(b) = q$. Use the intermediate value theorem on the composition $g \circ \alpha$.]

5.3. Show by example that if S is not connected then Theorem 2 of this section fails.

5.4. Show that the two orientations on the n-sphere $x_1^2 + \cdots + x_{n+1}^2 = r^2$ of radius $r > 0$ are given by $\mathbf{N}_1(p) = (p, p/r)$ and $\mathbf{N}_2(p) = (p, -p/r)$.

5.5. $\mathbb{R}^n$ may be viewed as the n-surface $x_{n+1} = 0$ in $\mathbb{R}^{n+1}$. Let $\mathbf{N}$ be the orientation on $\mathbb{R}^n \subset \mathbb{R}^{n+1}$ defined by $\mathbf{N}(p) = (p, 0, \ldots, 0, 1)$ for each $p \in \mathbb{R}^n$. (This $\mathbf{N}$ is called the *natural orientation* on $\mathbb{R}^n$.) Show that, given this orientation for each n,

(a) the positive tangent direction at $p \in \mathbb{R}^1$ is the direction $(p, 1, 0)$,
(b) the positive θ-rotation in $\mathbb{R}_p^2$, $p \in \mathbb{R}^2$, is counterclockwise rotation through the angle θ, and

(c) the ordered orthonormal basis $\{(p, 1, 0, 0, 0), (p, 0, 1, 0, 0), (p, 0, 0, 1, 0)\}$ for $\mathbb{R}^3_p$, $p \in \mathbb{R}^3$, is right-handed.

5.6. Let C be an oriented plane curve and let $\mathbf{v}$ be a nonzero vector tangent to C at $p \in C$. Show that the basis $\{\mathbf{v}\}$ for C_p is consistent with the orientation on C if and only if the positive tangent direction at p is $\mathbf{v}/\|\mathbf{v}\|$. [*Hint*: Let θ denote the angle measured counterclockwise from $(p, 1, 0)$ to the orientation direction $\mathbf{N}(p)$, so that $\mathbf{N}(p) = (p, \cos\theta, \sin\theta)$. Express both $\mathbf{v}$ and the positive tangent direction at p in terms of θ.]

5.7. Recall that the *cross product* $\mathbf{v} \times \mathbf{w}$ of two vectors $\mathbf{v} = (p, v_1, v_2, v_3)$ and $\mathbf{w} = (p, w_1, w_2, w_3)$ in $\mathbb{R}^3_p$ $(p \in \mathbb{R}^3)$ is defined by

$$\mathbf{v} \times \mathbf{w} = (p, v_2 w_3 - v_3 w_2, v_3 w_1 - v_1 w_3, v_1 w_2 - v_2 w_1).$$

(a) Show that $\mathbf{v} \times \mathbf{w}$ is orthogonal to both $\mathbf{v}$ and $\mathbf{w}$ and that $\|\mathbf{v} \times \mathbf{w}\| = \|\mathbf{v}\|\,\|\mathbf{w}\| \sin\theta$, where $\theta = \cos^{-1}(\mathbf{v} \cdot \mathbf{w}/\|\mathbf{v}\|\,\|\mathbf{w}\|)$ is the angle between $\mathbf{v}$ and $\mathbf{w}$.
(b) Show that if $\mathbf{u} = (p, u_1, u_2, u_3)$ then

$$\mathbf{u} \cdot (\mathbf{v} \times \mathbf{w}) = \mathbf{v} \cdot (\mathbf{w} \times \mathbf{u}) = \mathbf{w} \cdot (\mathbf{u} \times \mathbf{v}) = \begin{vmatrix} u_1 & u_2 & u_3 \\ v_1 & v_2 & v_3 \\ w_1 & w_2 & w_3 \end{vmatrix}.$$

(c) Show that the only vector $\mathbf{x}$ in $\mathbb{R}^3_p$ such that $\mathbf{u} \cdot \mathbf{x}$ is equal to the determinant above (part b) for all $\mathbf{u} \in \mathbb{R}^3_p$ is $\mathbf{x} = \mathbf{v} \times \mathbf{w}$.

5.8. Let S be an oriented 2-surface in $\mathbb{R}^3$ and let $\{\mathbf{v}, \mathbf{w}\}$ be an ordered basis for the tangent space S_p to S at $p \in S$. Show that the consistency of $\{\mathbf{v}, \mathbf{w}\}$ with the orientation $\mathbf{N}$ of S is equivalent with each of the following conditions:

(a) $\mathbf{N}(p) \cdot (\mathbf{v} \times \mathbf{w}) > 0$
(b) $\mathbf{w}/\|\mathbf{w}\| = R_\theta(\mathbf{v}/\|\mathbf{v}\|)$ for some θ with $0 < \theta < \pi$, where R_θ is the positive θ-rotation in S_p.

5.9. Let S be an oriented 3-surface in $\mathbb{R}^4$ and let $p \in S$.

(a) Show that, given vectors $\mathbf{v} = (p, v)$ and $\mathbf{w} = (p, w)$ in S_p, there is a unique vector $\mathbf{v} \times \mathbf{w} \in S_p$ such that

$$\mathbf{u} \cdot (\mathbf{v} \times \mathbf{w}) = \det\begin{pmatrix} u \\ v \\ w \\ N(p) \end{pmatrix}$$

for all $\mathbf{u} = (p, u) \in S_p$, where $\mathbf{N}(p) = (p, N(p))$ is the orientation direction at p. This vector $\mathbf{v} \times \mathbf{w}$ is the *cross product* of $\mathbf{v}$ and $\mathbf{w}$.
(b) Check that the cross product in S_p has the following properties:

(i) $(\mathbf{v} + \mathbf{w}) \times \mathbf{x} = \mathbf{v} \times \mathbf{x} + \mathbf{w} \times \mathbf{x}$
(ii) $\mathbf{v} \times (\mathbf{w} + \mathbf{x}) = \mathbf{v} \times \mathbf{w} + \mathbf{v} \times \mathbf{x}$
(iii) $(c\mathbf{v}) \times \mathbf{w} = c(\mathbf{v} \times \mathbf{w})$
(iv) $\mathbf{v} \times (c\mathbf{w}) = c(\mathbf{v} \times \mathbf{w})$
(v) $\mathbf{v} \times \mathbf{w} = -\mathbf{w} \times \mathbf{v}$
(vi) $\mathbf{u} \cdot (\mathbf{v} \times \mathbf{w}) = \mathbf{v} \cdot (\mathbf{w} \times \mathbf{u}) = \mathbf{w} \cdot (\mathbf{u} \times \mathbf{v})$
(vii) $\mathbf{v} \times \mathbf{w}$ is orthogonal to both $\mathbf{v}$ and $\mathbf{w}$

(viii) $\mathbf{u} \cdot (\mathbf{v} \times \mathbf{w}) = \mathbf{0}$ if and only if $\{\mathbf{u}, \mathbf{v}, \mathbf{w}\}$ is linearly dependent
(ix) An ordered orthonormal basis $\{\mathbf{e}_1, \mathbf{e}_2, \mathbf{e}_3\}$ for S_p is right-handed if and only if $\mathbf{e}_3 \cdot (\mathbf{e}_1 \times \mathbf{e}_2) > 0$.

5.10. Let S be an oriented n-surface in $\mathbb{R}^{n+1}$, with orientation $\mathbf{N}$, and let $p \in S$.

(a) Show that an ordered basis for S_p is inconsistent with $\mathbf{N}$ if and only if it is consistent with $-\mathbf{N}$.

(b) Suppose $\{\mathbf{v}_1, \ldots, \mathbf{v}_n\}$ is an ordered basis for S_p which is consistent with $\mathbf{N}$ and suppose $\{\mathbf{w}_1, \ldots, \mathbf{w}_n\}$ is another ordered basis for S_p. Show that $\{\mathbf{w}, \ldots, \mathbf{w}_n\}$ is also consistent with $\mathbf{N}$ if and only if the matrix (a_{ij}), where $\mathbf{w}_i = \sum_j a_{ij} \mathbf{v}_j$, has positive determinant. [*Hint*: Complete each basis to a basis for $\mathbb{R}^{n+1}_p$ by adjoining $\mathbf{N}(p)$. What is the relationship between (a_{ij}) and the two matrices which determine the consistency of the given bases with $\mathbf{N}$?]

The Gauss Map 6

An oriented n-surface in $\mathbb{R}^{n+1}$ is more than just an n-surface S, it is an n-surface S together with a smooth unit normal vector field $\mathbf{N}$ on S. The function $N: S \to \mathbb{R}^{n+1}$ associated with the vector field $\mathbf{N}$ by $\mathbf{N}(p) = (p, N(p))$, $p \in S$, actually maps S into the unit n-sphere $\mathbf{S}^n \subset \mathbb{R}^{n+1}$, since $\|N(p)\| = 1$ for all $p \in S$. Thus, associated to each oriented n-surface S is a smooth map $N: S \to \mathbf{S}^n$, called the *Gauss map*. N may be thought of as the map which assigns to each point $p \in S$ the point in $\mathbb{R}^{n+1}$ obtained by "translating" the unit normal vector $N(p)$ to the origin (see Figure 6.1).

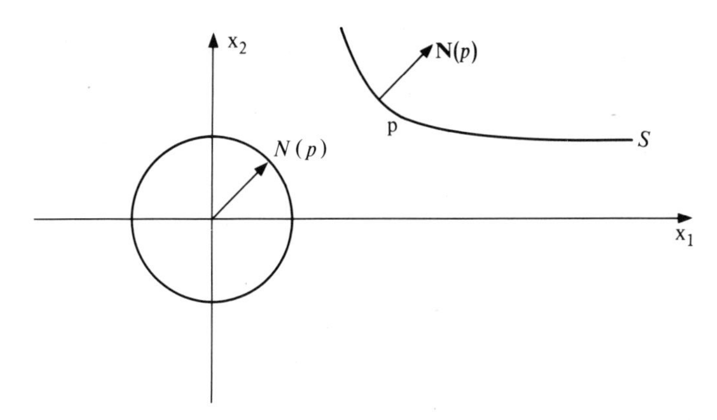

Figure 6.1 The Gauss map of a 1-surface in $\mathbb{R}^2$.

The image of the Gauss map,

$$N(S) = \{q \in \mathbf{S}^n : q = N(p) \text{ for some } p \in S\}$$

is called the *spherical image* of the oriented n-surface S (see Figure 6.2).

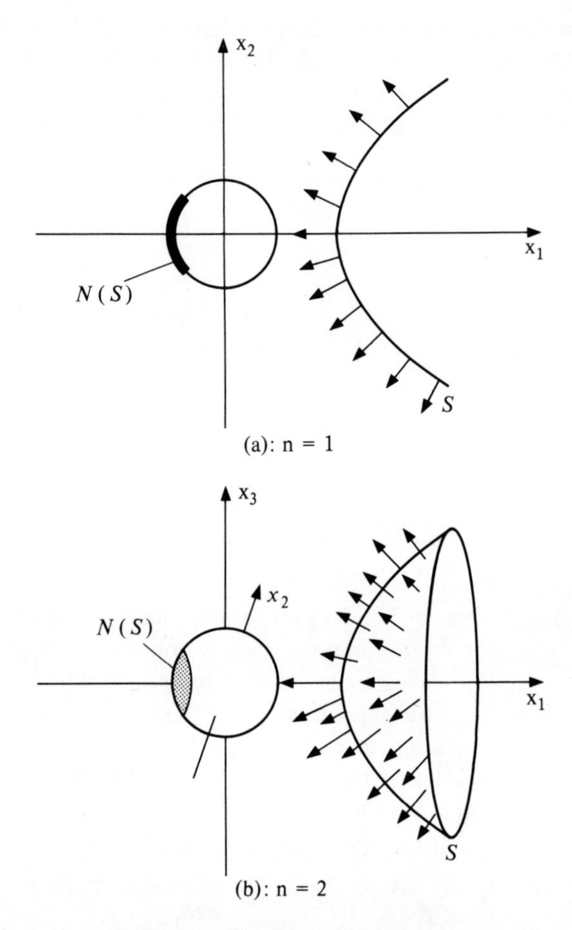

Figure 6.2 The spherical image of one sheet of a 2-sheeted hyperboloid $x_1^2 - x_2^2 - \cdots - x_{n+1}^2 = 4$, $x_1 > 0$, oriented by $\mathbf{N} = -\nabla f/\|\nabla f\|$ where $f(x_1, \ldots, x_{n+1}) = x_1^2 - x_2^2 - \cdots - x_{n+1}^2$.

The spherical image of an oriented n-surface S records the set of directions which occur as normal directions to S. Hence its size is a measure of how much the surface curves around in $\mathbb{R}^{n+1}$. For an n-plane, which doesn't curve around at all, the spherical image is a single point. If an n-surface is compact (closed and bounded) then it must curve all the way around: the spherical image will be all of $\mathbf{S}^n$. Although we do not yet have enough machinery to prove this theorem in full generality, we can already prove an important special case, namely the case in which S is a level set of a smooth function defined on all of $\mathbb{R}^{n+1}$.

Theorem. *Let S be a compact connected oriented n-surface in $\mathbb{R}^{n+1}$ exhibited as a level set $f^{-1}(c)$ of a smooth function $f\colon \mathbb{R}^{n+1} \to \mathbb{R}$ with $\nabla f(p) \neq \mathbf{0}$ for all $p \in S$. Then the Gauss map maps S onto the unit sphere $\mathbf{S}^n$.*

Proof. The idea of the proof is as follows. Given $v \in S^n$, consider the n-plane $v^\perp$. By moving this n-plane far enough in the v-direction, it will have null intersection with S. Bringing it back in until it just touches S at some point p, it will be tangent there (see Figure 6.3). Hence at this point, $N(p) = \pm v$. If

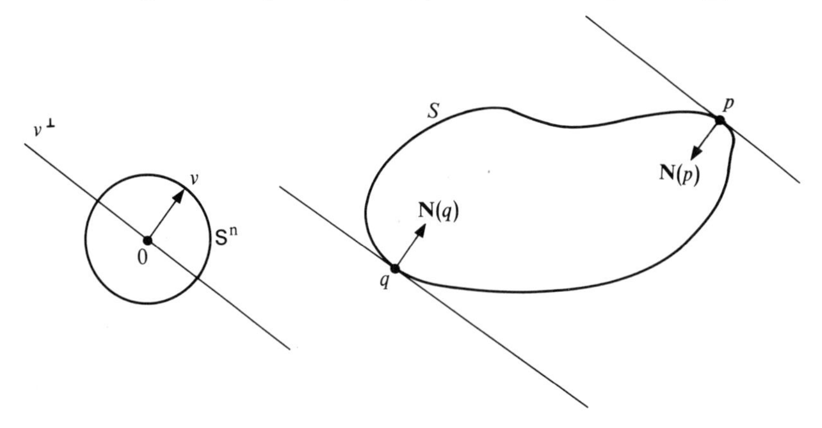

Figure 6.3 The Gauss map of a compact oriented n-surface is onto.

$N(p) = -v$, then $N(q) = v$ where q is obtained similarly, by moving the n-plane in from the opposite direction.

More precisely, consider the function $g: \mathbb{R}^{n+1} \to \mathbb{R}$ defined by $g(p) = p \cdot v$; i.e., $g(x_1, \ldots, x_{n+1}) = a_1 x_1 + \cdots + a_{n+1} x_{n+1}$ where

$$v = (a_1, \ldots, a_{n+1}).$$

The level sets of g are the n-planes parallel to $v^\perp$. Since S is compact, the restriction to S of the function g attains its maximum and its minimum, say at p and q respectively. By the Lagrange multiplier theorem (Chapter 4),

$$(p, v) = \nabla g(p) = \lambda \nabla f(p) = \lambda \|\nabla f(p)\| \mathbf{N}(p)$$

for some $\lambda \in \mathbb{R}$. Hence v and $N(p)$ are multiples of one another. Since both have unit length, it follows that $N(p) = \pm v$. Similarly, $N(q) = \pm v$.

It remains only to check that $N(q) \neq N(p)$. For this, it suffices to construct a continuous function $\alpha: [a, b] \to \mathbb{R}^{n+1}$, differentiable at a and b, such that

(i) $\alpha(a) = p, \quad \alpha(b) = q, \quad \dot{\alpha}(a) = (p, v), \quad \dot{\alpha}(b) = (q, v),$ and
(ii) $\alpha(t) \notin S$ for $a < t < b$.

For then, by (i), if $\mathbf{N} = \nabla f / \|\nabla f\|$,

$$\begin{aligned}(f \circ \alpha)'(a) &= \nabla f(\alpha(a)) \cdot \dot{\alpha}(a) \\ &= \|\nabla f(p)\| \mathbf{N}(p) \cdot (p, v) = \|\nabla f(p)\| N(p) \cdot v\end{aligned}$$

and similarly

$$(f \circ \alpha)'(b) = \|\nabla f(q)\| N(q) \cdot v$$

so the derivative $(f \circ \alpha)'$ has the same sign at both endpoints. If $\mathbf{N} = -\nabla f/\|\nabla f\|$, the same conclusion holds. Thus, if $N(p)$ were equal to $N(q)$ $(= \pm v)$ then $f \circ \alpha$ would either be increasing at both end points or be decreasing at both end points. Since $f \circ \alpha(a) = f \circ \alpha(b) = c$, this would imply that there exist t_1 and t_2 between a and b with $f \circ \alpha(t_1) > c$ and $f \circ \alpha(t_2) < c$. But then, by the intermediate value theorem, there would exist t_3 between t_1 and t_2 such that $f \circ \alpha(t_3) = c$, contradicting (ii).

To construct α, enclose S in the interior of a large sphere S_1. This is possible since S is compact. Set (see Figure 6.4) $\alpha_1(t) = p + tv$ $(0 \le t \le a_1)$,

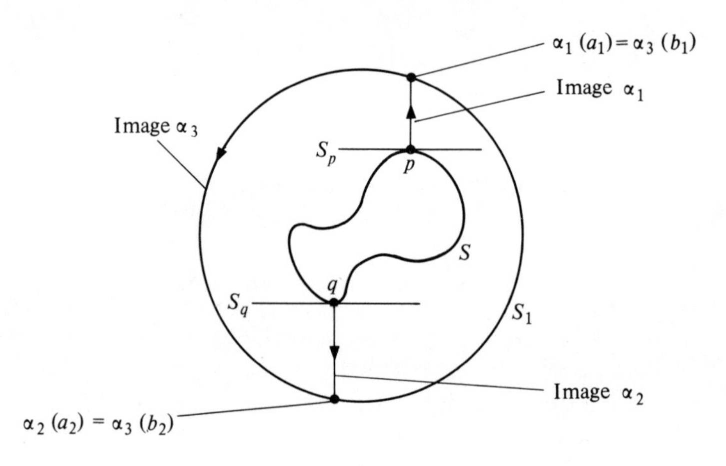

Figure 6.4 $N(q) = -N(p)$.

where a_1 is such that $\alpha_1(a_1) \in S_1$, and set $\alpha_2(t) = q - tv$ $(0 \le t \le a_2)$, where a_2 is such that $\alpha_2(a_2) \in S_1$. Let $\alpha_3 \colon [b_1, b_2] \to S_1$ be such that $\alpha_3(b_1) = \alpha_1(a_1)$ and $\alpha_3(b_2) = \alpha_2(a_2)$. Such an α_3 exists because the n-sphere S_1 is connected for $n \ge 1$. Then the function α defined by

$$\alpha(t) = \begin{cases} \alpha_1(t) & (0 \le t \le a_1) \\ \alpha_3(t + b_1 - a_1) & (a_1 \le t \le a_1 + b_2 - b_1) \\ \alpha_2(a_1 + a_2 + b_2 - b_1 - t) & (a_1 + b_2 - b_1 \le t \\ & \le a_1 + a_2 + b_2 - b_1) \end{cases}$$

has the required properties, where $a = 0$, $b = a_1 + a_2 + b_2 - b_1$. Continuity and condition (i) are easy to check, and condition (ii) is satisfied because

(1) $\alpha_1(t) \notin S$ for $t > 0$ since $(g \circ \alpha_1)'(t) = \nabla g(\alpha_1(t)) \cdot \dot{\alpha}_1(t) = \mathbf{v} \cdot \mathbf{v} > 0$ so g is increasing along α_1, and the maximum value of g on S is attained at $\alpha_1(0) = p$;

(2) $\alpha_2(t) \notin S$ for $t > 0$ since $(g \circ \alpha_2)'(t) = \mathbf{v} \cdot (-\mathbf{v}) < 0$ so g is decreasing along α_2, and the minimum value of g on S is attained at $\alpha_2(0) = q$; and

(3) $\alpha_3(t) \notin S$ for $t \in [b_1, b_2]$ since $\alpha_3(t) \in S_1$ and $S_1 \cap S = \varnothing$. □

Remark. Some insight into the general case of this theorem can be gained from the following intuitive argument. Suppose S is compact and connected. Then S divides $\mathbb{R}^{n+1}$ into two parts, a part inside S and a part outside S. Let $S = f^{-1}(c)$ where $f: U \to \mathbb{R}$. Then, for small enough $\varepsilon > 0$, the level set $f^{-1}(c + \varepsilon)$ will be an n-surface S_+ gotten by pushing each point of S a short distance out from S along ∇f (see Figure 6.5).

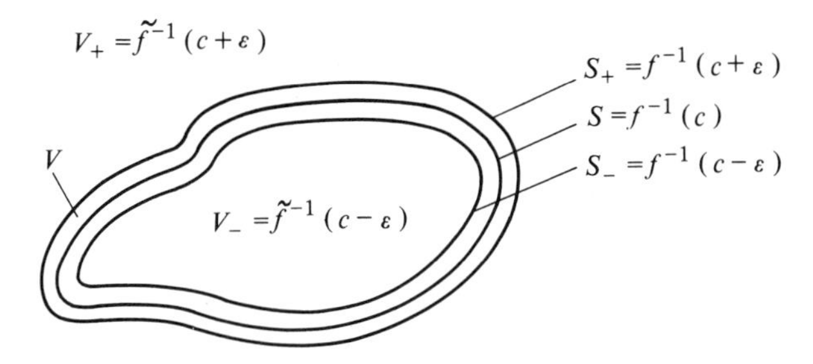

Figure 6.5 Given a compact connected n-surface $S = f^{-1}(c)$, the nearby level sets $f^{-1}(c - \varepsilon)$ and $f^{-1}(c + \varepsilon)$ are slightly inside and slightly outside S.

(Possibly, $f^{-1}(c + \varepsilon)$ might also contain some points far away from S but we can ignore such points in the present argument.) Similarly, for small enough $\varepsilon > 0$, the level set $f^{-1}(c - \varepsilon)$ will be an n-surface S_- on the other side of S, gotten by pushing each point of S a short distance out along $-\nabla f$. Denoting by V the set of points between S_- and S_+, by V_+ the set of points in $\mathbb{R}^{n+1} - V$ which lie on the same side of S as S_+, and by V_- the set of points in $\mathbb{R}^{n+1} - V$ which lie on the other side of S, we can define $\tilde{f}: \mathbb{R}^{n+1} \to \mathbb{R}$ by

$$\tilde{f}(p) = \begin{cases} f(p) & \text{for } p \in V \\ c + \varepsilon & \text{for } p \in V_+ \\ c - \varepsilon & \text{for } p \in V_-. \end{cases}$$

Then $\tilde{f}$ is continuous on $\mathbb{R}^{n+1}$, smooth on the open set V about S, and $\tilde{f}^{-1}(c) = S$. The above proof can now be applied, with f replaced everywhere by $\tilde{f}$, to show that the Gauss map is onto.

Exercises

In Exercises 6.1–6.5, describe the spherical image, when $n = 1$ and when $n = 2$, of the given n-surface, oriented by $\nabla f / \|\nabla f\|$ where f is the function defined by the left hand side of each equation.

6.1. The cylinder $x_2^2 + \cdots + x_{n+1}^2 = 1$.

6.2. The cone $-x_1^2 + x_2^2 + \cdots + x_{n+1}^2 = 0$, $x_1 > 0$.

6.3. The sphere $x_1^2 + x_2^2 + \cdots + x_{n+1}^2 = r^2$ $(r > 0)$.

6.4. The paraboloid $-x_1 + x_2^2 + \cdots + x_{n+1}^2 = 0$.

6.5. The 1-sheeted hyperboloid

$$-(x_1^2/a^2) + x_2^2 + \cdots + x_{n+1}^2 = 1 \ (a > 0).$$

What happens to the spherical image when $a \to \infty$? When $a \to 0$?

6.6. Show that the spherical image of an n-surface with orientation $\mathbf{N}$ is the reflection through the origin of the spherical image of the same n-surface with orientation $-\mathbf{N}$.

6.7. Let $a = (a_1, \ldots, a_{n+1}) \in \mathbb{R}^{n+1}$, $a \neq 0$. Show that the spherical image of an n-surface S is contained in the n-plane $a_1 x_1 + \cdots + a_{n+1} x_{n+1} = 0$ if and only if for every $p \in S$ there is an open interval I about 0 such that $p + ta \in S$ for all $t \in I$. [*Hint*: For the "only if" part, apply the corollary to Theorem 1, Chapter 5, to the constant vector field $\mathbf{X}(q) = (q, a)$.]

6.8. Show that if the spherical image of a connected n-surface S is a single point then S is part or all of an n-plane. [*Hint*: First show, by applying the corollary to Theorem 1, Chapter 5, to the constant vector fields $\mathbf{W}(q) = (q, w)$, where $w \perp v$, $\{v\} = N(S)$, that if B is an open ball which is contained in U and $p \in S \cap B$ then $H \cap B \subset S$ where H is the n-plane $\{x \in \mathbb{R}^{n+1}: x \cdot v = p \cdot v\}$. Then show that if $\alpha: [a, b] \to S$ is continuous and $\alpha(t) \in B$ for $t_1 \leq t \leq t_2$ then $\alpha(t_1) \cdot v = \alpha(t_2) \cdot v$ by showing that if, e.g., $\alpha(t_1) \cdot v < \alpha(t_2) \cdot v$ then S contains the open set $\{x \in B: \alpha(t_1) \cdot v < x \cdot v < \alpha(t_2) \cdot v\}$, which is impossible (why?).]

6.9. Let $S = f^{-1}(c)$, where $f: \mathbb{R}^{n+1} \to R$ is a smooth function such that $\nabla f(p) \neq \mathbf{0}$ for all $p \in S$. Suppose $\alpha: \mathbb{R} \to \mathbb{R}^{n+1}$ is a parametrized curve which is nowhere tangent to S (i.e., $\nabla f(\alpha(t)) \cdot \dot{\alpha}(t) \neq 0$ for all t with $\alpha(t) \in S$; see Figure 6.6).

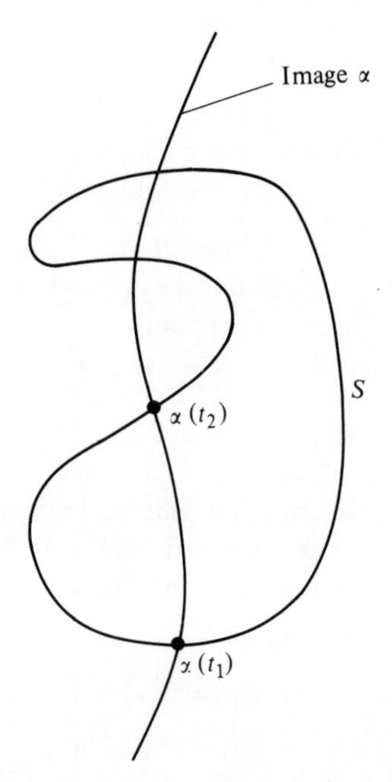

Figure 6.6 The curve α must cross the compact n-surface S an even number of times.

(a) Show that at each pair of consecutive crossings of S by α, the direction of the orientation $\nabla f/\|\nabla f\|$ on S reverses relative to the direction of α [i.e., show that if $\alpha(t_1) \in S$ and $\alpha(t_2) \in S$, where $t_1 < t_2$, and $\alpha(t) \notin S$ for $t_1 < t < t_2$, then $\nabla f(\alpha(t_1)) \cdot \dot{\alpha}(t_1) > 0$ if and only if $\nabla f(\alpha(t_2)) \cdot \dot{\alpha}(t_2) < 0$.]

(b) Show that if S is compact and α goes to ∞ in both directions [i.e., $\lim_{t \to -\infty} \|\alpha(t)\| = \lim_{t \to +\infty} \|\alpha(t)\| = \infty$] then α crosses S an even number of times.

6.10. Let S be a compact n-surface in $\mathbb{R}^{n+1}$. A point $p \in \mathbb{R}^{n+1} - S$ is *outside* S if there exists a continuous map $\alpha\colon [0, \infty) \to \mathbb{R}^{n+1} - S$ such that $\alpha(0) = p$ and $\lim_{t \to \infty} \|\alpha(t)\| = \infty$. Let $\mathcal{O}(S)$ denote the set of all points outside S.

(a) Show that if $\beta\colon [a, b] \to \mathbb{R}^{n+1} - S$ is continuous and $\beta(a) \in \mathcal{O}(S)$ then $\beta(t) \in \mathcal{O}(S)$ for all $t \in [a, b]$.

(b) Show that $\mathcal{O}(S)$ is a connected open subset of $\mathbb{R}^{n+1}$.

7 Geodesics

Geodesics are curves in n-surfaces which play the same role as do straight lines in $\mathbb{R}^n$. Before formulating a precise definition, we must introduce the process of differentiation of vector fields and functions defined along parametrized curves. In order to allow the possibility that such vector fields and functions may take on different values at a point where a parametrized curve crosses itself, it is convenient to regard these fields and functions to be defined on the parameter interval rather than on the image of the curve.

A *vector field* $\mathbf{X}$ *along the parametrized curve* $\alpha: I \to \mathbb{R}^{n+1}$ is a function which assigns to each $t \in I$ a vector $\mathbf{X}(t)$ at $\alpha(t)$; i.e., $\mathbf{X}(t) \in \mathbb{R}^{n+1}_{\alpha(t)}$ for all $t \in I$. A *function* f *along* α is simply a function $f: I \to \mathbb{R}$. Thus, for example, the velocity $\dot{\alpha}$ of the parametrized curve $\alpha: I \to \mathbb{R}^{n+1}$ is a vector field along α (Figure 7.1); its length $\|\dot{\alpha}\|: I \to \mathbb{R}$, defined by $\|\dot{\alpha}\|(t) = \|\dot{\alpha}(t)\|$ for all $t \in I$, is a function along α. $\|\dot{\alpha}\|$ is called the *speed* of α.

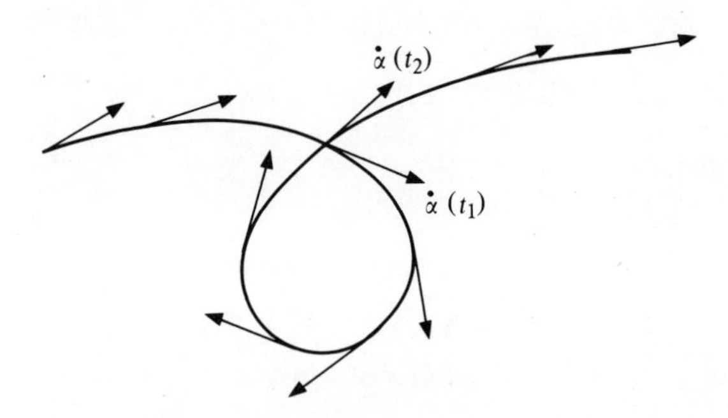

Figure 7.1 The velocity field along a parametrized curve α. Note that $\alpha(t_1) = \alpha(t_2)$ does not imply that $\dot{\alpha}(t_1) = \dot{\alpha}(t_2)$.

Vector fields and functions along parametrized curves frequently occur as restrictions. Thus if $\mathbf{X}$ is a vector field on U, where U is an open subset of $\mathbb{R}^{n+1}$ containing Image α, then $\mathbf{X} \circ \alpha$ is a vector field along α. Similarly $f \circ \alpha$ is a function along α whenever $f\colon U \to \mathbb{R}$, where $U \supset$ Image α.

Each vector field $\mathbf{X}$ along α is of the form

$$\mathbf{X}(t) = (\alpha(t), X_1(t), \ldots, X_{n+1}(t))$$

where each component X_i is a function along α. $\mathbf{X}$ is *smooth* if each $X_i\colon I \to \mathbb{R}$ is smooth. The *derivative* of a smooth vector field $\mathbf{X}$ along α is the vector field $\dot{\mathbf{X}}$ along α defined by

$$\dot{\mathbf{X}}(t) = \left(\alpha(t), \frac{dX_1}{dt}(t), \ldots, \frac{dX_{n+1}}{dt}(t)\right).$$

$\dot{\mathbf{X}}(t)$ measures the rate of change of the vector part $(X_1(t), \ldots, X_{n+1}(t))$ of $\mathbf{X}(t)$ along α. Thus, for example, the *acceleration* $\ddot{\alpha}$ of a parametrized curve α is the vector field along α obtained by differentiating the velocity field $\dot{\alpha}$ $[\ddot{\alpha} = \dot{(\dot{\alpha})}]$ (see Figure 7.2).

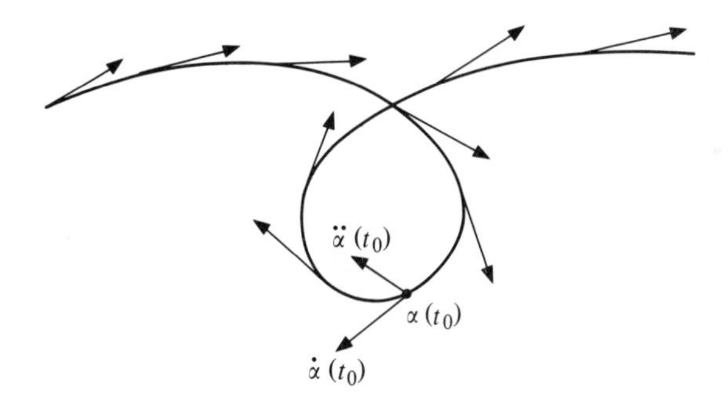

Figure 7.2 The acceleration $\ddot{\alpha}(t_0)$ is the derivative at t_0 of the velocity vector field $\dot{\alpha}$.

It is easy to check (Exercise 7.4) that differentiation of vector fields along parametrized curves has the following properties. For $\mathbf{X}$ and $\mathbf{Y}$ smooth vector fields along the parametrized curve $\alpha\colon I \to \mathbb{R}^{n+1}$ and f a smooth function along α,

(i) $\dot{(\mathbf{X} + \mathbf{Y})} = \dot{\mathbf{X}} + \dot{\mathbf{Y}}$
(ii) $\dot{(f\mathbf{X})} = f'\mathbf{X} + f\dot{\mathbf{X}}$
(iii) $(\mathbf{X} \cdot \mathbf{Y})' = \dot{\mathbf{X}} \cdot \mathbf{Y} + \mathbf{X} \cdot \dot{\mathbf{Y}}$

where $\mathbf{X} + \mathbf{Y}$, $f\mathbf{X}$, and $\mathbf{X} \cdot \mathbf{Y}$ are defined along α by

$$(\mathbf{X} + \mathbf{Y})(t) = \mathbf{X}(t) + \mathbf{Y}(t)$$

$$(f\mathbf{X})(t) = f(t)\mathbf{X}(t)$$

$$(\mathbf{X} \cdot \mathbf{Y})(t) = \mathbf{X}(t) \cdot \mathbf{Y}(t)$$

for all $t \in I$.

A *geodesic* in an n-surface $S \subset \mathbb{R}^{n+1}$ is a parametrized curve $\alpha\colon I \to S$ whose acceleration is everywhere orthogonal to S; that is, $\ddot{\alpha}(t) \in S_{\alpha(t)}^{\perp}$ for all $t \in I$. Thus a geodesic is a curve in S which always goes "straight ahead" in the surface. Its acceleration serves only to keep it in the surface. It has no component of acceleration tangent to the surface.

Note that geodesics have constant speed, because $\dot{\alpha}(t) \in S_{\alpha(t)}$ and $\ddot{\alpha}(t) \in S_{\alpha(t)}^{\perp}$ for all $t \in I$ implies that

$$\frac{d}{dt}\|\dot{\alpha}(t)\|^2 = \frac{d}{dt}(\dot{\alpha}(t) \cdot \dot{\alpha}(t)) = 2\dot{\alpha}(t) \cdot \ddot{\alpha}(t) = 0.$$

EXAMPLE 1. If an n-surface S contains a straight line segment $\alpha(t) = p + tv$ ($t \in I$) then that segment is a geodesic in S. Indeed, $\ddot{\alpha}(t) = 0$ for all $t \in I$ so in particular $\ddot{\alpha}(t) \perp S_{\alpha(t)}$ for all $t \in I$.

EXAMPLE 2. For each a, b, c, $d \in \mathbb{R}$, the parametrized curve $\alpha(t) = (\cos(at + b), \sin(at + b), ct + d)$ is a geodesic in the cylinder $x_1^2 + x_2^2 = 1$ in $\mathbb{R}^3$, because

$$\ddot{\alpha}(t) = (\alpha(t), -a^2\cos(at + b), -a^2\sin(at + b), 0) = \pm a^2\mathbf{N}(\alpha(t))$$

for all $t \in \mathbb{R}$ (see Figure 7.3).

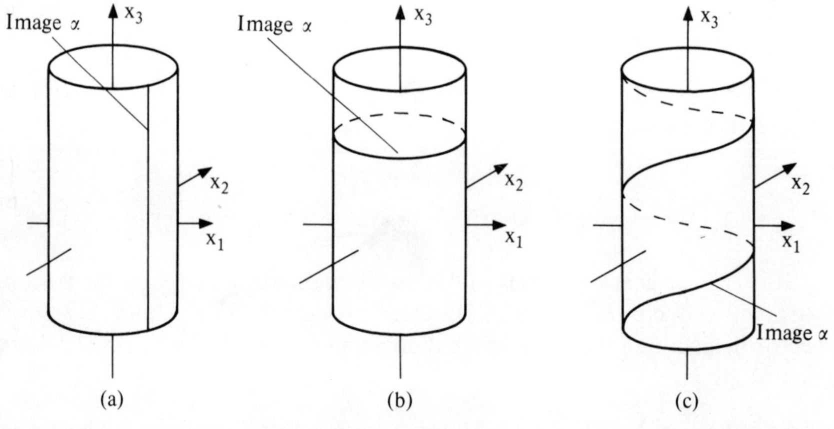

Image α x_3 x_2 x_1 (a) (b) (c)

Figure 7.3 Geodesics $\alpha(t) = (\cos(at + b), \sin(at + b), ct + d)$ in the cylinder $x_1^2 + x_2^2 = 1$. (a) $a = 0$, (b) $c = 0$, (c) $a \neq 0$, $c \neq 0$.

EXAMPLE 3. For each pair of orthogonal unit vectors $\{e_1, e_2\}$ in $\mathbb{R}^3$ and each $a \in \mathbb{R}$, the great circle (or point if $a = 0$) $\alpha(t) = (\cos at)e_1 + (\sin at)e_2$ is a geodesic in the 2-sphere $x_1^2 + x_2^2 + x_3^2 = 1$ in $\mathbb{R}^3$, because $\ddot{\alpha}(t) = (\alpha(t), -a^2\alpha(t)) = \pm a^2\mathbf{N}(\alpha(t))$ for all $t \in \mathbb{R}$ (see Figure 7.4).

Intuitively, it seems clear that given any point p in an n-surface S and any initial velocity $\mathbf{v}$ at p ($\mathbf{v} \in S_p$) there should be a geodesic in S passing through

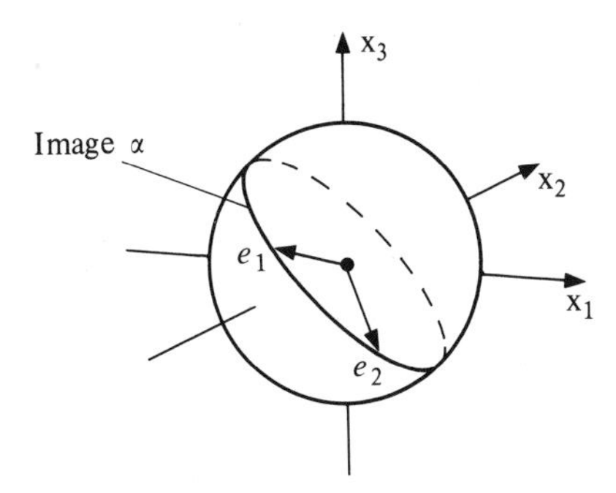

Figure 7.4 Great circles are geodesics in the 2-sphere.

p with initial velocity $\mathbf{v}$. After all, a racing car travelling on S, passing through p with velocity $\mathbf{v}$, should be able to continue travelling "straight ahead" on S at constant speed $\|\mathbf{v}\|$, thereby tracing out a geodesic in S. The following theorem shows that this is the case, and that the geodesic with these properties is essentially unique.

Theorem. *Let S be an n-surface in $\mathbb{R}^{n+1}$, let $p \in S$, and let $\mathbf{v} \in S_p$. Then there exists an open interval I containing* 0 *and a geodesic $\alpha: I \to S$ such that*

(i) $\alpha(0) = p$ *and* $\dot{\alpha}(0) = \mathbf{v}$.
(ii) *If $\beta: \tilde{I} \to S$ is any other geodesic in S with $\beta(0) = p$ and $\dot{\beta}(0) = \mathbf{v}$, then $\tilde{I} \subset I$ and $\beta(t) = \alpha(t)$ for all $t \in \tilde{I}$.*

Remark. The geodesic α is called the *maximal geodesic in S passing through p with initial velocity* $\mathbf{v}$.

Proof. $S = f^{-1}(c)$ for some $c \in \mathbb{R}$ and some smooth function $f: U \to \mathbb{R}$ (U open in $\mathbb{R}^{n+1}$) with $\nabla f(p) \neq \mathbf{0}$ for all $p \in S$. Since $\nabla f(p) \neq \mathbf{0}$ for all p in some open set containing S, we may assume (by shrinking U if necessary) that $\nabla f(p) \neq 0$ for all $p \in U$. Set $\mathbf{N} = \nabla f/\|\nabla f\|$.

By definition, a parametrized curve $\alpha: I \to S$ is a geodesic of S if and only if its acceleration is everywhere perpendicular to S; that is, if and only if $\ddot{\alpha}(t)$ is a multiple of $\mathbf{N}(\alpha(t))$ for all $t \in I$:

$$\ddot{\alpha}(t) = g(t)\mathbf{N}(\alpha(t))$$

for all $t \in I$, where $g: I \to \mathbb{R}$. Taking the dot product of both sides of this equation with $\mathbf{N}(\alpha(t))$ we find

$$\begin{aligned} g = \ddot{\alpha} \cdot \mathbf{N} \circ \alpha &= (\dot{\alpha} \cdot \mathbf{N} \circ \alpha)' - \dot{\alpha} \cdot \mathbf{N} \dot{\circ} \alpha \\ &= -\dot{\alpha} \cdot \mathbf{N} \dot{\circ} \alpha, \end{aligned}$$

since $\dot{\alpha} \cdot \mathbf{N} \circ \alpha = 0$. Thus $\alpha: I \to S$ is a geodesic if and only if it satisfies the differential equation

$$\text{(G)} \qquad \ddot{\alpha} + (\dot{\alpha} \cdot \mathbf{N} \dot{\circ} \alpha)(\mathbf{N} \circ \alpha) = \mathbf{0}.$$

This is a second order differential equation in α. (If we write $\alpha(t) = (x_1(t), \ldots, x_{n+1}(t))$, this vector differential equation becomes the system of second order differential equations

$$\frac{d^2 x_i}{dt^2} + \sum_{j,k=1}^{n+1} N_i(x_1, \ldots, x_{n+1}) \frac{\partial N_j}{\partial x_k}(x_1, \ldots, x_{n+1}) \frac{dx_j}{dt}\frac{dx_k}{dt} = 0$$

where the N_j ($j \in \{1, \ldots, n+1\}$) are the components of $\mathbf{N}$.) By the existence theorem for the solutions of such equations,† there exists an open interval I_1 about 0 and a solution $\beta_1 : I_1 \to U$ of this differential equation satisfying the initial conditions $\beta_1(0) = p$ and $\dot{\beta}_1(0) = \mathbf{v}$ (that is, satisfying $x_i(0) = p_i$ and $(dx_i/dt)(0) = v_i$ for $i \in \{1, \ldots, n+1\}$, where $p = (p_1, \ldots, p_{n+1})$ and $\mathbf{v} = (p, v_1, \ldots, v_{n+1})$). Moreover, this solution is unique in the sense that if $\beta_2 : I_2 \to U$ is another solution of (G), with $\beta_2(0) = p$ and $\dot{\beta}_2(0) = \mathbf{v}$, then $\beta_1(t) = \beta_2(t)$ for all $t \in I_1 \cap I_2$. It follows that there exists a maximal open interval I (I is the union of the domains of all solutions to (G) which map 0 into p and have initial velocity $\mathbf{v}$) and a unique solution $\alpha : I \to U$ of (G) satisfying $\alpha(0) = p$ and $\dot{\alpha}(0) = \mathbf{v}$. Furthermore, if $\beta : \tilde{I} \to U$ is any solution of (G) with $\beta(0) = p$ and $\dot{\beta}(0) = \mathbf{v}$ then $\tilde{I} \subset I$ and β is the restriction of α to $\tilde{I}$.

To complete the proof, it remains only to show that the solution α to (G) is actually a curve in S. For, if so, it must be a geodesic since it satisfies the geodesic equation (G), and the rest of the theorem follows from the uniqueness statements above.

To see that α is in fact a curve in S, note first that for every solution $\alpha : I \to U$ of (G), $\dot{\alpha} \cdot \mathbf{N} \circ \alpha = 0$. Indeed,

$$(\dot{\alpha} \cdot \mathbf{N} \circ \alpha)' = \ddot{\alpha} \cdot \mathbf{N} \circ \alpha + \dot{\alpha} \cdot \mathbf{N} \dot{\circ} \alpha = 0$$

by (G), so $\dot{\alpha} \cdot \mathbf{N} \circ \alpha$ is constant along α, and

$$(\dot{\alpha} \cdot \mathbf{N} \circ \alpha)(0) = \mathbf{v} \cdot \mathbf{N}(p) = 0$$

since $\mathbf{v} \in S_p$ and $\mathbf{N}(p) \perp S_p$. It follows then that

$$(f \circ \alpha)'(t) = \nabla f(\alpha(t)) \cdot \dot{\alpha}(t) = \|\nabla f(\alpha(t))\| \mathbf{N}(\alpha(t)) \cdot \dot{\alpha}(t) = 0$$

for all $t \in I$ so $f \circ \alpha$ is constant, and $f(\alpha(0)) = f(p) = c$ so $f(\alpha(t)) = c$ for all $t \in I$; that is, Image $\alpha \subset f^{-1}(c) = S$. □

It follows from the theorem just proved that each maximal geodesic on the unit 2-sphere in $\mathbb{R}^3$ (Example 3) is either a great circle (parametrized by a constant speed parametrization) or is constant ($\alpha(t) = p$ for all t, some p) since such a curve can be found through each point p with any given initial velocity. Similarly, each maximal geodesic on the cylinder $x_1^2 + x_2^2 = 1$ in $\mathbb{R}^3$ (Example 2) is either a vertical line, a horizontal circle, a helix (spiral), or is constant.

† See e.g. W. Hurewicz, *Lectures on Ordinary Differential Equations*, Cambridge, Mass.: MIT Press (1958), pp. 32–33. See also Exercise 9.15.

Exercises

7.1. Find the velocity, the acceleration, and the speed of each of the following parametrized curves:

(a) $\alpha(t) = (t, t^2)$
(b) $\alpha(t) = (\cos t, \sin t)$
(c) $\alpha(t) = (\cos 3t, \sin 3t)$
(d) $\alpha(t) = (\cos t, \sin t, t)$
(e) $\alpha(t) = (\cos t, \sin t, 2 \cos t, 2 \sin t)$.

7.2. Show that if $\alpha: I \to \mathbb{R}^{n+1}$ is a parametrized curve with constant speed then $\ddot{\alpha}(t) \perp \dot{\alpha}(t)$ for all $t \in I$.

7.3. Let $\alpha: I \to \mathbb{R}^{n+1}$ be a parametrized curve with $\dot{\alpha}(t) \neq \mathbf{0}$ for all $t \in I$. Show that there exists a unit speed reparametrization β of α; i.e., show that there exists an interval J and a smooth function $h: J \to I$ (onto) such that $h' > 0$ and such that $\beta = \alpha \circ h$ has unit speed. [*Hint*: Set $h = s^{-1}$ where $s(t) = \int_{t_0}^{t} \|\dot{\alpha}(t)\|\, dt$ for some $t_0 \in I$.]

7.4. Let $\mathbf{X}$ and $\mathbf{Y}$ be smooth vector fields along the parametrized curve $\alpha: I \to \mathbb{R}^{n+1}$ and let $f: I \to \mathbb{R}$ be a smooth function along α. Verify that

(a) $(\mathbf{X} \dot{+} \mathbf{Y}) = \dot{\mathbf{X}} + \dot{\mathbf{Y}}$
(b) $(\dot{f\mathbf{X}}) = f'\mathbf{X} + f\dot{\mathbf{X}}$
(c) $(\mathbf{X} \cdot \mathbf{Y})' = \dot{\mathbf{X}} \cdot \mathbf{Y} + \mathbf{X} \cdot \dot{\mathbf{Y}}$.

7.5. Let S denote the cylinder $x_1^2 + x_2^2 = r^2$ of radius $r > 0$ in $\mathbb{R}^3$. Show that α is a geodesic of S if and only if α is of the form

$$\alpha(t) = (r \cos(at + b), r \sin(at + b), ct + d)$$

for some $a, b, c, d \in \mathbb{R}$.

7.6. Show that a parametrized curve α in the unit n-sphere $x_1^2 + \cdots + x_{n+1}^2 = 1$ is a geodesic if and only if it is of the form

$$\alpha(t) = (\cos at)e_1 + (\sin at)e_2$$

for some orthogonal pair of unit vectors $\{e_1, e_2\}$ in $\mathbb{R}^{n+1}$ and some $a \in \mathbb{R}$. (For $a \neq 0$, these curves are "great circles" on the n-sphere.)

7.7. Show that if $\alpha: I \to S$ is a geodesic in an n-surface S and if $\beta = \alpha \circ h$ is a reparametrization of $\alpha(h: \tilde{I} \to I)$ then β is a geodesic in S if and only if there exist $a, b \in \mathbb{R}$ with $a > 0$ such that $h(t) = at + b$ for all $t \in \tilde{I}$.

7.8. Let C be a plane curve in the upper half plane $x_2 > 0$ and let S be the surface of revolution obtained by rotating C about the x_1-axis (see Example 5, Chapter 4). Let $\alpha: I \to C$, $\alpha(t) = (x_1(t), x_2(t))$, be a constant speed parametrized curve in C. For each $\theta \in \mathbb{R}$, define $\alpha_\theta: I \to S$ by

$$\alpha_\theta(t) = (x_1(t), x_2(t) \cos \theta, x_2(t) \sin \theta)$$

and, for each $t \in I$, define $\beta_t: \mathbb{R} \to S$ by the same formula:

$$\beta_t(\theta) = (x_1(t), x_2(t) \cos \theta, x_2(t) \sin \theta).$$

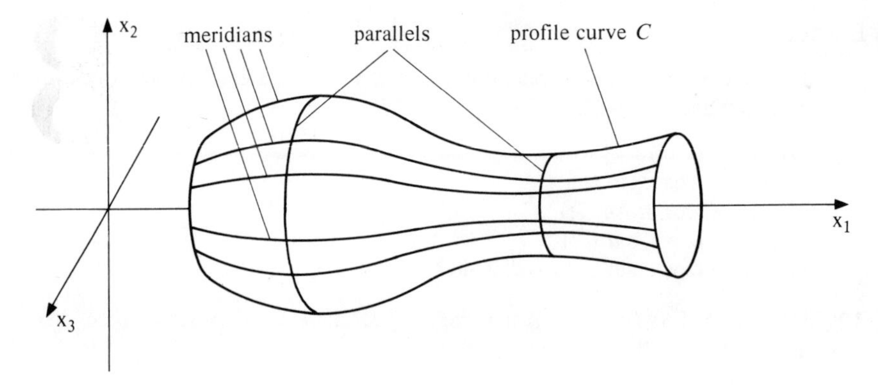

Figure 7.5 Geodesics on a surface of revolution: all meridians are geodesics; a parallel is a geodesic if it cuts the "profile curve" C at a point where the slope of C is zero.

The curves α_θ are called *meridians* of S, and the circles β_t are *parallels* of S (see Figure 7.5).

(a) Show that meridians and parallels always meet orthogonally; i.e., $\dot{\alpha}_\theta(t) \cdot \dot{\beta}_t(\theta) = 0$ for all $t \in I$, $\theta \in \mathbb{R}$.
(b) Show that each meridian α_θ is a geodesic of S. [*Hint*: Note that $\{\dot{\alpha}_\theta(t), \dot{\beta}_t(\theta)\}$ spans S_p, where $p = \alpha_\theta(t)$. Hence it suffices to check that $\ddot{\alpha}_\theta(t)$ is perpendicular to both $\dot{\alpha}_\theta(t)$ and $\dot{\beta}_t(\theta)$.]
(c) Show that a parallel β_t is a geodesic of S if and only if the slope $x_2'(t)/x_1'(t)$ of the tangent line to C at $\alpha(t)$ is zero.

7.9. Let S be an n-surface in $\mathbb{R}^{n+1}$, let $\mathbf{v} \in S_p$, $p \in S$, and let $\alpha\colon I \to S$ be the maximal geodesic in S with initial velocity $\mathbf{v}$. Show that the maximal geodesic β in S with initial velocity $c\mathbf{v}$ ($c \in \mathbb{R}$) is given by the formula $\beta(t) = \alpha(ct)$.

7.10. Let S be an n-surface in $\mathbb{R}^{n+1}$, let $p \in S$, let $\mathbf{v} \in S_p$, and let $\alpha\colon I \to S$ be the maximal geodesic in S passing through p with velocity $\mathbf{v}$. Show that if $\beta\colon \tilde{I} \to S$ is any geodesic in S with $\beta(t_0) = p$ and $\dot{\beta}(t_0) = \mathbf{v}$ for some $t_0 \in \tilde{I}$ then $\beta(t) = \alpha(t - t_0)$ for all $t \in \tilde{I}$.

7.11. Let S be an n-surface in $\mathbb{R}^{n+1}$ and let $\beta\colon I \to S$ be a geodesic in S with $\beta(t_0) = \beta(0)$ and $\dot{\beta}(t_0) = \dot{\beta}(0)$ for some $t_0 \in I$, $t_0 \neq 0$. Show that β is periodic by showing that $\beta(t + t_0) = \beta(t)$ for all t such that both t and $t + t_0 \in I$. [*Hint*: Use Exercise 7.10.]

7.12. An n-surface S in $\mathbb{R}^{n+1}$ is said to be *geodesically complete* if every maximal geodesic in S has domain $\mathbb{R}$. Which of the following n-surfaces are geodesically complete?

(a) The n-sphere $x_1^2 + \cdots + x_{n+1}^2 = 1$.
(b) The n-sphere with the north pole deleted: $x_1^2 + \cdots + x_{n+1}^2 = 1$, $x_{n+1} \neq 1$.
(c) The cone $x_1^2 + x_2^2 - x_3^2 = 0$, $x_3 > 0$ in $\mathbb{R}^3$.
(d) The cylinder $x_1^2 + x_2^2 = 1$ in $\mathbb{R}^3$.
(e) The cylinder in $\mathbb{R}^3$ with a straight line deleted: $x_1^2 + x_2^2 = 1$, $x_1 \neq 1$.

Parallel Transport 8

A vector field $\mathbf{X}$ along a parametrized curve $\alpha\colon I \to S$ in an n-surface S is *tangent to S along α* if $\mathbf{X}(t) \in S_{\alpha(t)}$ for all $t \in I$. The derivative $\dot{\mathbf{X}}$ of such a vector field is, however, generally not tangent to S. We can, nevertheless, obtain a vector field tangent to S by projecting $\dot{\mathbf{X}}(t)$ orthogonally onto $S_{\alpha(t)}$ for each $t \in I$ (see Figure 8.1). This process of differentiating and then

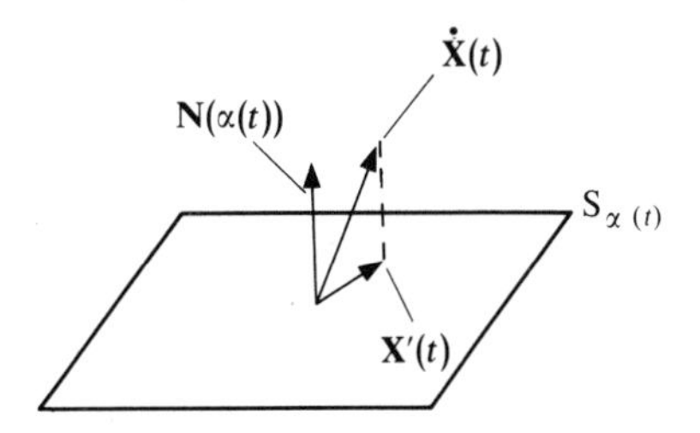

Figure 8.1 The covariant derivative $\mathbf{X}'(t)$ is the orthogonal projection onto the tangent space of the ordinary derivative $\dot{\mathbf{X}}(t)$.

projecting onto the tangent space to S defines an operation with the same properties as differentiation, except that now differentiation of vector fields tangent to S yields vector fields tangent to S. This operation is called covariant differentiation.

Let S be an n-surface in $\mathbb{R}^{n+1}$, let $\alpha\colon I \to S$ be a parametrized curve in S, and let $\mathbf{X}$ be a smooth vector field tangent to S along α. The *covariant derivative* of $\mathbf{X}$ is the vector field $\mathbf{X}'$ tangent to S along α defined by

$$\mathbf{X}'(t) = \dot{\mathbf{X}}(t) - [\dot{\mathbf{X}}(t) \cdot \mathbf{N}(\alpha(t))]\mathbf{N}(\alpha(t)),$$

where $\mathbf{N}$ is an orientation on S. Note that $\mathbf{X}'(t)$ is independent of the choice of $\mathbf{N}$ since replacing $\mathbf{N}$ by $-\mathbf{N}$ has no effect on the above formula.

It is easy to check that covariant differentiation has the following properties: for $\mathbf{X}$ and $\mathbf{Y}$ smooth vector fields tangent to S along a parameterized curve $\alpha: I \to S$ and f a smooth function along α,

(i) $$(\mathbf{X} + \mathbf{Y})' = \mathbf{X}' + \mathbf{Y}'$$
(ii) $$(f\mathbf{X})' = f'\mathbf{X} + f\mathbf{X}'$$
(iii) $$(\mathbf{X} \cdot \mathbf{Y})' = \mathbf{X}' \cdot \mathbf{Y} + \mathbf{X} \cdot \mathbf{Y}'.$$

These properties follow immediately from the corresponding properties of ordinary differentiation. For example, the following computation verifies (iii):

$$\begin{aligned}(\mathbf{X} \cdot \mathbf{Y})' &= \dot{\mathbf{X}} \cdot \mathbf{Y} + \mathbf{X} \cdot \dot{\mathbf{Y}} \\ &= [\mathbf{X}' + (\dot{\mathbf{X}} \cdot \mathbf{N} \circ \alpha)\mathbf{N} \circ \alpha] \cdot \mathbf{Y} + \mathbf{X} \cdot [\mathbf{Y}' + (\dot{\mathbf{Y}} \cdot \mathbf{N} \circ \alpha)\mathbf{N} \circ \alpha] \\ &= \mathbf{X}' \cdot \mathbf{Y} + \mathbf{X} \cdot \mathbf{Y}',\end{aligned}$$

since $\mathbf{N}$ is perpendicular to S and $\mathbf{X}$ and $\mathbf{Y}$ are tangent to S.

Intuitively, the covariant derivative $\mathbf{X}'$ measures the rate of change of $\mathbf{X}$ along α *as seen from the surface* S (by ignoring the component of $\dot{\mathbf{X}}$ normal to S). Note that a parametrized curve $\alpha: I \to S$ is a geodesic in S if and only if its *covariant acceleration* $(\dot{\alpha})'$ is zero along α.

The covariant derivative leads naturally to a concept of parallelism on an n-surface. In $\mathbb{R}^{n+1}$, vectors $\mathbf{v} = (p, v) \in \mathbb{R}_p^{n+1}$ and $\mathbf{w} = (q, w) \in \mathbb{R}_q^{n+1}$ are said to be *Euclidean parallel* if $v = w$ (see Figure 8.2(a)). A vector field $\mathbf{X}$ along a

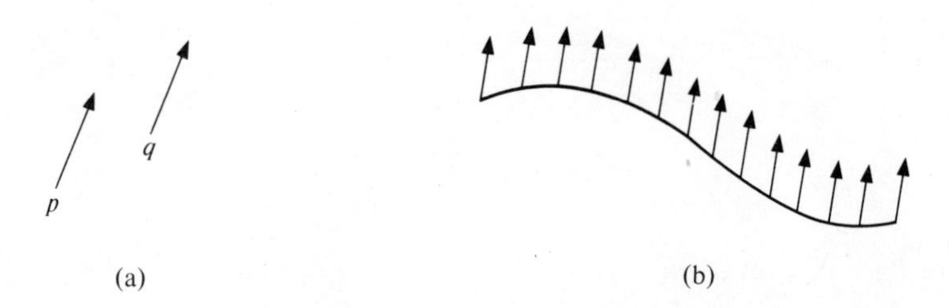

Figure 8.2 Euclidean parallelism in $\mathbb{R}^2$: (a) parallel vectors; (b) a parallel vector field.

parametrized curve $\alpha: I \to \mathbb{R}^{n+1}$ is Euclidean parallel if $X(t_1) = X(t_2)$ for all $t_1, t_2 \in I$, where $\mathbf{X}(t) = (\alpha(t), X(t))$ for $t \in I$ (see Figure 8.2(b)). Thus $\mathbf{X}$ is Euclidean parallel along α if and only if $\dot{\mathbf{X}} = \mathbf{0}$.

Given an n-surface S in $\mathbb{R}^{n+1}$ and a parametrized curve $\alpha: I \to S$, a smooth vector field $\mathbf{X}$ tangent to S along α is said to be *Levi-Civita parallel*, or simply *parallel*, if $\mathbf{X}' = \mathbf{0}$. Intuitively, $\mathbf{X}$ is parallel along α if $\mathbf{X}$ is a constant vector field along α, as seen from S.

Note that Levi-Civita parallelism has the following properties:

(i) If $\mathbf{X}$ is parallel along α, then $\mathbf{X}$ has constant length, since

$$\frac{d}{dt}\|\mathbf{X}\|^2 = \frac{d}{dt}(\mathbf{X}\cdot\mathbf{X}) = 2\mathbf{X}'\cdot\mathbf{X} = 0.$$

(ii) If $\mathbf{X}$ and $\mathbf{Y}$ are two parallel vector fields along α, then $\mathbf{X}\cdot\mathbf{Y}$ is constant along α, since

$$(\mathbf{X}\cdot\mathbf{Y})' = \mathbf{X}'\cdot\mathbf{Y} + \mathbf{X}\cdot\mathbf{Y}' = 0.$$

(iii) If $\mathbf{X}$ and $\mathbf{Y}$ are parallel along α, then the angle $\cos^{-1}(\mathbf{X}\cdot\mathbf{Y}/\|\mathbf{X}\|\,\|\mathbf{Y}\|)$ between $\mathbf{X}$ and $\mathbf{Y}$ is constant along α, since $\mathbf{X}\cdot\mathbf{Y}$, $\|\mathbf{X}\|$, and $\|\mathbf{Y}\|$ are each constant along α.

(iv) If $\mathbf{X}$ and $\mathbf{Y}$ are parallel along α then so are $\mathbf{X}+\mathbf{Y}$ and $c\mathbf{X}$, for all $c \in \mathbb{R}$.

(v) The velocity vector field along a parametrized curve α in S is parallel if and only if α is a geodesic.

Theorem 1. *Let S be an n-surface in $\mathbb{R}^{n+1}$, let $\alpha\colon I \to S$ be a parametrized curve in S, let $t_0 \in I$, and let $\mathbf{v} \in S_{\alpha(t_0)}$. Then there exists a unique vector field $\mathbf{V}$, tangent to S along α, which is parallel and has $\mathbf{V}(t_0) = \mathbf{v}$.*

PROOF. We require a vector field $\mathbf{V}$ tangent to S along α satisfying $\mathbf{V}' = \mathbf{0}$. But

$$\begin{aligned}\mathbf{V}' &= \dot{\mathbf{V}} - (\dot{\mathbf{V}}\cdot\mathbf{N}\circ\alpha)\mathbf{N}\circ\alpha\\ &= \dot{\mathbf{V}} - [(\mathbf{V}\cdot\mathbf{N}\circ\alpha)' - \mathbf{V}\cdot\mathbf{N}\,\dot{\circ}\,\alpha]\mathbf{N}\circ\alpha\\ &= \dot{\mathbf{V}} + (\mathbf{V}\cdot\mathbf{N}\,\dot{\circ}\,\alpha)\mathbf{N}\circ\alpha\end{aligned}$$

so $\mathbf{V}' = 0$ if and only if $\mathbf{V}$ satisfies the differential equation

$$\text{(P)}\qquad \dot{\mathbf{V}} + (\mathbf{V}\cdot\mathbf{N}\,\dot{\circ}\,\alpha)\mathbf{N}\circ\alpha = \mathbf{0}.$$

This is a first order differential equation in $\mathbf{V}$. (If we write $\mathbf{V}(t) = (\alpha(t), V_1(t), \ldots, V_{n+1}(t))$, this vector differential equation becomes the system of first order differential equations

$$\frac{dV_i}{dt} + \sum_{j=1}^{n+1}(N_i\circ\alpha)(N_j\circ\alpha)'V_j = 0$$

where the N_j ($j \in \{1, \ldots, n+1\}$) are the components of $\mathbf{N}$.) By the existence and uniqueness theorem for solutions of first order differential equations, there exists a unique vector field $\mathbf{V}$ along α satisfying equation (P) together with the initial condition $\mathbf{V}(t_0) = \mathbf{v}$ (that is, satisfying $V_i(t_0) = v_i$ for $i \in \{1, \ldots, n+1\}$, where $\mathbf{v} = (\alpha(t_0), v_1, \ldots, v_{n+1})$). The existence and uniqueness theorem does not guarantee, however, that $\mathbf{V}$ is tangent to S along α.

To see that $\mathbf{V}$ is indeed tangent to S, simply note that, by (P),

$$\begin{aligned}(\mathbf{V}\cdot\mathbf{N}\circ\alpha)' &= \dot{\mathbf{V}}\cdot\mathbf{N}\circ\alpha + \mathbf{V}\cdot\mathbf{N}\mathbin{\dot\circ}\alpha \\ &= [-(\mathbf{V}\cdot\mathbf{N}\mathbin{\dot\circ}\alpha)\mathbf{N}\circ\alpha]\cdot\mathbf{N}\circ\alpha + \mathbf{V}\cdot\mathbf{N}\mathbin{\dot\circ}\alpha \\ &= -\mathbf{V}\cdot\mathbf{N}\mathbin{\dot\circ}\alpha + \mathbf{V}\cdot\mathbf{N}\mathbin{\dot\circ}\alpha = 0,\end{aligned}$$

so $\mathbf{V}\cdot\mathbf{N}\circ\alpha$ is constant along α and, since $(\mathbf{V}\cdot\mathbf{N}\circ\alpha)(t_0) = \mathbf{v}\cdot\mathbf{N}(\alpha(t_0)) = 0$, this constant must be zero. Finally, this vector field $\mathbf{V}$, tangent to S along α, is parallel because it satisfies equation (P). □

Remark. We have implicitly assumed in the above proof that the solution $\mathbf{V}$ of (P) satisfying $\mathbf{V}(t_0) = \mathbf{v}$ is defined on the whole interval I and not just on some smaller interval containing t_0. That this is indeed the case can be seen from the following argument. Suppose $\tilde{I} \subset I$ is the maximal interval on which there exists a solution $\mathbf{V}$ of (P) satisfying $\mathbf{V}(t_0) = \mathbf{v}$. If $\tilde{I} \neq I$, there exists an endpoint b of $\tilde{I}$ with $b \in I$. Let $\{t_i\}$ be a sequence in $\tilde{I}$ with $\lim_{i\to\infty} t_i = b$. Since $\|\mathbf{V}\|$ is constant on $\tilde{I}$, $\|\mathbf{V}(t_i)\| = \|\mathbf{v}\|$ for all i, so the sequence $\{V(t_i)\}$ of vector parts of $\{\mathbf{V}(t_i)\}$ takes values in a compact set, the sphere of radius $\|\mathbf{v}\|$ about the origin in $\mathbb{R}^{n+1}$. It follows that $\{V(t_i)\}$ must have a convergent subsequence $\{V(t_{i_k})\}$. Let $w = \lim_{k\to\infty} V(t_{i_k})$, and let $\mathbf{W}$ be a solution of (P), on some interval J containing b, with $\mathbf{W}(b) = (\alpha(b), w)$. Then $\mathbf{W} - \mathbf{V}$ is also a solution of (P), on $\tilde{I} \cap J$, and in particular $\|\mathbf{W} - \mathbf{V}\|$ is constant on $\tilde{I} \cap J$. But

$$\lim_{k\to\infty} \|W(t_{i_k}) - V(t_{i_k})\| = \|w - w\| = 0$$

so $\|\mathbf{W} - \mathbf{V}\| = 0$ on $J \cap \tilde{I}$; that is, $\mathbf{W} = \mathbf{V}$ on $J \cap \tilde{I}$. Hence the vector field on $I \cup \tilde{I}$ which is equal to $\mathbf{V}$ on I and to $\mathbf{W}$ on $\tilde{I}$ extends $\mathbf{V}$ to a solution of (P) on an interval larger than $\tilde{I}$, contradicting the maximality of $\tilde{I}$. Hence $\tilde{I} = I$, as claimed.

Corollary. *Let S be a 2-surface in $\mathbb{R}^3$ and let $\alpha: I \to S$ be a geodesic in S with $\dot{\alpha} \neq \mathbf{0}$. Then a vector field $\mathbf{X}$ tangent to S along α is parallel along α if and only if both $\|\mathbf{X}\|$ and the angle between $\mathbf{X}$ and $\dot{\alpha}$ are constant along α* (see Figure 8.3).

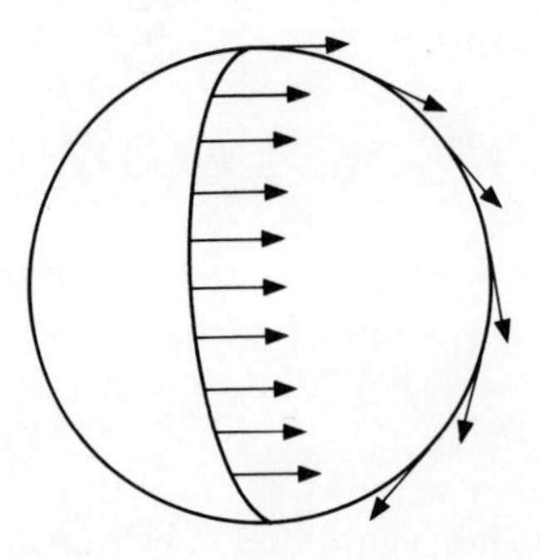

Figure 8.3 Levi-Civita parallel vector fields along geodesics in the 2-sphere.

PROOF. The "only if" statement is immediate from properties (i) and (iii) above. So suppose both $\|\mathbf{X}\|$ and the angle θ between $\mathbf{X}$ and $\dot{\alpha}$ are constant along α. Let $t_0 \in I$ and let $\mathbf{v} \in S_{\alpha(t_0)}$ be a unit vector orthogonal to $\dot{\alpha}(t_0)$. Let $\mathbf{V}$ be the unique parallel vector field along α such that $\mathbf{V}(t_0) = \mathbf{v}$. Then $\|\mathbf{V}\| = 1$ and $\mathbf{V} \cdot \dot{\alpha} = 0$ along α so $\{\dot{\alpha}(t), \mathbf{V}(t)\}$ is an orthogonal basis for $S_{\alpha(t)}$, for each $t \in I$. In particular, there exist smooth functions $f, g \colon I \to \mathbb{R}$ such that $\mathbf{X} = f\dot{\alpha} + g\mathbf{V}$. Since

$$\cos \theta = \mathbf{X} \cdot \dot{\alpha}/\|\mathbf{X}\|\,\|\dot{\alpha}\| = f\,\|\dot{\alpha}\|/\|\mathbf{X}\|$$

and

$$\|\mathbf{X}\|^2 = f^2\|\dot{\alpha}\|^2 + g^2,$$

the constancy of θ, $\|\mathbf{X}\|$, and $\|\dot{\alpha}\|$ along α implies that f and g are constant along α. Hence $\mathbf{X}$ is parallel along α, by property (iv) above. □

Parallelism can be used to transport tangent vectors from one point of an n-surface to another. Given two points p and q in an n-surface S, a *parametrized curve in S from p to q* is a smooth map $\alpha\colon [a, b] \to S$, from a closed interval $[a, b]$ into S, with $\alpha(a) = p$ and $\alpha(b) = q$. By *smoothness* of a map α defined on a closed interval we mean that α is the restriction to $[a, b]$ of a smooth map from some open interval containing $[a, b]$ into S. Each parametrized curve $\alpha\colon [a, b] \to S$ from p to q determines a map $P_\alpha\colon S_p \to S_q$ by

$$P_\alpha(\mathbf{v}) = \mathbf{V}(b)$$

where, for $\mathbf{v} \in S_p$, $\mathbf{V}$ is the unique parallel vector field along α with $\mathbf{V}(a) = \mathbf{v}$. $P_\alpha(\mathbf{v})$ is called the *parallel transport* (or *parallel translate*) of $\mathbf{v}$ along α to q.

EXAMPLE. For $\theta \in \mathbb{R}$, let $\alpha_\theta\colon [0, \pi] \to \mathbf{S}^2$ be the parametrized curve in the unit 2-sphere $\mathbf{S}^2$, from the north pole $p = (0, 0, 1)$ to the south pole $q = (0, 0, -1)$, defined by

$$\alpha_\theta(t) = (\cos\theta \sin t, \sin\theta \sin t, \cos t).$$

Thus, for each θ, α_θ is half of a great circle on $\mathbf{S}^2$ (see Figure 8.4). Let $\mathbf{v} = (p, 1, 0, 0) \in \mathbf{S}^2_p$. Since α_θ is a geodesic in $\mathbf{S}^2$, a vector field tangent to $\mathbf{S}^2$ along α_θ will be parallel if and only if it has constant length and keeps constant angle with $\dot{\alpha}_\theta$. The one with initial value $\mathbf{v}$ is

$$\mathbf{V}_\theta(t) = (\cos\theta)\dot{\alpha}_\theta(t) - (\sin\theta)\mathbf{N}(\alpha_\theta(t)) \times \dot{\alpha}_\theta(t),$$

where $\mathbf{N}$ is the outward orientation on $\mathbf{S}^2$. Hence

$$\begin{aligned} P_{\alpha_\theta}(\mathbf{v}) &= \mathbf{V}_\theta(\pi) \\ &= (\cos\theta)(q, -\cos\theta, -\sin\theta, 0) - (\sin\theta)(q, -\sin\theta, \cos\theta, 0) \\ &= -(q, \cos 2\theta, \sin 2\theta, 0). \end{aligned}$$

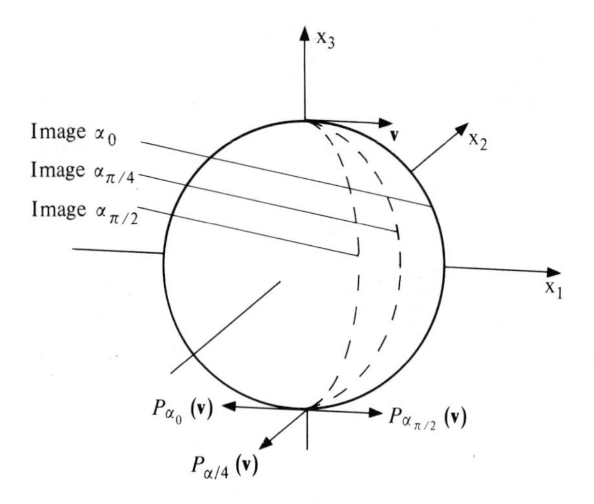

Figure 8.4 Parallel transport along geodesics in the 2-sphere.

Note that parallel transport from p to q is path dependent; that is, if α and β are two parametrized curves in S from p to q and $\mathbf{v} \in S_p$, then, in general, $P_\alpha(\mathbf{v}) \neq P_\beta(\mathbf{v})$.

Tangent vectors $\mathbf{v} \in S_p$, $p \in S$, may also be transported along piecewise smooth curves in S. A *piecewise smooth parametrized curve* α in S is a continuous map $\alpha\colon [a, b] \to S$ such that the restriction of α to $[t_i, t_{i+1}]$ is smooth for each $i \in \{0, 1, \ldots, k\}$, where $a = t_0 < t_1 < \cdots < t_{k+1} = b$ (see Figure 8.5).

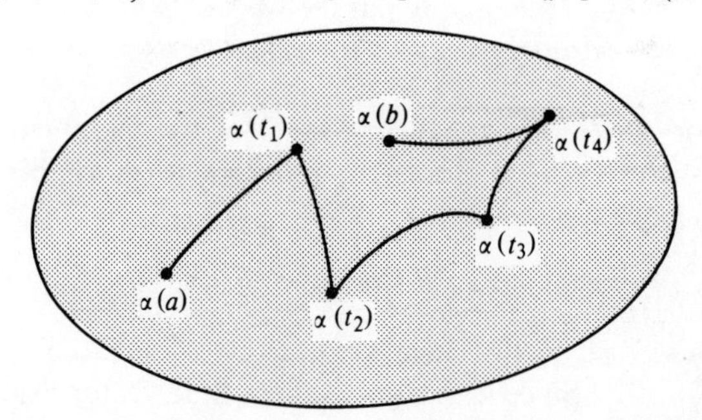

Figure 8.5 A piecewise smooth curve α in a 2-surface.

The *parallel transport* of $\mathbf{v} \in S_{\alpha(a)}$ along α to $\alpha(b)$ is obtained by transporting $\mathbf{v}$ along α to $\alpha(t_1)$ to get $\mathbf{v}_1 \in S_{\alpha(t_1)}$, then transporting $\mathbf{v}_1$ along α to $\alpha(t_2)$ to get $\mathbf{v}_2 \in S_{\alpha(t_2)}$, and so on, finally obtaining $P_\alpha(\mathbf{v})$ by transporting $\mathbf{v}_k \in S_{\alpha(t_k)}$ along α to $\alpha(b)$.

Theorem 2. *Let S be an n-surface in $\mathbb{R}^{n+1}$, let p, $q \in S$, and let α be a piecewise smooth parametrized curve from p to q. Then parallel transport $P_\alpha\colon S_p \to S_q$*

along α is a vector space isomorphism which preserves dot products; that is,

(i) P_α *is a linear map*
(ii) P_α *is one to one and onto*
(iii) $P_\alpha(\mathbf{v}) \cdot P_\alpha(\mathbf{w}) = \mathbf{v} \cdot \mathbf{w}$ *for all* $\mathbf{v}, \mathbf{w} \in S_p$.

PROOF. Property (i) is an immediate consequence of the fact that if $\mathbf{V}$ and $\mathbf{W}$ are parallel vector fields along a parametrized curve in S, then so are $\mathbf{V} + \mathbf{W}$ and $c\mathbf{V}$, for all $c \in \mathbb{R}$. Similarly, property (iii) follows from the fact that if $\mathbf{V}$ and $\mathbf{W}$ are parallel then $\mathbf{V} \cdot \mathbf{W}$ is constant. Finally, the kernel (null space) of P_α is zero because $\|P_\alpha(\mathbf{v})\| = 0$ implies $\|\mathbf{v}\| = 0$, by (iii), so P_α is a one to one linear map from one n-dimensional vector space to another. But all such maps are onto. □

EXERCISES

8.1. Let S be an n-surface in $\mathbb{R}^{n+1}$, let $\alpha: I \to S$ be a parametrized curve, and let $\mathbf{X}$ and $\mathbf{Y}$ be vector fields tangent to S along α. Verify that

(a) $(\mathbf{X} + \mathbf{Y})' = \mathbf{X}' + \mathbf{Y}'$, and
(b) $(f\mathbf{X})' = f'\mathbf{X} + f\mathbf{X}'$,

for all smooth functions f along α.

8.2. Let S be an n-plane $a_1 x_1 + \cdots + a_{n+1} x_{n+1} = b$ in $\mathbb{R}^{n+1}$, let $p, q \in S$, and let $\mathbf{v} = (p, v) \in S_p$. Show that if α is any parametrized curve in S from p to q then $P_\alpha(\mathbf{v}) = (q, v)$. Conclude that, in an n-plane, parallel transport is path independent.

8.3. Let $\alpha: [0, \pi] \to \mathbf{S}^2$ be the half great circle in $\mathbf{S}^2$, running from the north pole $p = (0, 0, 1)$ to the south pole $q = (0, 0, -1)$, defined by $\alpha(t) = (\sin t, 0, \cos t)$. Show that, for $\mathbf{v} = (p, v_1, v_2, 0) \in \mathbf{S}^2_p$, $P_\alpha(\mathbf{v}) = (q, -v_1, v_2, 0)$. [*Hint*: Check this first when $\mathbf{v} = (p, 1, 0, 0)$ and when $\mathbf{v} = (p, 0, 1, 0)$; then use the linearity of P_α.]

8.4. Let p be a point in the 2-sphere $\mathbf{S}^2$ and let $\mathbf{v}$ and $\mathbf{w} \in \mathbf{S}^2_p$ be such that $\|\mathbf{v}\| = \|\mathbf{w}\|$. Show that there is a piecewise smooth parametrized curve $\alpha: [a, b] \to \mathbf{S}^2$, with $\alpha(a) = \alpha(b) = p$, such that $P_\alpha(\mathbf{v}) = \mathbf{w}$. [*Hint*: Consider closed curves α, with $\dot\alpha(a) = \mathbf{v}$, which form geodesic triangles with $\alpha(t) \perp p$ for t in the "middle segment" of $[a, b]$.]

8.5. Let $\alpha: I \to \mathbb{R}^{n+1}$ be a parametrized curve with $\alpha(t) \in S_1 \cap S_2$ for all $t \in I$, where S_1 and S_2 are two n-surfaces in $\mathbb{R}^{n+1}$. Suppose $\mathbf{X}$ is a vector field along α which is tangent both to S_1 and to S_2 along α.

(a) Show by example that $\mathbf{X}$ may be parallel along α viewed as a curve in S_1 but not parallel along α viewed as a curve in S_2.
(b) Show that if S_1 is tangent to S_2 along α (that is, $(S_1)_{\alpha(t)} = (S_2)_{\alpha(t)}$ for all $t \in I$) then $\mathbf{X}$ is parallel along α in S_1 if and only if $\mathbf{X}$ is parallel along α in S_2.
(c) Show that, if S_1 and S_2 are n-surfaces which are tangent along a parametrized curve $\alpha: I \to S_1 \cap S_2$, then α is a geodesic in S_1 if and only if α is a geodesic in S_2.

8.6. Let S be an n-surface and let $\alpha: I \to S$ be a parametrized curve in S. Let $\beta: \tilde{I} \to S$ be defined by $\beta = \alpha \circ h$ where $h: \tilde{I} \to I$ is a smooth function with $h'(t) \neq 0$ for all $t \in \tilde{I}$. Show that a vector field $\mathbf{X}$ tangent to S along α is parallel if and only if $X \circ h$ is parallel along β. Conclude that parallel transport from $p \in S$ to $q \in S$ along a parametrized curve α in S is the same as parallel transport from p to q along any reparametrization of α, and that parallel transport from q to p along $\alpha \circ h$, where $h(t) = -t$, is the inverse of parallel transport from p to q along α.

8.7. Let S be an n-surface in $\mathbb{R}^{n+1}$, let $p \in S$, and let G_p denote the group of non-singular linear transformations from S_p to itself. Let

$$H_p = \{T \in G_p\colon T = P_\alpha \text{ for some piecewise smooth } \alpha\colon [a, b] \to S \text{ with } \alpha(a) = \alpha(b) = p\}.$$

Show that H_p is a subgroup of G_p by showing that

(i) for each pair of piecewise smooth curves α and β in S from p to p there is a piecewise smooth curve γ from p to p such that

$$P_\gamma = P_\beta \circ P_\alpha, \text{ and}$$

(ii) for each α in S from p to p there is a β in S from p to p such that $P_\beta = P_\alpha^{-1}$.

(The subgroup H_p is called the *holonomy group* of S at p.)

8.8. Let $\alpha: I \to S$ be a unit speed curve in an n-surface S, and let $\mathbf{X}$ be a smooth vector field, tangent to S along α, which is everywhere orthogonal to α ($\mathbf{X}(t) \cdot \dot{\alpha}(t) = 0$ for all $t \in I$). Define the *Fermi derivative* $\mathbf{X}'$ of $\mathbf{X}$ by

$$\mathbf{X}'(t) = \mathbf{X}'(t) - [\mathbf{X}'(t) \cdot \dot{\alpha}(t)]\dot{\alpha}(t).$$

(a) Show that if $\mathbf{X}$ and $\mathbf{Y}$ are smooth vector fields along α which are tangent to S and orthogonal to α then

(i) $(\mathbf{X} + \mathbf{Y})' = \mathbf{X}' + \mathbf{Y}'$
(ii) $(f\mathbf{X})' = f'\mathbf{X} + f\mathbf{X}'$ for all smooth functions f along α, and
(iii) $(\mathbf{X} \cdot \mathbf{Y})' = \mathbf{X}' \cdot \mathbf{Y} + \mathbf{X} \cdot \mathbf{Y}'$.

(b) Show that if $a \in I$ and $\mathbf{v} \in S_{\alpha(a)}$ is orthogonal to $\dot{\alpha}(a)$ then there exists a unique vector field $\mathbf{V}$ along α, tangent to S and orthogonal to α, such that $\mathbf{V}' = 0$ and $\mathbf{V}(a) = \mathbf{v}$. ($\mathbf{V}$ is said to be *Fermi parallel* along α.)

(c) For $\alpha: [a, b] \to S$ a parametrized curve in S and $\mathbf{v} \in S_{\alpha(a)}$, with $\mathbf{v} \perp \dot{\alpha}(0)$, let $F_\alpha(\mathbf{v}) = \mathbf{V}(b)$ where $\mathbf{V}$ is as in part b). Show that F_α is a vector space isomorphism from $\dot{\alpha}(a)^\perp$ onto $\dot{\alpha}(b)^\perp$, where $\dot{\alpha}(t)^\perp$ is the orthogonal complement of $\dot{\alpha}(t)$ in $S_{\alpha(t)}$. Also show that F_α preserves dot products. ($F_\alpha(\mathbf{v})$ is the *Fermi transport* of $\mathbf{v}$ along α to $\alpha(b)$.)

The Weingarten Map 9

We shall now consider the local behavior of curvature on an n-surface. The way in which an n-surface curves around in $\mathbb{R}^{n+1}$ is measured by the way the normal direction changes as we move from point to point on the surface. In order to measure the rate of change of the normal direction, we need to be able to differentiate vector fields on n-surfaces.

Recall that, given a smooth function f defined on an open set U in $\mathbb{R}^{n+1}$ and a vector $\mathbf{v} \in \mathbb{R}_p^{n+1}$, $p \in U$, the *derivative* of f with respect to $\mathbf{v}$ is the real number

$$\nabla_{\mathbf{v}} f = (f \circ \alpha)'(t_0)$$

where $\alpha: I \to U$ is any parametrized curve in U with $\dot{\alpha}(t_0) = \mathbf{v}$. Note that, although the curve α appears in the formula defining $\nabla_{\mathbf{v}} f$, the value of the derivative does not depend on the choice of α. Indeed, by the chain rule,

$$\nabla_{\mathbf{v}} f = (f \circ \alpha)'(t_0) = \nabla f(\alpha(t_0)) \cdot \dot{\alpha}(t_0) = \nabla f(p) \cdot \mathbf{v}.$$

This formula, expressing $\nabla_{\mathbf{v}} f$ in terms of the gradient of f, shows that the value of $\nabla_{\mathbf{v}} f$ is independent of the choice of curve α passing through p with velocity $\mathbf{v}$. It is frequently the most useful formula to use in computations. This formula also shows that the function which sends $\mathbf{v}$ into $\nabla_{\mathbf{v}} f$ is a linear map from $\mathbb{R}_p^{n+1}$ to $\mathbb{R}$; that is,

$$\nabla_{\mathbf{v}+\mathbf{w}} f = \nabla_{\mathbf{v}} f + \nabla_{\mathbf{w}} f$$

and

$$\nabla_{c\mathbf{v}} f = c\nabla_{\mathbf{v}} f$$

for all $\mathbf{v}, \mathbf{w} \in \mathbb{R}_p^{n+1}$ and $c \in \mathbb{R}$.

Note that $\nabla_{\mathbf{v}} f$ depends on the magnitude of $\mathbf{v}$ as well as on the direction of $\mathbf{v}$. The formula $\nabla_{2\mathbf{v}} f = 2\nabla_{\mathbf{v}} f$, for example, expresses the fact that if we move twice as fast through p, the observed rate of change of f will double.

When $\|\mathbf{v}\| = 1$, the derivative $\nabla_{\mathbf{v}} f$ is called the *directional derivative* of f at p in the direction $\mathbf{v}$.

Given an n-surface S in $\mathbb{R}^{n+1}$ and a smooth function $f\colon S \to \mathbb{R}$, its *derivative* with respect to a vector $\mathbf{v}$ tangent to S is defined similarly, by

$$\nabla_{\mathbf{v}} f = (f \circ \alpha)'(t_0)$$

where $\alpha\colon I \to S$ is any parametrized curve in S with $\dot{\alpha}(t_0) = \mathbf{v}$. Note that the value of $\nabla_{\mathbf{v}} f$ is independent of the curve α in S passing through p with velocity $\mathbf{v}$, since

$$\nabla_{\mathbf{v}} f = (\tilde{f} \circ \alpha)'(t_0) = \nabla \tilde{f}(\alpha(t_0)) \cdot \dot{\alpha}(t_0) = \nabla \tilde{f}(p) \cdot \mathbf{v}$$

where $\tilde{f}\colon U \to \mathbb{R}$ is any smooth function, defined on an open set U containing S, whose restriction to S is f. It also follows from this last formula that the function which sends $\mathbf{v}$ into $\nabla_{\mathbf{v}} f$ is a linear map from S_p to $\mathbb{R}$.

The *derivative of a smooth vector field* $\mathbf{X}$ on an open set U in $\mathbb{R}^{n+1}$ with respect to a vector $\mathbf{v} \in \mathbb{R}^{n+1}_p$, $p \in U$, is defined by

$$\nabla_{\mathbf{v}} \mathbf{X} = (\mathbf{X} \mathbin{\dot{\circ}} \alpha)(t_0)$$

where $\alpha\colon I \to U$ is any parametrized curve in U such that $\dot{\alpha}(t_0) = \mathbf{v}$. For $\mathbf{X}$ a smooth vector field on an n-surface S in $\mathbb{R}^{n+1}$ and $\mathbf{v}$ a vector tangent to S at $p \in S$, the derivative $\nabla_{\mathbf{v}} \mathbf{X}$ is defined by the same formula, where now α is required to be a parametrized curve in S with $\dot{\alpha}(t_0) = \mathbf{v}$. Note that, in both situations, $\nabla_{\mathbf{v}} \mathbf{X} \in \mathbb{R}^{n+1}_p$ and that

$$\begin{aligned}\nabla_{\mathbf{v}} \mathbf{X} &= (\alpha(t_0), (X_1 \circ \alpha)'(t_0), \ldots, (X_{n+1} \circ \alpha)'(t_0)) \\ &= (p, \nabla_{\mathbf{v}} X_1, \ldots, \nabla_{\mathbf{v}} X_{n+1})\end{aligned}$$

where the X_i are the components of $\mathbf{X}$. In particular, the value of $\nabla_{\mathbf{v}} \mathbf{X}$ does not depend on the choice of α.

It is easy to check (Exercise 9.4) that differentiation of vector fields has the following properties:

(i) $\nabla_{\mathbf{v}}(\mathbf{X} + \mathbf{Y}) = \nabla_{\mathbf{v}} \mathbf{X} + \nabla_{\mathbf{v}} \mathbf{Y}$
(ii) $\nabla_{\mathbf{v}}(f\mathbf{X}) = (\nabla_{\mathbf{v}} f)\mathbf{X}(p) + f(p)(\nabla_{\mathbf{v}} \mathbf{X})$
(iii) $\nabla_{\mathbf{v}}(\mathbf{X} \cdot \mathbf{Y}) = (\nabla_{\mathbf{v}} \mathbf{X}) \cdot \mathbf{Y}(p) + \mathbf{X}(p) \cdot (\nabla_{\mathbf{v}} \mathbf{Y})$

for all smooth vector fields $\mathbf{X}$ and $\mathbf{Y}$ on U (or on S) and all smooth functions $f\colon U \to \mathbb{R}$ (or $f\colon S \to \mathbb{R}$). Here, the sum $\mathbf{X} + \mathbf{Y}$ of two vector fields $\mathbf{X}$ and $\mathbf{Y}$ is the vector field defined by $(\mathbf{X} + \mathbf{Y})(q) = \mathbf{X}(q) + \mathbf{Y}(q)$, the product of a function f and a vector field $\mathbf{X}$ is the vector field defined by $(f\mathbf{X})(q) = f(q)\mathbf{X}(q)$, and the dot product of vector fields $\mathbf{X}$ and $\mathbf{Y}$ is the function defined by $(\mathbf{X} \cdot \mathbf{Y})(q) = \mathbf{X}(q) \cdot \mathbf{Y}(q)$, for all $q \in U$ (or for all $q \in S$). Moreover, for each smooth vector field $\mathbf{X}$, the function which sends $\mathbf{v}$ into $\nabla_{\mathbf{v}} \mathbf{X}$ is a linear map, from $\mathbb{R}_p^{n+1}$ into $\mathbb{R}_p^{n+1}$ if $\mathbf{X}$ is a vector field on an open set U, and from S_p into $\mathbb{R}_p^{n+1}$ if $\mathbf{X}$ is a vector field on an n-surface S.

Note that the derivative $\nabla_{\mathbf{v}}\mathbf{X}$ of a tangent vector field $\mathbf{X}$ on an n-surface S with respect to a vector $\mathbf{v}$ tangent to S at $p \in S$ will not in general be tangent to S. In later chapters we will find it useful to consider the tangential component $D_{\mathbf{v}}\mathbf{X}$ of $\nabla_{\mathbf{v}}\mathbf{X}$:

$$D_{\mathbf{v}}\mathbf{X} = \nabla_{\mathbf{v}}\mathbf{X} - (\nabla_{\mathbf{v}}\mathbf{X} \cdot \mathbf{N}(p))\mathbf{N}(p),$$

where $\mathbf{N}$ is an orientation on S. $D_{\mathbf{v}}\mathbf{X}$ is called the *covariant derivative* of the tangent vector field $\mathbf{X}$ with respect to $\mathbf{v} \in S_p$. Note that $D_{\mathbf{v}}\mathbf{X} = (\mathbf{X} \circ \alpha)'(t_0)$ where $\alpha\colon I \to S$ is any parametrized curve in S with $\dot{\alpha}(t_0) = \mathbf{v}$. Covariant differentiation has the same properties ((i)–(iii) above, with ∇ replaced by D) as ordinary differentiation (see Exercise 9.5). Moreover, for each smooth tangent vector field $\mathbf{X}$ on S, the function which sends $\mathbf{v}$ into $D_{\mathbf{v}}\mathbf{X}$ is a linear map from S_p into S_p.

We are now ready to study the rate of change of the normal direction $\mathbf{N}$ on an oriented n-surface S in $\mathbb{R}^{n+1}$. Note that, for $p \in S$ and $\mathbf{v} \in S_p$, the derivative $\nabla_{\mathbf{v}}\mathbf{N}$ is tangent to S (i.e., $\nabla_{\mathbf{v}}\mathbf{N} \perp \mathbf{N}(p)$) since

$$\begin{aligned} 0 = \nabla_{\mathbf{v}}(1) = \nabla_{\mathbf{v}}(\mathbf{N} \cdot \mathbf{N}) &= (\nabla_{\mathbf{v}}\mathbf{N}) \cdot \mathbf{N}(p) + \mathbf{N}(p) \cdot (\nabla_{\mathbf{v}}\mathbf{N}) \\ &= 2(\nabla_{\mathbf{v}}\mathbf{N}) \cdot \mathbf{N}(p). \end{aligned}$$

The linear map $L_p\colon S_p \to S_p$ defined by

$$L_p(\mathbf{v}) = -\nabla_{\mathbf{v}}\mathbf{N}$$

is called the *Weingarten map* of S at p. The geometric meaning of L_p can be seen from the formula

$$\nabla_{\mathbf{v}}\mathbf{N} = -(\mathbf{N} \dot{\circ} \alpha)(t_0)$$

where $\alpha\colon I \to S$ is any parametrized curve in S with $\dot{\alpha}(t_0) = \mathbf{v}$: $L_p(\mathbf{v})$ measures (up to sign) the rate of change of $\mathbf{N}$ (i.e., the turning of $\mathbf{N}$ since $\mathbf{N}$ has constant length) as one passes through p along any such curve α. Since the tangent space $S_{\alpha(t)}$ to S at $\alpha(t)$ is just $[\mathbf{N}(\alpha(t))]^{\perp}$, the tangent space turns as the normal $\mathbf{N}$ turns and so $L_p(\mathbf{v})$ can be interpreted as a measure of the turning of the tangent space as one passes through p along α (see Figure 9.1). Thus L_p contains information about the shape of S; for this reason L_p is sometimes called the *shape operator* of S at p.

For computational purposes, it is important to note that $L_p(\mathbf{v})$ can be obtained from the formula

$$\begin{aligned} L_p(\mathbf{v}) = -\nabla_{\mathbf{v}}\mathbf{N} &= -(p, \nabla_{\mathbf{v}} N_1, \ldots, \nabla_{\mathbf{v}} N_{n+1}) \\ &= -(p, \nabla\tilde{N}_1(p) \cdot \mathbf{v}, \ldots, \nabla\tilde{N}_{n+1}(p) \cdot \mathbf{v}), \end{aligned}$$

where $\tilde{N}$ is *any* smooth vector field defined on an open set U containing S with $\tilde{\mathbf{N}}(q) = \mathbf{N}(q)$ for all $q \in S$. Note that $\tilde{\mathbf{N}}(q)$ need not be a unit vector for

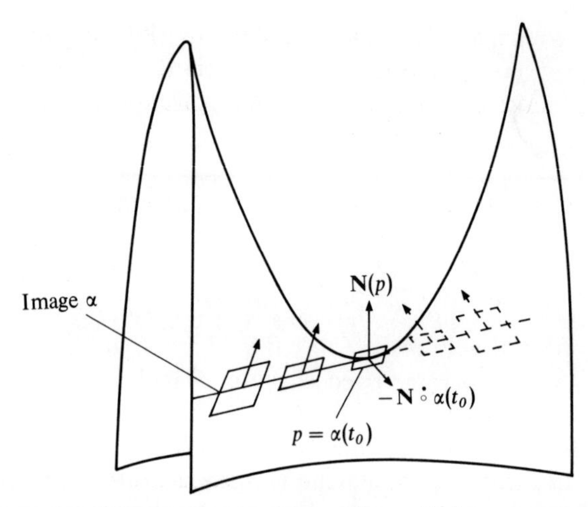

Figure 9.1 The Weingarten map. $L_p(\mathbf{v}) = -(\mathbf{N} \circ \alpha)^{\cdot}(t_0)$ measures the turning of the normal, hence the turning of the tangent space, as one passes through p along the curve α.

$q \notin S$. When $f\colon U \to \mathbb{R}$ is such that $S = f^{-1}(c)$ for some $c \in \mathbb{R}$ and $\mathbf{N}(q) = \nabla f(q)/\|\nabla f(q)\|$ for all $q \in S$, it is natural to take $\tilde{\mathbf{N}} = \nabla f/\|\nabla f\|$. Sometimes, however, another choice of $\tilde{\mathbf{N}}$ is more convenient, as the following example shows.

Example. Let S be the n-sphere $x_1^2 + \cdots + x_{n+1}^2 = r^2$ of radius $r > 0$, oriented by the inward unit normal vector field $\mathbf{N}$:

$$\mathbf{N}(q) = (q, -q/\|q\|) = (q, -q/r)$$

for $q \in S$. Setting $\tilde{\mathbf{N}}(q) = (q, -q/r)$ for $q \in \mathbb{R}^{n+1}$ (i.e.,

$$\tilde{\mathbf{N}}(x_1, \ldots, x_{n+1}) = \left(x_1, \ldots, x_{n+1}, -\frac{x_1}{r}, \ldots, -\frac{x_{n+1}}{r}\right),$$

we have, for $p \in S$ and $\mathbf{v} \in S_p$,

$$\begin{aligned} L_p(\mathbf{v}) &= -\nabla_{\mathbf{v}}\mathbf{N} = -(p, \nabla_{\mathbf{v}}\tilde{N}_1, \ldots, \nabla_{\mathbf{v}}\tilde{N}_{n+1}) \\ &= -\left(p, \nabla_{\mathbf{v}}\left(-\frac{x_1}{r}\right), \ldots, \nabla_{\mathbf{v}}\left(-\frac{x_{n+1}}{r}\right)\right) \\ &= \frac{1}{r}(p, \nabla_{\mathbf{v}}(x_1), \ldots, \nabla_{\mathbf{v}}(x_{n+1})). \end{aligned}$$

But, for each $i \in \{1, \ldots, n+1\}$,

$$\nabla_{\mathbf{v}} x_i = \nabla x_i(p) \cdot \mathbf{v} = (p, 0, \ldots, 1, \ldots, 0) \cdot (p, v_1, \ldots, v_{n+1}) = v_i$$

so

$$L_p(\mathbf{v}) = \frac{1}{r}(p, v_1, \ldots, v_{n+1}) = \frac{1}{r}\mathbf{v}.$$

Thus the Weingarten map of the n-sphere of radius r is simply multiplication by $1/r$. Note however the dependence on the choice of orientation: if S is oriented by the outward normal $-\mathbf{N}$, the Weingarten map will be multiplication by $-1/r$.

The following two theorems exhibit important properties of the Weingarten map.

Theorem 1. *Let S be an n-surface in $\mathbb{R}^{n+1}$, oriented by the unit normal vector field $\mathbf{N}$. Let $p \in S$ and $\mathbf{v} \in S_p$. Then for every parametrized curve $\alpha: I \to S$, with $\dot{\alpha}(t_0) = \mathbf{v}$ for some $t_0 \in I$,*

$$\ddot{\alpha}(t_0) \cdot \mathbf{N}(p) = L_p(\mathbf{v}) \cdot \mathbf{v}.$$

This theorem says that the normal component $\ddot{\alpha}(t_0) \cdot \mathbf{N}(p)$ of acceleration is the same for all parametrized curves α in S passing through p with velocity $\mathbf{v}$. In particular, if the normal component of acceleration is non-zero for some curve α with $\dot{\alpha}(t_0) = \mathbf{v}$ then it is non-zero for all curves in S passing through p with the same velocity (see Figure 9.2). This component of acceler-

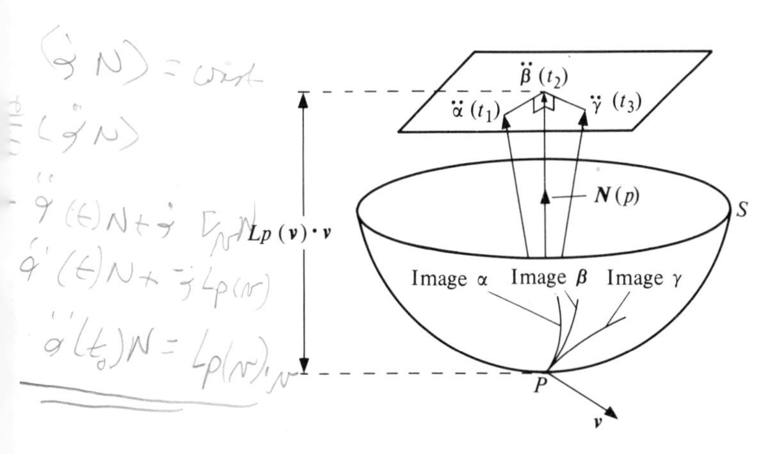

Figure 9.2 All parametrized curves in S passing through p with the same velocity will necessarily have the same normal component of acceleration at p. In the figure, $\dot{\alpha}(t_1) = \dot{\beta}(t_2) = \dot{\alpha}(t_3) = \mathbf{v}$; β is a geodesic.

ation is forced on every such curve in S by the shape of S at p and, according to the above formula, it can be computed directly from the value of the Weingarten map L_p on $\mathbf{v}$.

Note that when α is a geodesic, its only component of acceleration is normal to the surface, and this acceleration is forced on α by the shape of the surface.

For S the n-sphere of radius r, the computation in the example above shows that every unit speed curve α in S has normal component of acceleration pointing inward with magnitude $1/r$.

PROOF OF THEOREM 1. Since α is a parametrized curve in S, $\dot{\alpha}(t) \in S_{\alpha(t)} = [\mathbf{N}(\alpha(t))]^\perp$ for all $t \in I$; i.e., $\dot{\alpha} \cdot (\mathbf{N} \circ \alpha) = 0$ along α. Hence

$$\begin{aligned}
0 &= [\dot{\alpha} \cdot (\mathbf{N} \circ \alpha)]'(t_0) \\
&= \ddot{\alpha}(t_0) \cdot (\mathbf{N} \circ \alpha)(t_0) + \dot{\alpha}(t_0) \cdot (\mathbf{N} \dot{\circ} \alpha)(t_0) \\
&= \ddot{\alpha}(t_0) \cdot \mathbf{N}(\alpha(t_0)) + \mathbf{v} \cdot \nabla_{\mathbf{v}} \mathbf{N} \\
&= \ddot{\alpha}(t_0) \cdot \mathbf{N}(p) - \mathbf{v} \cdot L_p(\mathbf{v})
\end{aligned}$$

so $\ddot{\alpha}(t_0) \cdot \mathbf{N}(p) = L_p(\mathbf{v}) \cdot \mathbf{v}$ as claimed. □

Theorem 2. *The Weingarten map L_p is self-adjoint; that is,*

$$L_p(\mathbf{v}) \cdot \mathbf{w} = \mathbf{v} \cdot L_p(\mathbf{w})$$

for all $\mathbf{v}, \mathbf{w} \in S_p$.

PROOF. Let $f\colon U \to \mathbb{R}$ (U open in $\mathbb{R}^{n+1}$) be such that $S = f^{-1}(c)$ for some $c \in \mathbb{R}$ and such that $\mathbf{N}(p) = \nabla f(p)/\|\nabla f(p)\|$ for all $p \in S$. Then

$$\begin{aligned}
L_p(\mathbf{v}) \cdot \mathbf{w} = (-\nabla_{\mathbf{v}}\mathbf{N}) \cdot \mathbf{w} &= -\nabla_{\mathbf{v}}\left(\frac{\nabla f}{\|\nabla f\|}\right) \cdot \mathbf{w} \\
&= -\left[\nabla_{\mathbf{v}}\left(\frac{1}{\|\nabla f\|}\right)\nabla f(p) + \frac{1}{\|\nabla f(p)\|}\nabla_{\mathbf{v}}(\nabla f)\right] \cdot \mathbf{w} \\
&= -\nabla_{\mathbf{v}}\left(\frac{1}{\|\nabla f\|}\right)\nabla f(p) \cdot \mathbf{w} - \frac{1}{\|\nabla f(p)\|}[\nabla_{\mathbf{v}}(\nabla f)] \cdot \mathbf{w}.
\end{aligned}$$

Since $\nabla f(p) \cdot \mathbf{w} = 0$, the first term drops out. Thus

$$\begin{aligned}
L_p(\mathbf{v}) \cdot \mathbf{w} &= -\frac{1}{\|\nabla f(p)\|}[\nabla_{\mathbf{v}}(\nabla f)] \cdot \mathbf{w} \\
&= -\frac{1}{\|\nabla f(p)\|}\left(p, \nabla_{\mathbf{v}}\frac{\partial f}{\partial x_1}, \ldots, \nabla_{\mathbf{v}}\frac{\partial f}{\partial x_{n+1}}\right) \cdot \mathbf{w} \\
&= -\frac{1}{\|\nabla f(p)\|}\left(p, \nabla\left(\frac{\partial f}{\partial x_1}\right)(p) \cdot \mathbf{v}, \ldots, \nabla\left(\frac{\partial f}{\partial x_{n+1}}\right)(p) \cdot \mathbf{v}\right) \cdot \mathbf{w} \\
&= -\frac{1}{\|\nabla f(p)\|}\left(p, \sum_{i=1}^{n+1}\frac{\partial^2 f}{\partial x_i\,\partial x_1}(p)v_i, \ldots, \sum_{i=1}^{n+1}\frac{\partial^2 f}{\partial x_i\,\partial x_{n+1}}(p)v_i\right) \cdot \mathbf{w} \\
&= -\frac{1}{\|\nabla f(p)\|}\sum_{i,j=1}^{n+1}\frac{\partial^2 f}{\partial x_i\,\partial x_j}(p)v_i w_j,
\end{aligned}$$

where $\mathbf{v} = (p, v_1, \ldots, v_{n+1})$ and $\mathbf{w} = (p, w_1, \ldots, w_{n+1})$.

The same computation, with $\mathbf{v}$ and $\mathbf{w}$ interchanged, shows that

$$L_p(\mathbf{w}) \cdot \mathbf{v} = -\frac{1}{\|\nabla f(p)\|}\sum_{i,j=1}^{n+1}\frac{\partial^2 f}{\partial x_i\,\partial x_j}(p)w_i v_j.$$

Since $\partial^2 f/\partial x_i\,\partial x_j = \partial^2 f/\partial x_j\,\partial x_i$ for all (i, j), we can finally conclude that

$$L_p(\mathbf{v}) \cdot \mathbf{w} = -\frac{1}{\|\nabla f(p)\|}\sum_{i,j=1}^{n+1}\frac{\partial^2 f}{\partial x_i\,\partial x_j}(p)v_i w_j$$

$$= -\frac{1}{\|\nabla f(p)\|}\sum_{i,j=1}^{n+1}\frac{\partial^2 f}{\partial x_j\,\partial x_i}(p)v_i w_j$$

$$= -\frac{1}{\|\nabla f(p)\|}\sum_{i,j=1}^{n+1}\frac{\partial^2 f}{\partial x_j\,\partial x_i}(p)w_j v_i = L_p(\mathbf{w}) \cdot \mathbf{v}. \qquad \square$$

Exercises

9.1. Compute $\nabla_{\mathbf{v}} f$ where $f: \mathbb{R}^{n+1} \to \mathbb{R}$ and $\mathbf{v} \in \mathbb{R}^{n+1}_p$, $p \in \mathbb{R}^{n+1}$, are given by

(a) $f(x_1, x_2) = 2x_1^2 + 3x_2^2$, $\mathbf{v} = (1, 0, 2, 1)$ $(n = 1)$
(b) $f(x_1, x_2) = x_1^2 - x_2^2$, $\mathbf{v} = (1, 1, \cos\theta, \sin\theta)$ $(n = 1)$
(c) $f(x_1, x_2, x_3) = x_1 x_2 x_3^2$, $\mathbf{v} = (1, 1, 1, a, b, c)$ $(n = 2)$
(d) $f(q) = q \cdot q$, $\mathbf{v} = (p, v)$ (arbitrary n).

9.2. Let U be an open set in $\mathbb{R}^{n+1}$ and let $f: U \to \mathbb{R}$ be a smooth function. Show that if $\mathbf{e}_i = (p, 0, \ldots, 1, \ldots, 0)$ where $p \in U$ and the 1 is in the $(i + 1)$th spot (i spots after the p), then $\nabla_{\mathbf{e}_i} f = (\partial f/\partial x_i)(p)$.

9.3. Compute $\nabla_{\mathbf{v}}\mathbf{X}$ where $\mathbf{v} \in \mathbb{R}^{n+1}_p$, $p \in \mathbb{R}^{n+1}$, and $\mathbf{X}$ are given by

(a) $\mathbf{X}(x_1, x_2) = (x_1, x_2, x_1 x_2, x_2^2)$, $\mathbf{v} = (1, 0, 0, 1)$ $(n = 1)$
(b) $\mathbf{X}(x_1, x_2) = (x_1, x_2, -x_2, x_1)$, $\mathbf{v} = (\cos\theta, \sin\theta, -\sin\theta, \cos\theta)$ $(n = 1)$
(c) $\mathbf{X}(q) = (q, 2q)$, $\mathbf{v} = (0, \ldots, 0, 1, \ldots, 1)$ (arbitrary n).

9.4. Verify that differentiation of vector fields has the properties (i)–(iii) as stated on page 54.

9.5. Show that covariant differentiation of vector fields has the following properties: for $p \in S$ and $\mathbf{v} \in S_p$,

(i) $D_{\mathbf{v}}(\mathbf{X} + \mathbf{Y}) = D_{\mathbf{v}}\mathbf{X} + D_{\mathbf{v}}\mathbf{Y}$
(ii) $D_{\mathbf{v}}(f\mathbf{X}) = (\nabla_{\mathbf{v}} f)\mathbf{X}(p) + f(p)D_{\mathbf{v}}\mathbf{X}$
(iii) $\nabla_{\mathbf{v}}(\mathbf{X} \cdot \mathbf{Y}) = (D_{\mathbf{v}}\mathbf{X}) \cdot \mathbf{Y}(p) + \mathbf{X}(p) \cdot (D_{\mathbf{v}}\mathbf{Y})$

for all smooth tangent vector fields $\mathbf{X}$ and $\mathbf{Y}$ on S and all smooth functions $f: S \to \mathbb{R}$.

9.6. Suppose $\mathbf{X}$ is a smooth unit vector field on an n-surface S in $\mathbb{R}^{n+1}$; i.e., $\|\mathbf{X}(q)\| = 1$ for all $q \in S$. Show that $\nabla_{\mathbf{v}}\mathbf{X} \perp \mathbf{X}(p)$ for all $\mathbf{v} \in S_p$, $p \in S$. Show further that, if $\mathbf{X}$ is a unit tangent vector field on S, then $D_{\mathbf{v}}\mathbf{X} \perp \mathbf{X}(p)$.

9.7. A smooth tangent vector field $\mathbf{X}$ on an n-surface S is said to be a *geodesic vector field*, or *geodesic flow*, if all integral curves of $\mathbf{X}$ are geodesics of S.
(a) Show that a smooth tangent vector field $\mathbf{X}$ on S is a geodesic field if and only if $D_{\mathbf{X}(p)}\mathbf{X} = \mathbf{0}$ for all $p \in S$.
(b) Describe a geodesic flow on a 2-surface of revolution in $\mathbb{R}^3$.

9.8. Compute the Weingarten map for

(a) the hyperplane $a_1 x_1 + \cdots + a_{n+1} x_{n+1} = b$ $[(a_1, \ldots, a_{n+1}) \neq (0, \ldots, 0)]$
(b) the circular cylinder $x_2^2 + x_3^2 = a^2$ in $\mathbb{R}^3$ $(a \neq 0)$

(Choose your own orientations).

9.9. Show that if S is an n-surface and $\mathbf{N}$ is a unit normal vector field on S, then the Weingarten map of S oriented by $-\mathbf{N}$ is the negative of the Weingarten map of S oriented by $\mathbf{N}$.

9.10. Let V be a finite dimensional vector space with inner product (dot product). Let $L\colon V \to V$ be a linear map.
(a) Show that there exists a unique linear map $L^*\colon V \to V$ such that $L^*(v) \cdot w = v \cdot L(w)$ for all $v, w \in V$. [*Hint*: Choose an orthonormal basis $\{e_1, \ldots, e_n\}$ for V and compute $L^*(e_i)$ for each i]. (L^* is the *adjoint* of L.)
(b) Show that the matrix for L^* relative to an orthonormal basis for V is the transpose of the matrix for L relative to that basis. Conclude that L is self-adjoint ($L^* = L$) if and only if the matrix for L relative to any orthonormal basis for V is symmetric.

9.11. Let $S = f^{-1}(c)$ be an n-surface in $\mathbb{R}^{n+1}$, oriented by $\nabla f/\|\nabla f\|$. Suppose $p \in S$ is such that $\nabla f(p)/\|\nabla f(p)\| = \mathbf{e}_{n+1}$, where $\mathbf{e}_i = (p, 0, \ldots, 1, \ldots, 0)$ with the 1 in the $(i+1)$th spot (i spots after the p) for $i \in \{1, \ldots, n+1\}$. Show that the matrix for L_p with respect to the basis $\{\mathbf{e}_1, \mathbf{e}_2, \ldots, \mathbf{e}_n\}$ for S_p is

$$\left(-\frac{1}{\|\nabla f(p)\|}\frac{\partial^2 f}{\partial x_i\, \partial x_j}(p)\right).$$

9.12. Let S be an n-surface in $\mathbb{R}^{n+1}$, oriented by the unit normal vector field $\mathbf{N}$. Suppose $\mathbf{X}$ and $\mathbf{Y}$ are smooth tangent vector fields on S.

(a) Show that

$$(\nabla_{\mathbf{X}(p)}\mathbf{Y}) \cdot \mathbf{N}(p) = (\nabla_{\mathbf{Y}(p)}\mathbf{X}) \cdot \mathbf{N}(p)$$

for all $p \in S$. [*Hint*: Show that both sides are equal to $L_p(\mathbf{X}(p)) \cdot \mathbf{Y}(p)$.]
(b) Conclude that the vector field $[\mathbf{X}, \mathbf{Y}]$ defined on S by $[\mathbf{X}, \mathbf{Y}](p) = \nabla_{\mathbf{X}(p)}\mathbf{Y} - \nabla_{\mathbf{Y}(p)}\mathbf{X}$ is everywhere tangent to S. ($[\mathbf{X}, \mathbf{Y}]$ is called the *Lie bracket* of the vector fields $\mathbf{X}$ and $\mathbf{Y}$.)

9.13. The *derivative* at $p \in U$ (U open in $\mathbb{R}^{n+1}$) of a smooth map $F\colon U \to \mathbb{R}^{n+1}$ is the linear transformation $F'(p)\colon \mathbb{R}^{n+1} \to \mathbb{R}^{n+1}$ such that for every $\varepsilon > 0$ there exists a $\delta > 0$ guaranteeing that

$$\|F(p+v) - F(p) - F'(p)(v)\|/\|v\| < \varepsilon \quad \text{whenever } \|v\| < \delta.$$

Show that if $X\colon U \to \mathbb{R}^{n+1}$ is smooth and $\mathbf{X}$ is the vector field on U given by $\mathbf{X}(q) = (q, X(q))$ for all $q \in U$ then the derivative of $\mathbf{X}$ with respect to a vector $\mathbf{v} = (p, v) \in \mathbb{R}^{n+1}$, where $p \in U$, can be computed from the formula

$$\nabla_{\mathbf{v}}\mathbf{X} = (p, [X'(p)](v)).$$

9.14. Show that the Weingarten map at a point p of an n-surface S in $\mathbb{R}^{n+1}$ is essentially equal to the negative of the derivative at p of the Gauss map of S by showing that

$$L_p(p, v) = (p, -[\tilde{N}'(p)](v))$$

for $(p, v) \in S_p$, where $\tilde{N}: U \to \mathbb{R}^{n+1}$ is any smooth function, defined on an open set U containing S, whose restriction to S is the Gauss map N of S. [For the definition of the derivative, see Exercise 9.13.]

9.15. Let $f: U \to \mathbb{R}$, $S = f^{-1}(c)$, and $\mathbf{N} = \nabla f / \|\nabla f\|$ be as in the proof of the theorem of Chapter 7. Let $\mathbf{X}$ be the vector field on $U \times \mathbb{R}^{n+1}$ defined by

$$\mathbf{X}(\mathbf{w}) = \mathbf{X}(q, w) = (q, w, w, -(\mathbf{w} \cdot \nabla_{\mathbf{w}} \mathbf{N}) N(q))$$

where $N(q)$ is the vector part of $\mathbf{N}(q)$ ($\mathbf{N}(q) = (q, N(q))$. For each parametrized curve $\alpha: I \to U$, define the *natural lift* $\tilde{\alpha}$ of α to be the parametrized curve $\tilde{\alpha}: I \to U \times \mathbb{R}^{n+1}$ given by $\tilde{\alpha}(t) = \dot{\alpha}(t)$.

(a) Suppose $\alpha: I \to S$. Show that α is a geodesic of S if and only if its natural lift $\tilde{\alpha}$ is an integral curve of $\mathbf{X}$. [*Hint*: Show that $\dot{\tilde{\alpha}}(t) = \mathbf{X}(\tilde{\alpha}(t))$ for all $t \in I$ if and only if α satisfies the geodesic equation (G).] Conclude that if $\alpha: I \to S$ and $\beta: \tilde{I} \to S$ are geodesics in S with $\alpha(0) = \beta(0)$ and $\dot{\alpha}(0) = \dot{\beta}(0)$ then $\alpha(t) = \beta(t)$ for all $t \in I \cap \tilde{I}$.

(b) Given $\mathbf{v} = (p, v) \in U \times \mathbb{R}^{n+1}$, let $\beta: I \to U \times \mathbb{R}^{n+1}$ be the integral curve of $\mathbf{X}$ through $\mathbf{v}$. Then $\beta(t)$ is of the form $\beta(t) = (\beta_1(t), \beta_2(t))$ where $\beta_1: I \to U$ and $\beta_2: I \to \mathbb{R}^{n+1}$. Show that if $p \in S$ and $\mathbf{v} \in S_p$ then β_1 is a geodesic of S passing through p with initial velocity $\mathbf{v}$. [*Hint*: First check that β_1 satisfies the geodesic equation (G), then proceed as in the proof of the theorem of Chapter 7 to verify that β_1 actually maps I into S.]

Remark. Exercise 9.15 verifies the existence and uniqueness of a maximal geodesic α in S with initial conditions $\alpha(0) = p$ and $\dot{\alpha}(0) = \mathbf{v}$ using only the existence and uniqueness theorem for integral curves of vector fields. The introduction of the natural lift $\tilde{\alpha}$ of a curve α is the geometric analogue of the substitution $u_i = dx_i/dt$ which reduces the 2nd order differential system

$$\frac{d^2x_i}{dt^2} + \sum N_i \frac{\partial N_j}{\partial x_k} \frac{dx_j}{dt} \frac{dx_k}{dt} = 0$$

(in $n + 1$ variables x_i) to the first order differential system

$$\begin{cases} \dfrac{dx_i}{dt} = u_i \\[2ex] \dfrac{du_i}{dt} = -\sum N_i \dfrac{\partial N_j}{\partial x_k} u_j u_k \end{cases}$$

(in $2n + 2$ variables x_i and u_i). This first order system of differential equations is just the differential equation for the integral curves of $\mathbf{X}$ in $U \times \mathbb{R}^{n+1} \subset \mathbb{R}^{2n+2}$. The vector field $\mathbf{X}$ is called a *geodesic spray*.

10 Curvature of Plane Curves

Let $C = f^{-1}(c)$, where $f\colon U \to \mathbb{R}$, be a plane curve in the open set $U \subset \mathbb{R}^2$, oriented by $\mathbf{N} = \nabla f/\|\nabla f\|$. Then, for each $p \in C$, the Weingarten map L_p is a linear transformation on the 1-dimensional space C_p. Since every linear transformation from a 1-dimensional space to itself is multiplication by a real number, there exists, for each $p \in C$, a real number $\kappa(p)$ such that

$$L_p(\mathbf{v}) = \kappa(p)\mathbf{v} \text{ for all } \mathbf{v} \in C_p.$$

$\kappa(p)$ is called the *curvature* of C at p.

If $\mathbf{v}$ is any non-zero vector tangent to the plane curve C at $p \in C$ then

$$L_p(\mathbf{v}) \cdot \mathbf{v} = \kappa(p)\|\mathbf{v}\|^2$$

so the curvature of C at p is given by the formula

$$\kappa(p) = L_p(\mathbf{v}) \cdot \mathbf{v}/\|\mathbf{v}\|^2.$$

In particular, if $\alpha\colon I \to C$ is any parametrized curve in C with $\dot{\alpha}(t) \neq 0$ for all $t \in I$ then, by Theorem 1 of Chapter 9,

$$\kappa(\alpha(t)) = \frac{L_p(\dot{\alpha}(t)) \cdot \dot{\alpha}(t)}{\|\dot{\alpha}(t)\|^2} = \frac{\ddot{\alpha}(t) \cdot \mathbf{N}(\alpha(t))}{\|\dot{\alpha}(t)\|^2}.$$

If α is a unit speed parametrized curve in C, this formula reduces to

$$\kappa(\alpha(t)) = \ddot{\alpha}(t) \cdot \mathbf{N}(\alpha(t)).$$

Thus the curvature of C at $p \in C$ measures the normal component of acceleration of any unit speed parametrized curve in C passing through p.

Note in particular the significance of the sign of $\kappa(p)$: if $\kappa(p) > 0$ then the curve at p is turning toward its normal $\mathbf{N}(p)$, and if $\kappa(p) < 0$ the curve is turning away from $\mathbf{N}(p)$ (see Figure 10.1).

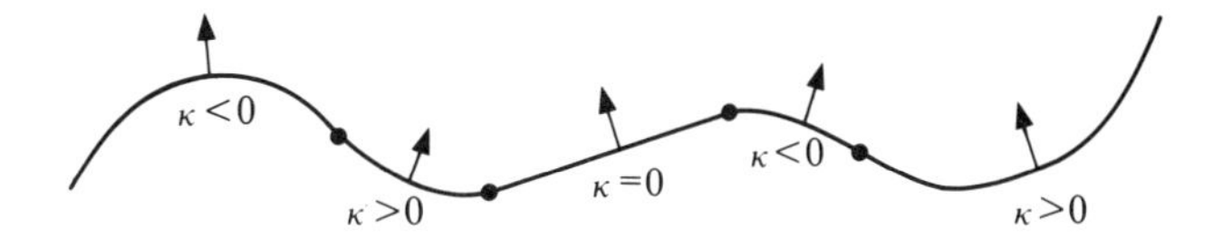

Figure 10.1 The curvature of C is positive at points where C is bending toward its normal and is negative where C is bending away from its normal.

One way to compute the curvature of a plane curve is to use the formula $\kappa \circ \alpha = (\ddot{\alpha} \cdot \mathbf{N} \circ \alpha)/\|\dot{\alpha}\|^2$ (or the equivalent formula in Exercise 10.1), where α is any parametrized curve in C whose velocity is nowhere zero. If such an α is oriented consistently with the orientation on C, it is called a local parametrization of C.

Given an oriented plane curve C and a point $p \in C$, a *parametrization of a segment of C containing p* is a parametrized curve $\alpha: I \to C$ which

(i) is regular; i.e., has $\dot{\alpha}(t) \neq \mathbf{0}$ for all $t \in I$,
(ii) is oriented consistently with C; i.e., is such that for each $t \in I$ the basis $\{\dot{\alpha}(t)\}$ for $C_{\alpha(t)}$ is consistent with the orientation $\mathbf{N}$ of C, and
(iii) has $p \in \text{Image } \alpha$.

If α is onto; i.e., if $\alpha(I) = C$, α is called a *global parametrization* of C. In general, α is called a *local parametrization* of C.

Local parametrizations of plane curves are, in principle, easy to obtain. If $C = f^{-1}(c)$ is oriented by $\mathbf{N} = \nabla f/\|\nabla f\|$, then $\nabla f(q) = (q, (\partial f/\partial x_1)(q), (\partial f/\partial x_2)(q))$ is orthogonal to C_q for each $q \in C$, and the vector field $\mathbf{X}$ given by $\mathbf{X}(q) = (q, (\partial f/\partial x_2)(q), -(\partial f/\partial x_1)(q))$ is everywhere orthogonal to ∇f ($\mathbf{X}(q)$ is obtained by rotating $\nabla f(q)$ through an angle of $-\pi/2$), so $\mathbf{X}$ is a tangent vector field on C. Further, $\mathbf{X}(q) \neq 0$ for $q \in C$ and $\{\mathbf{X}(q)\}$ is consistent with the orientation $\mathbf{N}$. Hence, given any point $p \in C$, the maximal integral curve $\alpha: I \to C$ of $\mathbf{X}$ through p will be a parametrization of a segment of C containing p.

Note that if, in this construction, the vector field ∇f is replaced by the vector field $\mathbf{N} = \nabla f/\|\nabla f\|$, then α becomes a unit speed parametrization of a segment of C containing p, since

$$\|\dot{\alpha}(t)\| = \|\mathbf{X}(\alpha(t))\| = \|\mathbf{N}(\alpha(t))\| = 1$$

for all $t \in I$.

Local parametrizations of plane curves are unique up to reparametrization: given any parametrization $\beta: \tilde{I} \to C$ of a segment of C containing p, there exists a smooth function $h: \tilde{I} \to \mathbb{R}$, with $h'(t) > 0$ for all $t \in \tilde{I}$, such that $\beta(t) = \alpha(h(t))$ for all $t \in \tilde{I}$, where α is the unit speed local parametrization constructed above. Indeed, since $\{\mathbf{X}(\beta(t))\}$ is a basis for the 1-dimensional vector space $C_{\beta(t)}$, $\dot{\beta}(t)$ is necessarily a multiple of $\mathbf{X}(\beta(t))$. In fact, $\dot{\beta}(t) = \|\dot{\beta}(t)\|\mathbf{X}(\beta(t))$ since $\|\mathbf{X}\| = 1$ and since $\{\dot{\beta}(t)\}$ and $\{\mathbf{X}(\beta(t))\}$ are both consistent with the orientation $\mathbf{N}$ of C. Setting

$$h(t) = \int_{t_0}^{t} \|\dot{\beta}(\tau)\| \, d\tau,$$

where t_0 is such that $\beta(t_0) = p$, we obtain a monotone increasing smooth function $h: \tilde{I} \to \mathbb{R}$ ($h'(t) = \|\dot{\beta}(t)\| > 0$ for all $t \in \tilde{I}$) which sends t_0 to 0. The parametrized curve $\beta \circ h^{-1}$ has velocity

$$\begin{aligned}(\beta \dot{\circ} h^{-1})(t) &= \dot{\beta}(h^{-1}(t))(h^{-1})'(t) \\ &= \dot{\beta}(h^{-1}(t))/h'(h^{-1}(t)) \\ &= \dot{\beta}(h^{-1}(t))/\|\dot{\beta}(h^{-1}(t))\| = \mathbf{X}(\beta(h^{-1}(t))\end{aligned}$$

and so is an integral curve of the vector field $\mathbf{X}$, with $\beta \circ h^{-1}(0) = p = \alpha(0)$. By the uniqueness of integral curves, domain $\beta \circ h^{-1} \subset I$ and $\beta \circ h^{-1}(t) = \alpha(t)$ for all $t \in$ domain $\beta \circ h^{-1}$. In other words, $\beta(t) = \alpha(h(t))$ for all $t \in \tilde{I}$, as claimed. Note in particular that *if $\beta: \tilde{I} \to C$ is a unit speed local parametrization of C with $\beta(t_0) = p$ then $h(t) = t - t_0$ and $\beta(t) = \alpha(t - t_0)$ for all $t \in \tilde{I}$.*

Example. Let C be the circle $f^{-1}(r^2)$, where $f(x_1, x_2) = (x_1 - a)^2 + (x_2 - b)^2$, oriented by the outward normal $\nabla f/\|\nabla f\|$. Since $\nabla f(p) = (p, 2(x_1 - a), 2(x_2 - b))$ for $p = (x_1, x_2) \in \mathbb{R}^2$, the integral curves of $\mathbf{X}(p) = (p, 2(x_2 - b), -2(x_1 - a))$ will be local parametrizations of C. The integral curve through $(a + r, b)$ gives the global parametrization $\alpha(t) = (a + r \cos 2t, b - r \sin 2t)$. Hence

$$\begin{aligned}\kappa(\alpha(t)) &= \frac{\ddot{\alpha}(t) \cdot \mathbf{N}(\alpha(t))}{\|\dot{\alpha}(t)\|^2} = \frac{\ddot{\alpha}(t)}{\|\dot{\alpha}(t)\|^2} \cdot \frac{\nabla f(\alpha(t))}{\|\nabla f(\alpha(t))\|} \\ &= \frac{(-4r \cos 2t, 4r \sin 2t) \cdot (2r \cos 2t, -2r \sin 2t)}{\|(-2r \sin 2t, -2r \cos 2t)\|^2 \|(2r \cos 2t, -2r \sin 2t)\|} \\ &= \frac{-8r^2}{(4r^2)(2r)} = -\frac{1}{r}.\end{aligned}$$

If C had been oriented by the inward normal, the curvature would have been $+1/r$ at each point.

For C an arbitrary oriented plane curve and $p \in C$ such that $\kappa(p) \neq 0$, there exists a unique oriented circle O, called the *circle of curvature* of C at p (see Figure 10.2), which

(i) is tangent to C at p (i.e., $C_p = \mathsf{O}_p$)
(ii) is oriented consistently with C (i.e., $N(p) = N_1(p)$ where N and N_1 are respectively the orientation normals of C and O), and
(iii) whose normal turns at the same rate at p as does the normal to C (i.e., $\nabla_{\mathbf{v}} \mathbf{N} = \nabla_{\mathbf{v}} \mathbf{N}_1$ for all $\mathbf{v} \in C_p = \mathsf{O}_p$).

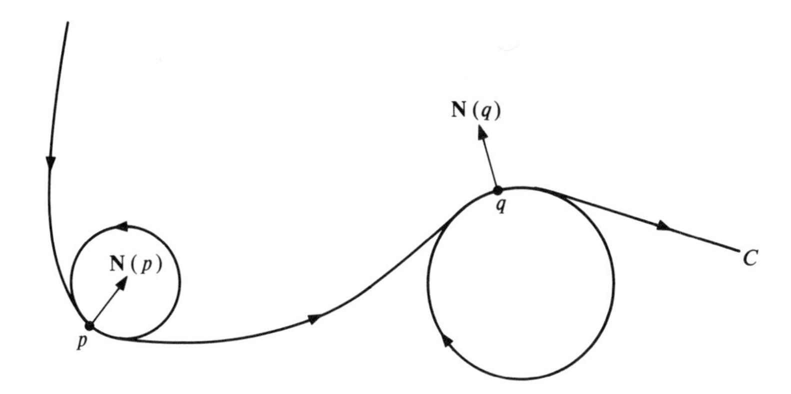

Figure 10.2 The circle of curvature at two points of an oriented plane curve C.

This circle of curvature is the circle which hugs the curve C closest among all circles containing p (see Exercises 10.8 and 10.9). Condition (i) says that the center of $\mathcal{O}$ is on the normal line to C at p, condition (iii) says that its radius r satisfies the equation $1/r = |\kappa(p)|$, where $\kappa(p)$ is the curvature of C at p, and condition (ii) says that $\mathbf{N}(p)$ points toward the center of $\mathcal{O}$ if $\kappa(p) > 0$ and away from the center of $\mathcal{O}$ if $\kappa(p) < 0$. The radius $r = 1/|\kappa(p)|$ of the circle of curvature is called the *radius of curvature* of C at p; its center is the *center of curvature* of C at p.

Exercises

10.1. Let $\alpha(t) = (x(t), y(t))$ $(t \in I)$ be a local parametrization of the oriented plane curve C. Show that

$$\kappa \circ \alpha = (x'y'' - y'x'')/(x'^2 + y'^2)^{3/2}.$$

10.2. Let $g\colon I \to \mathbb{R}$ be a smooth function and let C denote the graph of g. Show that the curvature of C at the point $(t, g(t))$ is $g''(t)/(1 + (g'(t))^2)^{3/2}$, for an appropriate choice of orientation.

10.3. Find global parametrizations of each of the following plane curves, oriented by $\nabla f/\|\nabla f\|$ where f is the function defined by the left side of each equation.

(a) $ax_1 + bx_2 = c$, $(a, b) \neq (0, 0)$.
(b) $x_1^2/a^2 + x_2^2/b^2 = 1$, $a \neq 0$, $b \neq 0$.
(c) $x_2 - ax_1^2 = c$, $a \neq 0$.
(d) $x_1^2 - x_2^2 = 1$, $x_1 > 0$.

10.4. Find the curvature κ of each of the oriented plane curves in Exercise 10.3.

10.5. Let C be an oriented plane curve. Let $p \in C$ and let $\mathbf{N}(p) = (p, N(p))$ denote the orientation unit normal vector at p. Show that if $\alpha\colon I \to C$ is any unit speed local parametrization of C with $\alpha(t_0) = p$, and $h(t) = (\alpha(t) - p) \cdot N(p)$ (see Figure 10.3), then $h(t_0) = h'(t_0) = 0$ and $h''(t_0) = \kappa(p)$.

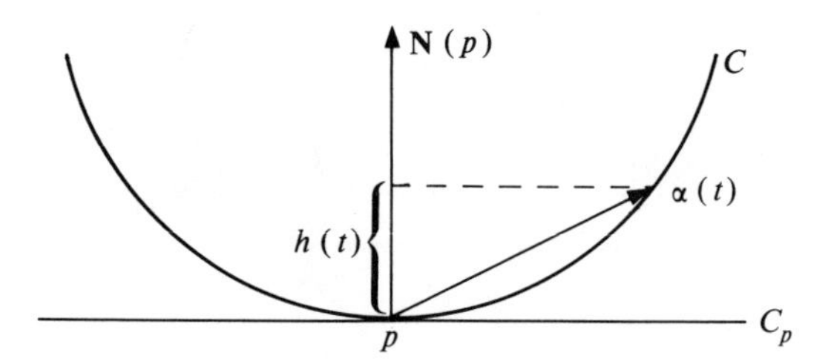

Figure 10.3 $h(t)$ is the projection of $\alpha(t) - p$ along $N(p)$. $h(t)$ may be thought of as the "height of $\alpha(t)$ above the line tangent to S at p."

10.6. Let C be a plane curve oriented by the unit normal vector field $\mathbf{N}$. Let $\alpha: I \to C$ be a unit speed local parametrization of C. For $t \in I$, let $\mathbf{T}(t) = \dot{\alpha}(t)$. Show that

$$\begin{cases} \dot{\mathbf{T}} = \kappa \mathbf{N} \\ \dot{\mathbf{N}} = -\kappa \mathbf{T} \end{cases}$$

or, more precisely,

$$\begin{cases} \dot{\mathbf{T}} = (\kappa \circ \alpha)(\mathbf{N} \circ \alpha) \\ (\mathbf{N} \circ \alpha)^{\cdot} = -(\kappa \circ \alpha)\mathbf{T}. \end{cases}$$

These formulas are called the *Frenet formulas* for a plane curve.

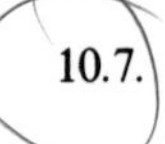
10.7. Let $\alpha: I \to \mathbb{R}^3$ be a unit speed parametrized curve in $\mathbb{R}^3$ such that $\dot{\alpha}(t) \times \ddot{\alpha}(t) \neq 0$ for all $t \in I$. Let $\mathbf{T}$, $\mathbf{N}$, and $\mathbf{B}$ denote the vector fields along α defined by $\mathbf{T}(t) = \dot{\alpha}(t)$, $\mathbf{N}(t) = \ddot{\alpha}(t)/\|\ddot{\alpha}(t)\|$, and $\mathbf{B}(t) = \mathbf{T}(t) \times \mathbf{N}(t)$ for all $t \in I$.

(a) Show that $\{\mathbf{T}(t), \mathbf{N}(t), \mathbf{B}(t)\}$ is orthonormal for each $t \in I$.
(b) Show that there exist smooth functions $\kappa: I \to \mathbb{R}$ and $\tau: I \to \mathbb{R}$ such that

$$\dot{\mathbf{T}} = \kappa \mathbf{N}$$

$$\dot{\mathbf{N}} = -\kappa \mathbf{T} + \tau \mathbf{B}$$

$$\dot{\mathbf{B}} = -\tau \mathbf{N}.$$

These formulas are called the *Frenet formulas* for parametrized curves in $\mathbb{R}^3$. The vector fields $\mathbf{N}$ and $\mathbf{B}$ are respectively the *principal normal* and the *binormal* vector fields along α. The functions κ and τ are called the *curvature* and *torsion* of α.

10.8. Show that the circle of curvature $\mathcal{O}$ at a point p of an oriented plane curve C, where $\kappa(p) \neq 0$, has second order contact with C at p; i.e., show that if α and β are unit speed local parametrizations of C and of $\mathcal{O}$, respectively, with $\alpha(0) = \beta(0) = p$, then $\dot{\alpha}(0) = \dot{\beta}(0)$ and $\ddot{\alpha}(0) = \ddot{\beta}(0)$.

10.9. Let C be an oriented plane curve, let $p \in C$, and let $\alpha: I \to C$ be a unit speed local parametrization of C with $\alpha(0) = p$. Assume $\kappa(p) \neq 0$. For $q \in \mathbb{R}^2$ and $r > 0$, define $f: I \to \mathbb{R}$ by $f(t) = \|\alpha(t) - q\|^2 - r^2$. Show that q is the center of curvature and r the radius of curvature of C at p if and only if $f(0) = f'(0) = f''(0) = 0$.

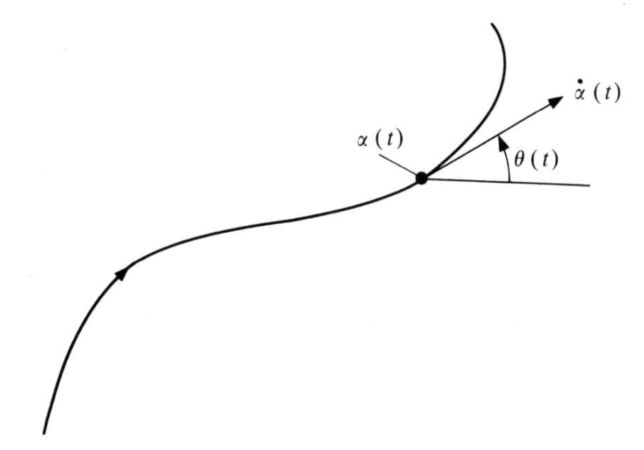

Figure 10.4 Inclination angle of a unit speed curve in $\mathbb{R}^2$.

10.10. Let $\alpha: I \to C$ be a unit speed local parametrization of the oriented plane curve C. Suppose $\theta: I \to \mathbb{R}$ is smooth and is such that

$$\dot{\alpha}(t) = (\alpha(t), \cos\theta(t), \sin\theta(t))$$

for all $t \in I$ (see Figure 10.4). (We shall, in the next chapter, be able to prove that such a function θ exists; see Exercise 11.15.) Show that $d\theta/dt = \kappa \circ \alpha$.

11 Arc Length and Line Integrals

Before analyzing the Weingarten map for n-surfaces ($n > 1$) we shall pause to see how parametrizations of plane curves can be used to evaluate integrals over the curve.

The *length* $l(\alpha)$ of a parametrized curve $\alpha: I \to \mathbb{R}^{n+1}$ is defined to be the integral of the speed of α:

$$l(\alpha) = \int_a^b \|\dot{\alpha}(t)\| \, dt,$$

where a and b are the endpoints of I (possibly $\pm\infty$). Note that $l(\alpha)$ may be $\pm\infty$. Also note that the length of a parametrized curve is the total "distance travelled." Whenever α retraces itself then that portion of the image which is covered more than once is counted more than once.

Note that if $\beta: \tilde{I} \to \mathbb{R}^{n+1}$ is a reparametrization of α, then $l(\beta) = l(\alpha)$. Indeed, if $\beta = \alpha \circ h$ where $h: \tilde{I} \to I$ is such that $h'(t) > 0$ for all $t \in \tilde{I}$, then

$$\begin{aligned} l(\beta) &= \int_c^d \|\dot{\beta}(t)\| \, dt = \int_c^d \|\dot{\alpha}(h(t))\| h'(t) \, dt \\ &= \int_a^b \|\dot{\alpha}(u)\| \, du = l(\alpha), \end{aligned}$$

where c and d are the endpoints of $\tilde{I}$.

If α is a unit speed curve, then for $t_1, t_2 \in I$ with $t_1 < t_2$,

$$\int_{t_1}^{t_2} \|\dot{\alpha}(t)\| \, dt = \int_{t_1}^{t_2} 1 \, dt = t_2 - t_1$$

so the length of any segment of α is just the length of the corresponding segment of the parameter interval. For this reason, unit speed curves are often said to be *parametrized by arc length.*

In order to apply the concept of length of a parametrized curve to define the length of an oriented plane curve, we need two preliminary results.

Theorem 1. *Let C be an oriented plane curve. Then there exists a global parametrization of C if and only if C is connected.*

PROOF. It is immediate from the definition of connectedness that any oriented plane curve which has a global parametrization must be connected.

Conversely, suppose C is connected. Let $p \in C$ and let $\alpha: I \to C$ be the local parametrization of C obtained as in the previous chapter. Recall that α is the maximal integral curve through p of the vector field $\mathbf{X}$ obtained by rotating through an angle of $-\pi/2$ the vector field ∇f, where $C = f^{-1}(c)$ and $\mathbf{N} = \nabla f/\|\nabla f\|$. Let $p_1 \in C$. We shall show that $p_1 \in$ Image α, hence that α is a global parametrization of C.

Since C is connected, there exists a continuous map $\beta: [a, b] \to C$ with $\beta(a) = p$ and $\beta(b) = p_1$. The proof will be complete if we can show that $\beta(t) \in$ Image α for all $t \in [a, b]$. To see that this is the case, let t_0 denote the least upper bound of the set $\{t \in [a, b]: \beta([a, t]) \subset \text{Image } \alpha\}$ and let γ be an integral curve of $\mathbf{X}$ with $\gamma(0) = \beta(t_0)$. We shall construct an open rectangle B about $p_0 = \beta(t_0)$ with the property that $C \cap B \subset$ Image γ. Assume for the moment that such a B exists. Then, by continuity of β, there exists a $\delta > 0$ such that $\beta(t) \in B$ (and hence $\beta(t) \in C \cap B \subset$ Image γ) for all $t \in [a, b]$ with $|t - t_0| < \delta$. Since $\beta(t) \in$ Image α for $a \leq t < t_0$ (and for $t = t_0$ if $t_0 = a$), $\beta(t) \in$ (Image γ) $\cap$ (Image α) for some $t \in [a, b]$ $(t \leq t_0)$. Therefore γ and α are integral curves of $\mathbf{X}$ passing through a common point. α being maximal, it follows that Image $\gamma \subset$ Image α (and, in fact, that there exists a $\tau \in \mathbb{R}$ such that $\gamma(t) = \alpha(t - \tau)$ for all t in the domain of γ.) Hence $\beta(t) \in C \cap B \subset$ Image $\gamma \subset$ Image α for all $t \in [a, b]$ with $|t - t_0| < \delta$. But this can only happen if $t_0 = b$ and $\beta(t_0) \in$ Image α, so $\beta(t) \in$ Image α for all $t \in [a, b]$ as claimed.

Now to complete the proof we need only to construct B. For this, let $u = (\partial f/\partial x_2)(p_0), -(\partial f/\partial x_1)(p_0))$ and $v = (\partial f/\partial x_1)(p_0), (\partial f/\partial x_2)(p_0))$ so that $(p_0, u) \in C_{p_0}$ and $(p_0, v) \perp C_{p_0}$, and let A denote the rectangle

$$A = \{p_0 + ru + sv: |r| < \varepsilon_1 \text{ and } |s| < \varepsilon_2\}$$

where $\varepsilon_1 > 0$ and $\varepsilon_2 > 0$ are chosen small enough so that A is contained in the domain of f and so that $\nabla f(q) \cdot (q, v) > 0$ for all $q \in A$ (see Figure 11.1). That this last condition can be satisfied is a consequence of the continuity of $(\partial f/\partial x_1, \partial f/\partial x_2) \cdot v$ and the fact that $((\partial f/\partial x_1, \partial f/\partial x_2) \cdot v)(p_0) = ((\partial f/\partial x_1)(p_0))^2 + ((\partial f/\partial x_2)(p_0))^2 > 0$. The condition $\nabla f(q) \cdot (q, v) > 0$ for all $q \in A$ guarantees that, for $|r| < \varepsilon_1$, the function $g_r(s) = f(p_0 + ru + sv)$ is strictly increasing on the interval $-\varepsilon_2 < s < \varepsilon_2$ $(g_r'(s) > 0)$ and hence that for each r with $|r| < \varepsilon_1$ there is at most one s with $|s| < \varepsilon_2$ such that $g_r(s) = f(p_0 + ru + sv) = c$. In other words, for each r with $|r| < \varepsilon_1$ there is at most one s with $|s| < \varepsilon_2$ such that $p_0 + ru + sv \in C$. Now $\gamma(t) = p_0 + h_1(t)u +$

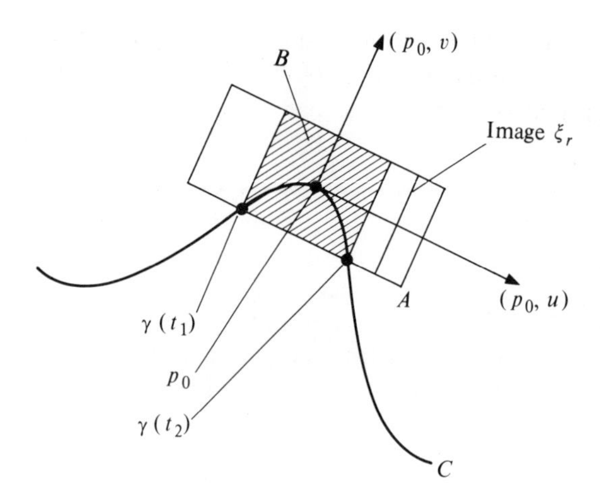

Figure 11.1 The rectangle A about p_0 is chosen so that each line segment $\xi_r(s) = (p_0 + ru) + sv$ (r fixed, $-\varepsilon_2 < s < \varepsilon_2$) meets C at most once. In the shaded rectangle B, each such line segment meets C exactly once.

$h_2(t)v$, where $h_1(t) = (\gamma(t) - p_0) \cdot u/\|u\|^2$ and $h_2(t) = (\gamma(t) - p_0) \cdot v/\|v\|^2$ are smooth functions of t with $h_1(0) = h_2(0) = 0$. Using the continuity of γ and of h_1', together with the fact that

$$h_1'(0) = \dot{\gamma}(0) \cdot (p_0, u/\|u\|^2) = \mathbf{X}(p_0) \cdot \mathbf{X}(p_0)/\|\mathbf{X}(p_0)\|^2 = 1,$$

we can choose $t_1 < 0$ and $t_2 > 0$ in the domain of γ so that both $\gamma(t) \in A$ and $h_1'(t) > 0$ for all $t \in (t_1, t_2)$ (Figure 11.1). Setting $r_1 = h_1(t_1)$ and $r_2 = h_1(t_2)$ it follows that for each $r \in (r_1, r_2)$ there is exactly one $t \in (t_1, t_2)$ with $h_1(t) = r$ (since $h_1(t)$ is continuous and strictly increasing) and that $s = h_2(t)$ for this t is an s (and therefore the unique s) with $|s| < \varepsilon_2$ such that $p_0 + ru + sv \in C$. In other words, if B is the rectangle

$$B = \{p_0 + ru + sv \colon r_1 < r < r_2,\ |s| < \varepsilon_2\},$$

then $p_0 + ru + sv \in B \cap C$ if and only if $r = h_1(t)$ and $s = h_2(t)$ for some $t \in (t_1, t_2)$; i.e., if and only if $p_0 + ru + sv \in \text{Image } \gamma$. Thus $C \cap B \subset \text{Image } \gamma$, as required. ☐

The proof of Theorem 1, with $\mathbf{X}$ replaced everywhere by the unit vector field $\mathbf{X}/\|\mathbf{X}\|$, also shows the existence of a global *unit speed* parametrization of any connected oriented plane curve.

Theorem 2. *Let C be a connected oriented plane curve and let $\beta\colon I \to C$ be a unit speed global parametrization of C. Then β is either one to one or periodic. Moreover, β is periodic if and only if C is compact.*

PROOF. Suppose $\beta(t_1) = \beta(t_2)$ for some $t_1, t_2 \in I$ with $t_1 \neq t_2$. Let $\mathbf{X}$ be the unit tangent vector field on C constructed as in the previous chapter and let α be the maximal integral curve of $\mathbf{X}$ with $\alpha(0) = \beta(t_1) = \beta(t_2)$. Then, since β is also an integral curve of $\mathbf{X}$, uniqueness of integral curves implies that

$$\beta(t) = \alpha(t - t_1)$$

and, at the same time,

$$\beta(t) = \alpha(t - t_2)$$

for all $t \in I$. Setting $\tau = t_2 - t_1$, we have

$$\beta(t) = \alpha(t - t_1) = \alpha((t\tau) - t_2) = \beta(t + \tau)$$

for all t such that both t and $t + \tau \in I$. Thus if β is not one to one then it is periodic.

If β is periodic then C must be compact because C is the image under the continuous map β of a closed interval $[t_0, t_0 + \tau]$. On the other hand, if β is not periodic, then β must be one to one so C cannot be compact because the function $\beta^{-1}: C \to \mathbb{R}$ is continuous on C yet attains no maximum value. The continuity of β^{-1} can be checked as follows. Given $t_0 \in I$ and $\varepsilon > 0$, set $\gamma(t) = \beta(t + t_0)$ for t such that $|t| < \varepsilon$ and $t + t_0 \in I$, and choose an open rectangle B about $p_0 = \beta(t_0) = \gamma(0)$ as in the proof of Theorem 1. Then $C \cap B \subset \text{Image } \gamma$ so $|\beta^{-1}(p) - t_0| = |\gamma^{-1}(p)| < \varepsilon$ whenever $p \in C \cap B$, as required for continuity. □

Recall that the *period* of a periodic function β is the smallest τ such that $\beta(t + \tau) = \beta(t)$ for all t such that both t and $t + \tau$ are in the domain of β. If τ is the period of β then any subset of the domain of β of the form $[t_0, t_0 + \tau)$ is called a *fundamental domain* of β. Note that the restriction of any periodic global parametrization β of a compact plane curve to a fundamental domain maps that fundamental domain one to one onto Image β. Hence, *if we allow half-open intervals as well as open intervals as domains of parametrized curves, every connected oriented plane curve admits a one to one unit speed global parametrization.* Moreover, any two such parametrizations $\alpha: I \to C$ and $\beta: \tilde{I} \to C$ have parameter intervals I and $\tilde{I}$ of the same length. Indeed, α and β are related by $\beta(t) = \alpha(t - t_0)$ for some $t_0 \in \mathbb{R}$ so $\tilde{I}$ is simply a translate of I. Hence we can define the *length* of a connected oriented plane curve C to be the length of I where $\alpha: I \to C$ is any one to one unit speed global parametrization of C.

Since the length of β is the same as the length of α where α is any reparametrization of β, it follows that the length of a connected oriented plane curve can be computed from the formula

$$l(C) = l(\alpha) = \int_a^b \|\dot{\alpha}(t)\| \, dt$$

where $\alpha: I \to C$ is any one to one global parametrization of C and a and b are the endpoints of I.

EXAMPLE. Let C denote the circle $(x_1 - a)^2 + (x_2 - b)^2 = r^2$ oriented by its outward normal. Then $\alpha\colon I \to C$, defined by $\alpha(t) = (a + r\cos 2t, b - r\sin 2t)$ is a global parametrization of C, as we saw in the previous chapter. α is periodic with period π so the restriction of α to the interval $[0, \pi)$ is a one to one global parametrization of C. Hence

$$l(C) = \int_0^\pi \|\dot{\alpha}(t)\|\, dt = \int_0^\pi \|(-2r\sin 2t, -2r\cos 2t)\|\, dt = 2\pi r.$$

The remainder of this chapter will be devoted to a discussion of differential 1-forms and their integrals.

A *differential 1-form*, usually called simply a *1-form*, on an open set $U \subset \mathbb{R}^{n+1}$ is a function $\omega\colon U \times \mathbb{R}^{n+1} \to \mathbb{R}$ such that, for each $p \in U$, the restriction of ω to $\mathbb{R}_p^{n+1} \subset U \times \mathbb{R}^{n+1}$ is linear.

EXAMPLE 1. Let $\mathbf{X}$ be a vector field on U and let $\omega_{\mathbf{X}}\colon U \times \mathbb{R}^{n+1} \to \mathbb{R}$ be defined by

$$\omega_{\mathbf{X}}(p, v) = \mathbf{X}(p) \cdot (p, v).$$

Then $\omega_{\mathbf{X}}$ is a 1-form on U, called the 1-form *dual* to $\mathbf{X}$.

EXAMPLE 2. For $f\colon U \to R$ a smooth function, define $df\colon U \times \mathbb{R}^{n+1} \to \mathbb{R}$ by

$$df(\mathbf{v}) = \nabla_{\mathbf{v}} f = \nabla f(p) \cdot \mathbf{v} \qquad (\mathbf{v} = (p, v) \in \mathbb{R}_p^{n+1},\ p \in U).$$

Then df is a 1-form on U, called the *differential* of f.

EXAMPLE 3. For each $i \in \{1, \ldots, n+1\}$, let $x_i\colon U \to \mathbb{R}$ $(U \subset \mathbb{R}^{n+1})$ be defined by

$$x_i(a_1, \ldots, a_{n+1}) = a_i.$$

The function x_i is called the ith *Cartesian coordinate function* on U. The 1-form dx_i simply picks off the ith component of each vector in its domain:

$$dx_i(\mathbf{v}) = \nabla x_i(p) \cdot \mathbf{v} = (p, 0, \ldots, 1, \ldots, 0) \cdot \mathbf{v} = v_i$$

for $\mathbf{v} = (p, v_1, \ldots, v_{n+1}) \in \mathbb{R}_p^{n+1}$, $p \in U$.

Remark. As is common in mathematics, we are using a single symbol in different situations to represent different quantities. We have used the symbol x_i to denote a real number when describing a point $(x_1, \ldots, x_{n+1})$ in $\mathbb{R}^{n+1}$, we have used x_i to denote a function with domain an interval when describing a parametrized curve $\alpha(t) = (x_1(t), \ldots, x_{n+1}(t))$ in $\mathbb{R}^{n+1}$, and now we are using x_i to denote a function whose domain is an open set in $\mathbb{R}^{n+1}$. These various uses of the symbol x_i are all standard. We could of course introduce extra notation to avoid overworking any given symbol but only at the cost of mushrooming symbology and non-conformity with common usage. We will continue to use the symbol x_i in each of these situations; the meaning in each given situation will always be clear in context.

A 1-form ω on $U \subset \mathbb{R}^{n+1}$ is *smooth* if it is smooth as a function $\omega\colon U \times \mathbb{R}^{n+1} \subset \mathbb{R}^{2n+2} \to \mathbb{R}$. Note that if $f\colon U \to \mathbb{R}$ is a smooth function on U then its differential df is a smooth 1-form on U.

The *sum* of two 1-forms ω_1 and ω_2 on an open set $U \subset \mathbb{R}^{n+1}$ is the 1-form $\omega_1 + \omega_2$ defined by

$$(\omega_1 + \omega_2)(\mathbf{v}) = \omega_1(\mathbf{v}) + \omega_2(\mathbf{v}).$$

The *product* of a function $f\colon U \to \mathbb{R}$ and a 1-form ω on U is the 1-form $f\omega$ on U defined by

$$(f\omega)(p, v) = f(p)\omega(p, v).$$

Note that the sum of two smooth 1-forms is smooth and that the product of a smooth function and a smooth 1-form is smooth.

Given a 1-form ω and a vector field $\mathbf{X}$ on $U \subset \mathbb{R}^{n+1}$, we can define a function $\omega(\mathbf{X})\colon U \to \mathbb{R}$ by

$$(\omega(\mathbf{X}))(p) = \omega(\mathbf{X}(p)).$$

Note that if ω and $\mathbf{X}$ are both smooth then so is $\omega(\mathbf{X})$.

Proposition. *For each 1-form ω on U (U open in $\mathbb{R}^{n+1}$) there exist unique functions $f_i\colon U \to \mathbb{R}$ ($i \in \{1, \ldots, n+1\}$) such that*

$$\omega = \sum_{i=1}^{n+1} f_i \, dx_i .$$

Moreover, ω is smooth if and only if each f_i is smooth.

PROOF. For each $j \in \{1, \ldots, n+1\}$, let $\mathbf{X}_j$ denote the smooth vector field on U defined by $\mathbf{X}_j(p) = (p, 0, \ldots, 1, \ldots, 0)$, with the 1 in the $(j+1)$th spot (j spots after the p). Then

$$dx_i(\mathbf{X}_j) = \begin{cases} 1 & \text{if } i = j \\ 0 & \text{if } i \neq j. \end{cases}$$

Thus if $\omega = \sum_{i=1}^{n+1} f_i \, dx_i$ then, for each $j \in \{1, \ldots, n+1\}$,

$$f_j = \left(\sum_{i=1}^{n+1} f_i \, dx_i\right)(\mathbf{X}_j) = \omega(\mathbf{X}_j).$$

This formula shows that the functions f_j, if they exist, are unique and also that they are smooth if ω is smooth. On the other hand, if we define functions f_j by the above formula, then the 1-forms ω and $\sum_{i=1}^{n+1} f_i \, dx_i$ have the same value on each of the basis vectors $\mathbf{X}_i(p)$ for $\mathbb{R}^{n+1}_p$ and hence by linearity they have the same value on all vectors in $\mathbb{R}^{n+1}_p$, $p \in U$, so $\omega = \sum_{i=1}^{n+1} f_i \, dx_i$. Clearly ω is smooth if each f_i is smooth. □

Corollary. *Let $f\colon U \to \mathbb{R}$ (U open in $\mathbb{R}^{n+1}$) be a smooth function. Then*

$$df = \sum_{i=1}^{n+1} \frac{\partial f}{\partial x_i}\, dx_i.$$

PROOF. $df(\mathbf{X}_j) = \nabla f \cdot \mathbf{X}_j = \partial f/\partial x_j$. □

Now let ω be a smooth 1-form on the open set $U \subset \mathbb{R}^{n+1}$ and let $\alpha\colon [a, b] \to U$ be a parametrized curve in U. The *integral* of ω over α is the real number

$$\int_\alpha \omega = \int_a^b \omega(\dot{\alpha}(t))\, dt.$$

Integrals of this type are called *line integrals.*

Note that if $\beta\colon [c, d] \to U$ is any reparametrization of α, $\beta = \alpha \circ h$ where $h\colon [c, d] \to [a, b]$ has derivative everywhere positive, then

$$\int_\beta \omega = \int_c^d \omega(\dot{\beta}(t))\, dt = \int_c^d \omega(\dot{\alpha}(h(t)h'(t)))\, dt$$
$$= \int_c^d \omega(\dot{\alpha}(h(t))h'(t))\, dt = \int_a^b \omega(\dot{\alpha}(u))\, du = \int_\alpha \omega.$$

In particular, if U is an open set in $\mathbb{R}^2$ and C is a compact connected oriented plane curve in U then we can define the *integral of ω over C* by

$$\int_C \omega = \int_\alpha \omega,$$

where $\alpha\colon [a, b] \to C$ is any parametrized curve whose restriction to $[a, b)$ is a one to one global parametrization of C; the result will be independent of the choice of α.

Also note that the line integral $\int_\alpha \omega$ can be defined for α a piecewise smooth parametrized curve. If $\alpha\colon [a, b] \to U \subset \mathbb{R}^{n+1}$ is continuous and such that the restriction of α to $[t_i, t_{i+1}]$ is smooth for each $i \in \{0, 1, \ldots, k\}$, where $a = t_0 < t_1 < \cdots < t_{k+1} = b$, then the integral of the smooth 1-form ω on U over α is

$$\int_\alpha \omega = \sum_{i=0}^{k} \int_{\alpha_i} \omega$$

where α_i is the restriction of α to $[t_i, t_{i+1}]$.

Remark. We have insisted, in defining the line integrals $\int_\alpha \omega$ and $\int_C \omega$, that the parametrized curve α have domain a closed interval and that the plane curve C be compact. This is done to assure the existence of the integrals. Note that it is not necessary to make these assumptions in defining the length integral because, the integrand being non-negative, this integral always either exists or diverges to $+\infty$.

EXAMPLE 1. Let U be open in $\mathbb{R}^{n+1}$ and let $f: U \to \mathbb{R}$ be smooth. Then, for $\alpha: [a, b] \to U$ any parametrized curve in U,

$$\int_\alpha df = \int_a^b df(\dot{\alpha}(t))\, dt = \int_a^b (f \circ \alpha)'(t)\, dt = f(\alpha(b)) - f(\alpha(a)).$$

In particular, if $\alpha(a) = \alpha(b)$ then $\int_\alpha df = 0$.

A 1-form which is the differential of a smooth function is said to be *exact*. A parametrized curve $\alpha: [a, b] \to \mathbb{R}^{n+1}$ with $\alpha(a) = \alpha(b)$ is said to be *closed*. The above computation shows that the integral of an exact 1-form over a closed curve is always zero. In particular, *the integral of an exact 1-form over a compact connected oriented plane curve is always zero.*

EXAMPLE 2. Let η denote the 1-form on $\mathbb{R}^2 - \{0\}$ defined by

$$\eta = -\frac{x_2}{x_1^2 + x_2^2}\, dx_1 + \frac{x_1}{x_1^2 + x_2^2}\, dx_2$$

and let C denote the ellipse $(x_1^2/a^2) + (x_2^2/b^2) = 1$, oriented by its inward normal. The parametrized curve $\alpha: [0, 2\pi] \to C$ defined by $\alpha(t) = (a \cos t, b \sin t)$ restricts to a one to one global parametrization of C on the interval $[0, 2\pi)$ so

$$\begin{aligned}
\int_C \eta = \int_\alpha \eta &= \int_0^{2\pi} \eta(\dot{\alpha}(t))\, dt \\
&= \int_0^{2\pi} \left[\frac{-x_2}{x_1^2 + x_2^2}(\alpha(t))\, dx_1(\dot{\alpha}(t)) + \frac{x_1}{x_1^2 + x_2^2}(\alpha(t))\, dx_2(\dot{\alpha}(t)) \right] dt \\
&= \int_0^{2\pi} \left[\frac{-b \sin t}{a^2 \cos^2 t + b^2 \sin^2 t}(-a \sin t) \right. \\
&\qquad\qquad \left. + \frac{a \cos t}{a^2 \cos^2 t + b^2 \sin^2 t}(b \cos t) \right] dt \\
&= \int_0^{2\pi} \frac{ab}{a^2 \cos^2 t + b^2 \sin^2 t}\, dt = 4 \int_0^{\pi/2} \frac{(b/a) \sec^2 t}{1 + (b/a)^2 \tan^2 t}\, dt \\
&= 4 \int_0^{\infty} \frac{du}{1 + u^2} = 2\pi.
\end{aligned}$$

Since its integral over the compact curve C is not zero, the 1-form η of Example 2 cannot be exact. However, its restriction to V (or, more precisely, to $V \times \mathbb{R}^2$), where V is the complement in $\mathbb{R}^2$ of any ray through the origin, *is* exact. Indeed, for v any unit vector in $\mathbb{R}^2$ and

$$V = \mathbb{R}^2 - \{rv: r \geq 0\},$$

$\eta = d\theta_V$ where $\theta_V: V \to \mathbb{R}$ is defined as follows. Let θ_v denote the unique real number with $0 \leq \theta_v < 2\pi$ such that $v = (\cos \theta_v, \sin \theta_v)$. Then, for each

$(x, y) \in V$, define $\theta_V(x, y)$ to be the unique real number with $\theta_v \le \theta_V(x, y) < \theta_v + 2\pi$ such that

$$\left(\frac{x}{(x^2 + y^2)^{1/2}}, \frac{y}{(x^2 + y^2)^{1/2}}\right) = (\cos \theta_V(x, y), \sin \theta_V(x, y))$$

(see Figure 11.2). In order to verify that $d\theta_V = \eta|_V$ simply note that $\tan \theta_V(x, y) = y/x$ and $\cot \theta_V(x, y) = x/y$ so in each sufficiently small open set we can solve one or the other of these equations for θ_V and compute

$$d\theta_V = \frac{\partial \theta_V}{\partial x_1}\, dx_1 + \frac{\partial \theta_V}{\partial x_2}\, dx_2 = -\frac{x_2}{x_1^2 + x_2^2}\, dx_1 + \frac{x_1}{x_1^2 + x_2^2}\, dx_2 .$$

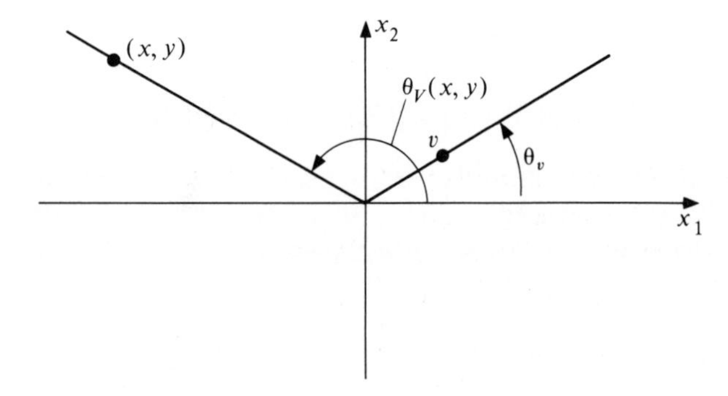

Figure 11.2 $\theta_V(x, y)$ is the inclination angle of the line segment from the origin to (x, y), $\theta_v \le \theta_V < \theta_v + 2\pi$.

Theorem 3. *Let η be the 1-form on $\mathbb{R}^2 - \{0\}$ defined by*

$$\eta = -\frac{x_2}{x_1^2 + x_2^2}\, dx_1 + \frac{x_1}{x_1^2 + x_2^2}\, dx_2 .$$

Then for $\alpha\colon [a, b] \to \mathbb{R}^2 - \{0\}$ any closed piecewise smooth parametrized curve in $\mathbb{R}^2 - \{0\}$,

$$\int_\alpha \eta = 2\pi k$$

for some integer k.

Proof. Define $\varphi\colon [a, b] \to \mathbb{R}$ by

$$\varphi(t) = \varphi(a) + \int_{\alpha_t} \eta$$

where α_t is the restriction of α to the interval $[a, t]$ and $\varphi(a)$ is chosen so that

$$\alpha(a)/\|\alpha(a)\| = (\cos \varphi(a), \sin \varphi(a)).$$

We claim that

$$(*) \qquad \alpha(t)/\|\alpha(t)\| = (\cos \varphi(t), \sin \varphi(t))$$

for all $t \in [a, b]$. Indeed, let t_0 denote the least upper bound of the set $\{\tau \in [a, b]\colon (*) \text{ holds for } a \le t \le \tau\}$. By continuity, (*) must hold at $t = t_0$. Setting $v = -\alpha(t_0)/\|\alpha(t_0)\|$ and defining θ_V as above we find that $(\cos \theta_V(\alpha(t_0)), \sin \theta_V(\alpha(t_0))) = \alpha(t_0)/\|\alpha(t_0)\| = (\cos \varphi(t_0), \sin \varphi(t_0))$ so $\varphi(t_0) - \theta_V(\alpha(t_0)) = 2\pi m$ for some integer m. Choosing $\delta > 0$ so that $\alpha(t) \in V$ for all $t \in [a, b]$ with $|t - t_0| < \delta$ we find further that, for $t \in [a, b]$, $|t - t_0| < \delta$, and $t \ne t_i$ (t_i a point where α fails to be smooth),

$$\frac{d}{dt}(\varphi(t) - \theta_V(\alpha(t))) = \eta(\dot{\alpha}(t)) - d\theta_V(\dot{\alpha}(t)) = 0,$$

so $\varphi(t) - \theta_V(\alpha(t)) = 2\pi m$ for all $t \in [a, b]$ with $|t - t_0| < \delta$. But then

$$(\cos \varphi(t), \sin \varphi(t)) = (\cos \theta_V(\alpha(t)), \sin \theta_V(\alpha(t))) = \alpha(t)/\|\alpha(t)\|$$

for all such t, which is possible only if $t_0 = b$. Thus $t_0 = b$ and (*) holds for all $t \in [a, b]$, as claimed.

Finally, since $\alpha(a) = \alpha(b)$, (*) implies that

$$(\cos \varphi(a), \sin \varphi(a)) = (\cos \varphi(b), \sin \varphi(b)),$$

so $\varphi(b) - \varphi(a) = 2\pi k$ for some integer k, and

$$\int_\alpha \eta = \int_a^b \eta(\dot{\alpha}(t))\, dt = \varphi(b) - \varphi(a) = 2\pi k. \qquad \square$$

The integer $k(\alpha) = (1/2\pi) \int_\alpha \eta$ is called the *winding number* of α since it counts the number of times the closed curve α winds around the origin.

Exercises

In Exercises 11.1–11.4, find the length of the given parametrized curve $\alpha\colon I \to \mathbb{R}^{n+1}$.

11.1. $\alpha(t) = (t^2, t^3)$, $I = [0, 2]$, $n = 1$.

11.2. $\alpha(t) = (\cos 3t, \sin 3t, 4t)$, $I = [-1, 1]$, $n = 2$.

11.3. $\alpha(t) = (\sqrt{2} \cos 2t, \sin 2t, \sin 2t)$, $I = [0, 2\pi]$, $n = 2$.

11.4. $\alpha(t) = (\cos t, \sin t, \cos t, \sin t)$, $I = [0, 2\pi]$, $n = 3$.

In Exercises 11.5–11.8, find the length of the connected oriented plane curve $f^{-1}(c)$, oriented by $\nabla f/\|\nabla f\|$, where $f\colon U \to \mathbb{R}$ and c are as given.

11.5. $f(x_1, x_2) = 5x_1 + 12x_2$, $U = \{(x_1, x_2)\colon x_1^2 + x_2^2 < 169\}$, $c = 0$.

11.6. $f(x_1, x_2) = \frac{1}{2}x_1^2 + \frac{1}{2}(x_2 - 1)^2$, $U = \mathbb{R}^2$, $c = 2$.

11.7. $f(x_1, x_2) = x_1^2 - x_2^2$, $U = \{(x_1, x_2)\colon 0 < x_1 < 2\}$, $c = 1$. [Set up, but do not evaluate, the integral.]

11.8. $f(x_1, x_2) = -9x_1^2 + 4x_2^3$, $U = \{(x_1, x_2)\colon x_1 > 0,\ 0 < x_2 < 3\}$, $c = 0$. [*Hint:* Note that there is a parametrization $\alpha(t) = (x_1(t), x_2(t))$ of $f^{-1}(0)$ with $x_2(t) = t$.]

11.9. Show that if C is a connected oriented plane curve and $\tilde{C}$ is the same curve with the opposite orientation, then $l(C) = l(\tilde{C})$.

11.10. Let C be a connected oriented plane curve, let $\alpha\colon I \to C$ be a one to one unit speed parametrization of C, and let $\kappa\colon C \to \mathbb{R}$ denote the curvature of C.

(a) Show that $\int_a^b |\kappa \circ \alpha(t)|\, dt$, where a and b are the endpoints of I, is independent of the choice of one to one unit speed parametrization α of C.
(b) Show that $\int_a^b |\kappa \circ \alpha(t)|\, dt = l(N \circ \alpha)$, where $N\colon C \to \mathbb{R}^2$ is the Gauss map of C.

[$\int_a^b |\kappa \circ \alpha(t)|\, dt$ is called the *total curvature* of C.]

11.11. Let f and g be smooth functions on the open set $U \subset \mathbb{R}^{n+1}$. Show that

(a) $d(f + g) = df + dg$.
(b) $d(fg) = g\,df + f\,dg$.
(c) If $h\colon \mathbb{R} \to \mathbb{R}$ is smooth then $d(h \circ f) = (h' \circ f)\, df$.

11.12. Compute the following line integrals.

(a) $\int_\alpha (x_2\, dx_1 - x_1\, dx_2)$ where $\alpha(t) = (2 \cos t, 2 \sin t)$, $0 \le t \le 2\pi$.
(b) $\int_C (-x_2\, dx_1 + x_1\, dx_2)$ where C is the ellipse $(x_1^2/a^2) + (x_2^2/b^2) = 1$, oriented by its inward normal.
(c) $\int_\alpha \sum_{i=1}^{n+1} x_i\, dx_i$ where $\alpha\colon [0, 1] \to \mathbb{R}^{n+1}$ is such that $\alpha(0) = (0, 0, \ldots, 0)$ and $\alpha(1) = (1, 1, \ldots, 1)$. [*Hint*: Find an $f\colon \mathbb{R}^{n+1} \to \mathbb{R}$ such that $df = \sum_{i=1}^{n+1} x_i\, dx_i$.]

11.13. Let $\omega = \sum_{i=1}^{n+1} f_i\, dx_i$ be a smooth 1-form on $\mathbb{R}^{n+1}$ and let $\alpha(t) = (x_1(t), \ldots, x_{n+1}(t))$, where the x_i are smooth real valued functions on $[a, b]$. Show that

$$\int_\alpha \omega = \int_a^b \sum_{i=1}^{n+1} (f_i \circ \alpha) \frac{dx_i}{dt}\, dt.$$

11.14. Let $C = f^{-1}(c)$ be a compact plane curve oriented by $\nabla f/\|\nabla f\|$, let $\mathbf{X}$ be the unit vector field on $U = \text{domain}\,(f)$ obtained by rotating $\nabla f/\|\nabla f\|$ through the angle $-\pi/2$, and let $\omega_{\mathbf{X}}$ be the 1-form on U dual to $\mathbf{X}$. Show that $\int_C \omega_{\mathbf{X}} = l(C)$.

11.15. Let $\alpha\colon I \to \mathbb{R}^2$ be a unit speed curve, let $t_0 \in I$, and let $\theta_0 \in \mathbb{R}$ be such that $\dot{\alpha}(t_0) = (\alpha(t_0), \cos \theta_0, \sin \theta_0)$. Show that there exists a unique smooth function $\theta\colon I \to \mathbb{R}$ with $\theta = \theta_0$ such that $\dot{\alpha}(t) = (\alpha(t), \cos \theta(t), \sin \theta(t))$ for all $t \in I$. [*Hint*: Set $\theta(t) = \theta_0 + \int_{t_0}^t \eta(\dot{\beta}(\tau))\, d\tau$ where η is the 1-form of Theorem 3 and $\beta = d\alpha/dt$.]

11.16. Let $\alpha\colon [a, b] \to \mathbb{R}^2 - \{0\}$ be closed piecewise smooth parametrized curve. Show that the winding number of α is the same as the winding number of $f\alpha$ where $f\colon [a, b] \to \mathbb{R}$ is any piecewise smooth function along α with $f(a) = f(b)$

and $f(t) > 0$ for all $t \in [a, b]$. Conclude that α and $\alpha/\|\alpha\|$ have the same winding number.

11.17. Let $a = t_0 < t_1 < \cdots < t_{k+1} = b$ and let $\varphi: [a, b] \times [0, 1] \to \mathbb{R}^2 - \{0\}$ be a continuous map such that, for each $u \in [0, 1]$, the map $\varphi_u: [a, b] \to \mathbb{R}$ defined by $\varphi_u(t) = \varphi(t, u)$ is smooth on each interval $[t_i, t_{i+1}]$. Assume that $\varphi_u(a) = \varphi_u(b)$ for all $u \in [0, 1]$. Show that the winding number $k(\varphi_u)$ is a continuous function of u and hence that $k(\varphi_u)$ is constant and, in particular, $k(\varphi_0) = k(\varphi_1)$. [The map φ is called a *homotopy* between φ_0 and φ_1.]

11.18. The winding number of $d\alpha/dt$ (which is the same, by Exercise 11.16, as the winding number of $d\alpha/dt/\|d\alpha/dt\|$), where $\alpha: [a, b] \to \mathbb{R}^2$ is a regular ($\dot{\alpha}(t) \neq 0$ for all t) parametrized curve with $\dot{\alpha}(a) = \dot{\alpha}(b)$, is called the *rotation index* of α.

(a) Show by example that for each integer k there is an α with rotation index k.

(b) Show that if α is the restriction to $[a, b]$ of a periodic regular parametrized curve with period $\tau = b - a$ and if α is one to one on $[a, b)$ then the rotation index of α is ± 1. [*Hint*: See Figure 11.3. Let $u \in \mathbb{R}^2$, $u \neq 0$, and choose t_0 so that $h(t) = \alpha(t) \cdot u$ has an absolute minimum at t_0. For

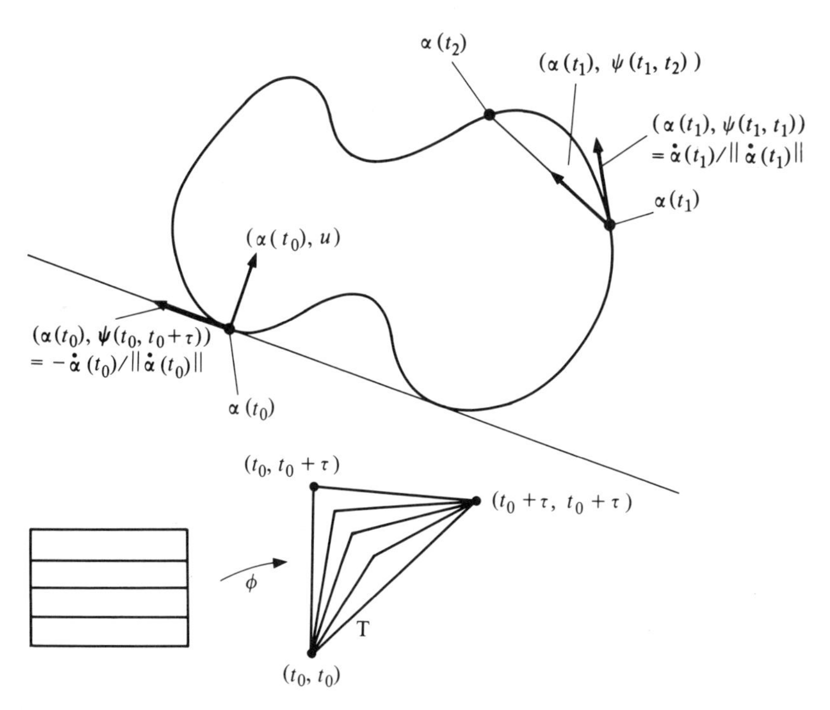

Figure 11.3 To show that the rotation index of α is ± 1, t_0 is chosen so that Image α lies completely on one side of the tangent line at $\alpha(t_0)$, ψ is the normalized secant map $\psi(t_1, t_2) = (\alpha(t_2) - \alpha(t_1))/\|\alpha(t_2) - \alpha(t_1)\|$ extended continuously to the closed triangle T, and $\phi: [t_0, t_0 + \tau] \times [0, 1] \to T$ is the homotopy which maps horizontal line segments to piecewise smooth curves in T, as indicated.

$t_0 \le t_1 \le t_2 \le t_0 + \tau$, let

$$\psi(t_1, t_2) = \begin{cases} \dfrac{d\alpha}{dt}(t_1) \Big/ \left\| \dfrac{d\alpha}{dt}(t_1) \right\| & \text{if } t_1 = t_2 \\ -\dfrac{d\alpha}{dt}(t_0) \Big/ \left\| \dfrac{d\alpha}{dt}(t_0) \right\| & \text{if } t_1 = t_0 \text{ and } t_2 = t_0 + \tau \\ (\alpha(t_2) - \alpha(t_1)) / \|\alpha(t_2) - \alpha(t_1)\| & \text{otherwise,} \end{cases}$$

define a homotopy φ: $[t_0, t_0 + \tau] \times [0, 1] \to \mathbb{R}^2 - \{0\}$ by $\varphi = \psi \circ \phi$ where

$$\phi(t, u) = \begin{cases} (t_0, t_0) + (t - t_0)(1 - u, 1 + u) \\ \qquad \text{if } t_0 \le t \le t_0 + \frac{1}{2}\tau \\ (t_0 + \tau, t_0 + \tau) - (t_0 + \tau - t)(1 + u, 1 - u) \\ \qquad \text{if } t_0 + \frac{1}{2}\tau \le t \le t_0 + \tau, \end{cases}$$

calculate the winding number of φ_1, and apply Exercise 11.17. In calculating the winding number of φ_1 note that Image $\varphi_1|_{[t_0, t_0 + \tau/2]}$ and Image $\varphi_1|_{[t_0 + \tau/2, t_0 + \tau]}$ are each contained in a region where the 1-form η is exact.]

11.19. Let C be a compact connected oriented plane curve with curvature κ everywhere positive and let α: $\mathbb{R} \to C$ be a unit speed global parametrization of C.

(a) Show that if $u \in S^1$ and h: $\mathbb{R} \to \mathbb{R}$ is defined by $h(t) = \alpha(t) \cdot u$ then $h'(t_0) = 0$ if and only if $N(\alpha(t_0)) = \pm u$, and $h''(t_0) = \kappa(\alpha(t_0))u \cdot N(\alpha(t_0))$ for all such t_0. Conclude that the Gauss map N of C is onto.

(b) Show that if $[t_0, t_0 + \tau)$ is a fundamental domain of α then the rotation index (Exercise 11.18) of $\alpha|_{[t_0, t_0 + \tau]}$ is equal to $(1/2\pi) \int_{t_0}^{t_0 + \tau} (\kappa \circ \alpha)(t)\, dt$. [*Hint*: Use Exercises 11.15 and 10.10.]

(c) Show that if $c \in (t_0, t_0 + \tau]$ is such that $N(c) = N(t_0)$ and $N(t) \ne N(t_0)$ for $t_0 < t < c$ then $\int_{t_0}^{c} (\kappa \circ \alpha)(t)\, dt = 2\pi$. [Use Exercise 10.10.] Conclude that $c = t_0 + \tau$ and that the Gauss map of C is one to one. [Use Exercise 11.18(*b*).]

11.20. A function f: $\mathbb{C} \to \mathbb{C}$ ($\mathbb{C}$ = {complex numbers}) may be viewed as a function from $\mathbb{R}^2$ to $\mathbb{R}^2$ by identifying each complex number $a + bi$ with the point $(a, b) \in \mathbb{R}^2$. In particular, we may view each polynomial function $f(z) = a_n z^n + \cdots + a_1 z + a_0$ ($a_0, \ldots, a_n \in \mathbb{C}$) as a smooth map from $\mathbb{R}^2$ into itself. Given such a polynomial f, let α_f: $[0, 2\pi] \to \mathbb{R}^2$ be defined by

$$\alpha_f(t) = f(\cos t, \sin t) = f(\cos t + i \sin t)$$

and let $k(f)$ denote the winding number of α_f.

(a) Show that if $f(z) = a_0 \ne 0$ for all z then $k(f) = 0$.

(b) Show that if $f(z) = a_n z^n$ with $a_n \ne 0$ then $k(f) = n$.

(c) Show that if f is any polynomial with $f(z) \ne 0$ for all $z \in \mathbb{C}$ with $|z| \le 1$ then $k(f) = 0$. [Apply Exercise 11.17 to φ: $[0, 2\pi] \times [0, 1] \to \mathbb{R}^2 - \{0\}$, $\varphi(t, u) = f(u(\cos t + i \sin t))$.]

(d) Show that if f is any polynomial with $f(z) \neq 0$ for all $z \in \mathbb{C}$ with $|z| \geq 1$ then $k(f) = n$. [Apply Exercise 11.17 to φ: $[0, 2\pi] \times [0, 1] \to \mathbb{R}^2 - \{0\}$,

$$\varphi(t, u) = \begin{cases} u^n f\left(\dfrac{1}{u}(\cos t + i \sin t)\right) & \text{if } u \neq 0 \\ a_n(\cos t + i \sin t)^n & \text{if } u = 0, \end{cases}$$

where a_n is the leading coefficient of $f(z)$.]

(e) Conclude that if f is a polynomial with $f(z) \neq 0$ for all $z \in \mathbb{C}$ then the degree of f must be zero. (This exercise proves the *fundamental theorem of algebra*: every non-constant polynomial with complex coefficients must have a root in $\mathbb{C}$.)

11.21. Let α: $[a, b] \to \mathbb{R}^2 - \{0\}$ be smooth and such that $\alpha(a) = \alpha(b)$. Suppose α hits the positive x_1-axis only finitely many times. Show that the winding number $k(\alpha)$ is equal to the algebraic number of crossings of the positive x_1-axis by α, where each upward crossing is counted positively and each downward crossing is counted negatively. [*Hint*: Let $t_1 < t_2 < \cdots < t_m$ be the set of all $t \in [a, b]$ such that $\alpha(t)$ lies on the positive x_1-axis. Let V and θ_V be as in our discussion of the winding number, with $v = (1, 0)$. Then

$$k(\alpha) = \sum_{i=0}^{m} \lim_{\varepsilon \to 0} \int_{t_i + \varepsilon}^{t_{i+1} - \varepsilon} d\theta_V(\dot{\alpha}(t))\, dt$$

where $t_0 = a$ and $t_{m+1} = b$. (If $t_1 = a$, the sum will range only from 1 to $m - 1$.)]

11.22. Let α: $I \to \mathbb{R}^2$ be a piecewise smooth closed parametrized curve. For $p = (a, b) \in \mathbb{R}^2 - \text{Image } \alpha$, define

$$k_p(\alpha) = \int_{\alpha} \frac{-(x_2 - b)\, dx_1 + (x_1 - a)\, dx_2}{(x_1 - a)^2 + (x_2 - b)^2}.$$

(a) Show that $k_p(\alpha)$ is an integer. [*Hint*: Show that $k_p(\alpha)$ is the winding number of β: $I \to \mathbb{R}^2 - \{0\}$ where $\beta(t) = \alpha(t) - p$.]

(b) Show that if p and $q \in \mathbb{R}^2 - \text{Image } \alpha$ can be joined by a continuous curve in $\mathbb{R}^2 - \text{Image } \alpha$ then $k_p(\alpha) = k_q(\alpha)$.

(The integer $k_p(\alpha)$ is the *winding number of* α *about* p.)

12 Curvature of Surfaces

Let S be an n-surface in $\mathbb{R}^{n+1}$, oriented by the unit normal vector field $\mathbf{N}$, and let $p \in S$. The Weingarten map $L_p: S_p \to S_p$, defined by $L_p(\mathbf{v}) = -\nabla_{\mathbf{v}}\mathbf{N}$ for $\mathbf{v} \in S_p$, measures the turning of the normal as one moves in S through p with various velocities $\mathbf{v}$. Thus L_p measures the way S curves in $\mathbb{R}^{n+1}$ at p. For $n = 1$, we have seen that L_p is just multiplication by a number $\kappa(p)$, the curvature of S at p. We shall now analyze L_p when $n > 1$.

Recall that, for $\mathbf{v} \in S_p$, $L_p(\mathbf{v}) \cdot \mathbf{v}$ is equal to the normal component of acceleration at p of every parametrized curve α in S passing through p with velocity $\mathbf{v}$. This component of acceleration is thus forced on α by the curvature of S in $\mathbb{R}^{n+1}$. When $\|\mathbf{v}\| = 1$, this number

$$k(\mathbf{v}) = L_p(\mathbf{v}) \cdot \mathbf{v}$$

is called the *normal curvature* of S at p in the direction $\mathbf{v}$. Note that if $k(\mathbf{v}) > 0$ then the surface S bends toward $\mathbf{N}$ in the direction $\mathbf{v}$, and if $k(\mathbf{v}) < 0$ it bends away from $\mathbf{N}$ in the direction $\mathbf{v}$ (see Figure 12.1). When $n = 1$, $k(\mathbf{v}) = \kappa(p)$ for both unit vectors $\mathbf{v} \in S_p$.

Example 1. Let S be the sphere $x_1^2 + \cdots + x_{n+1}^2 = r^2$ of radius r, oriented by the inward normal $\mathbf{N}(p) = (p, -p/\|p\|)$. Then, as we saw in Chapter 9, L_p is simply multiplication by $1/r$. Hence $k(\mathbf{v}) = 1/r$ has the same value for all tangent directions $\mathbf{v}$ at all points $p \in S$.

Example 2. Let S be the hyperboloid $-x_1^2 + x_2^2 + x_3^2 = 1$ in $\mathbb{R}^3$, oriented by the unit normal vector field $\mathbf{N}(p) = (p, -x_1/\|p\|, x_2/\|p\|, x_3/\|p\|)$ for $p = (x_1, x_2, x_3) \in S$ (Figure 12.1). Then for $p = (0, 0, 1)$, each unit vector $\mathbf{v} \in S_p$ is of the form $(p, v_1, v_2, 0)$ where $v_1^2 + v_2^2 = 1$, $L_p(\mathbf{v}) = -\nabla_{\mathbf{v}}\mathbf{N} = (p, v_1, -v_2, 0)$, and $k(\mathbf{v}) = v_1^2 - v_2^2$. In particular, $k(\mathbf{v}) = 1$ when

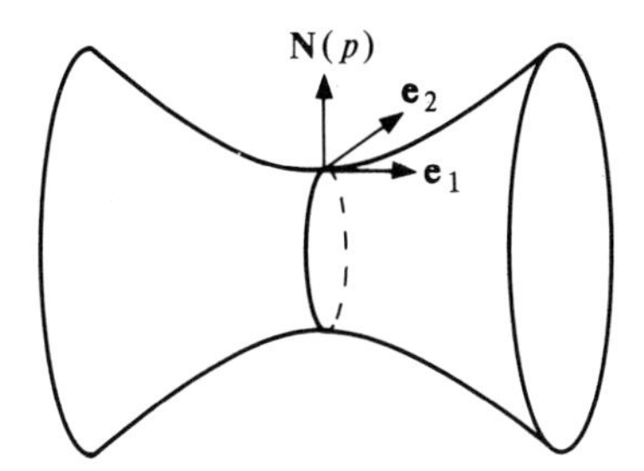

Figure 12.1 Normal curvature at $p = (0, 0, 1)$ is positive in the direction $\mathbf{e}_1 = (p, 1, 0, 0)$ and negative in the direction $\mathbf{e}_2 = (p, 0, 1, 0)$.

$\mathbf{v} = (p, 1, 0, 0)$ and $k(\mathbf{v}) = -1$ when $\mathbf{v} = (p, 0, 1, 0)$. Moreover, $k(\mathbf{v}) = 0$ when $\mathbf{v} = \pm(p, 1/\sqrt{2}, 1/\sqrt{2}, 0)$ and when $\mathbf{v} = \pm(p, 1/\sqrt{2}, -1/\sqrt{2}, 0)$ which is not surprising since the straight lines $\alpha(t) = (t/\sqrt{2}, t/\sqrt{2}, 1)$ and $\beta(t) = (t/\sqrt{2}, -t/\sqrt{2}, 1)$ both lie completely in S so S does not force any acceleration on parametrized curves in these directions through p (see Figure 9.1).

Further insight into the meaning of normal curvature can be gained from normal sections. Given an n-surface $S = f^{-1}(c)$ in $\mathbb{R}^{n+1}$, oriented by the unit normal vector field $\mathbf{N}$, the *normal section* determined by the unit vector $\mathbf{v} = (p, v) \in S_p$, $p \in S$, is the subset $\mathcal{N}(\mathbf{v})$ of $\mathbb{R}^{n+1}$ defined by

$$\mathcal{N}(\mathbf{v}) = \{q \in \mathbb{R}^{n+1} : q = p + xv + yN(p) \text{ for some } (x, y) \in \mathbb{R}^2\}$$

where N is the Gauss map $[\mathbf{N}(p) = (p, N(p))]$ (see Figure 12.2). $\mathcal{N}(\mathbf{v})$ is just a copy of $\mathbb{R}^2$, with p corresponding to the origin, $p + v$ corresponding to $(1, 0)$ and $p + N(p)$ corresponding to $(0, 1)$, so we shall identify $\mathcal{N}(\mathbf{v})$ with $\mathbb{R}^2$ and shall view the intersection $S \cap \mathcal{N}(\mathbf{v})$ as a subset of $\mathbb{R}^2$. More precisely,

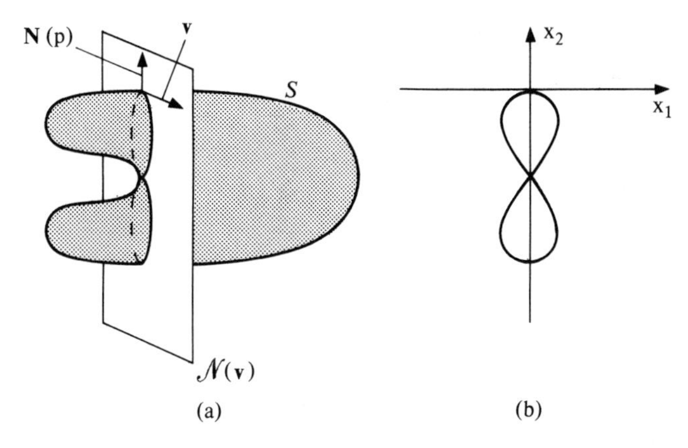

Figure 12.2 (a) The normal section $\mathcal{N}(\mathbf{v})$, $\mathbf{v} \in S_p$, $p \in S$. (b) $S \cap \mathcal{N}(\mathbf{v})$, viewed as a subset of $\mathbb{R}^2$.

define $i: \mathbb{R}^2 \to \mathbb{R}^{n+1}$ by $i(x, y) = p + xv + yN(p)$, so that $\mathcal{N}(\mathbf{v}) = i(\mathbb{R}^2)$. Then $i(x, y) \in S \cap \mathcal{N}(\mathbf{v}) \Leftrightarrow i(x, y) \in S \Leftrightarrow f \circ i(x, y) = c$ so, under i, the level set $(f \circ i)^{-1}(c)$ is identified with $S \cap \mathcal{N}(\mathbf{v})$. $(f \circ i)^{-1}(c)$ is not necessarily a (simple) plane curve, as Figure 12.2(b) shows. It is, however, if the points where ∇f is orthogonal to $\mathcal{N}(\mathbf{v})$ are deleted:

Theorem 1. *Let S be an oriented n-surface in $\mathbb{R}^{n+1}$ and let $\mathbf{v}$ be a unit vector in S_p, $p \in S$. Then there exists an open set $V \subset \mathbb{R}^{n+1}$ containing p such that $S \cap \mathcal{N}(\mathbf{v}) \cap V$ is a plane curve. Moreover, the curvature at p of this curve (suitably oriented) is equal to the normal curvature $k(\mathbf{v})$.*

PROOF. Let $f: U \to \mathbb{R}$ be such that $S = f^{-1}(c)$ and $\nabla f(q) \neq 0$ for all $q \in S$. Given $\mathbf{v} = (p, v) \in S_p$, $p \in S$, let $i: \mathbb{R}^2 \to \mathbb{R}^{n+1}$ be as above and let

$$V = \{q \in U: \text{either } \widetilde{\nabla f}(q) \cdot v \neq 0 \text{ or } \widetilde{\nabla f}(q) \cdot N(p) \neq 0\},$$

where $\widetilde{\nabla f}(q)$ is the vector part of $\nabla f(q)$ $(\nabla f(q) = (q, \widetilde{\nabla f}(q)))$. Then $p \in V$ and

$$\nabla(f \circ i)(x, y) = (x, y, \widetilde{\nabla f}(i(x, y)) \cdot v, \widetilde{\nabla f}(i(x, y)) \cdot N(p))$$

is never zero for $(x, y) \in i^{-1}(V)$ so

$$C = i^{-1}(S \cap \mathcal{N}(\mathbf{v}) \cap V) = (f \circ i)^{-1}(c) \cap i^{-1}(V)$$

is a plane curve (i.e., $S \cap \mathcal{N}(\mathbf{v}) \cap V$ is a plane curve) as required.

Moreover, if $\alpha(t) = (x(t), y(t))$ is a unit speed curve in C with $\dot{\alpha}(t_0) = (0, 0, 1, 0)$ (this vector is tangent to C since it is orthogonal to $\nabla(f \circ i)(0, 0)$), then $i \circ \alpha$ is a unit speed curve in $S \cap \mathcal{N}(\mathbf{v})$, since

$$\begin{aligned} \|(i \mathbin{\dot{\circ}} \alpha)(t)\|^2 &= \|(i \circ \alpha(t), x'(t)v + y'(t)N(p))\|^2 \\ &= (x'(t))^2 + (y'(t))^2 = \|\dot{\alpha}(t)\|^2 = 1, \end{aligned}$$

and $(i \mathbin{\dot{\circ}} \alpha)(t_0) = \mathbf{v}$. Now, if we orient C so that the orientation normal at $(0, 0)$ is $(0, 0, 0, 1)$, then the curvature of C at $(0, 0)$ is

$$\kappa(\alpha(t_0)) = \ddot{\alpha}(t_0) \cdot (\alpha(t_0), 0, 1) = y''(t_0)$$

whereas the normal curvature of S in the direction $\mathbf{v}$ is

$$k(\mathbf{v}) = (i \mathbin{\ddot{\circ}} \alpha)(t_0) \cdot \mathbf{N}(p) = (p, x''(t_0)v + y''(t_0)N(p)) \cdot (p, N(p)) = y''(t_0)$$

so $k(\mathbf{v}) = \kappa(\alpha(t_0))$, as was to be shown. □

For p a point in an oriented n-surface S, the normal curvature $k(\mathbf{v})$ is defined for each unit vector $\mathbf{v}$ in the tangent space S_p to S at p. Thus the normal curvature at p is a real valued function with domain the unit sphere in S_p. Since k is continuous and the sphere is compact, this function attains its maximum and its minimum. The following lemma shows that these extrema are eigenvalues of the Weingarten map L_p.

Lemma. *Let V be a finite dimensional vector space with dot product and* *let $L\colon V \to V$ be a self-adjoint linear transformation on V. Let $S = \{v \in V\colon v \cdot v = 1\}$ and define $f\colon S \to \mathbb{R}$ by $f(v) = L(v) \cdot v$. Suppose f is stationary at $v_0 \in S$ (that is, suppose $(f \circ \alpha)'(t_0) = 0$ for all parametrized curves $\alpha\colon I \to S$ with $\alpha(t_0) = v_0$). Then $L(v_0) = f(v_0)v_0$ (that is, v_0 is an eigenvector of L with eigenvalue $f(v_0)$).*

PROOF. Since f is stationary at v_0, $(f \circ \alpha)'(0) = 0$ for all parametrized curves α in S with $\alpha(0) = v_0$. For v any unit vector with $v \cdot v_0 = 0$, let $\alpha(t) = (\cos t)v_0 + (\sin t)v$. Then

$$\begin{aligned}
0 = (f \circ \alpha)'(0) &= \frac{d}{dt}\bigg|_0 L(\alpha(t)) \cdot \alpha(t) \\
&= \frac{d}{dt}\bigg|_0 [(\cos^2 t)L(v_0) \cdot v_0 + 2 \sin t \cos t\, L(v_0) \cdot v + (\sin^2 t)L(v) \cdot v] \\
&= 2L(v_0) \cdot v.
\end{aligned}$$

Thus $L(v_0) \perp v$ for all unit vectors $v \in v_0^\perp$. It follows then that $L(v_0) \perp v_0^\perp$; that is, $L(v_0) = \lambda v_0$ for some $\lambda \in \mathbb{R}$. Thus v_0 is an eigenvector of L. The eigenvalue λ is given by

$$\lambda = \lambda v_0 \cdot v_0 = L(v_0) \cdot v_0 = f(v_0). \qquad \square$$

Remark. In this lemma we have used the concept of a (smooth) parametrized curve $\alpha\colon I \to V$ where V is an arbitrary finite dimensional vector space with dot product. Smoothness makes sense in this setting because limits and derivatives can be defined in the usual way: $\lim_{t \to t_0} \alpha(t) = v$ means for every $\varepsilon > 0$ there is a $\delta > 0$ such that $\|\alpha(t) - v\| < \varepsilon$ whenever $0 < |t - t_0| < \delta$, and $(d\alpha/dt)(t_0) = \lim_{t \to t_0} (\alpha(t) - \alpha(t_0))/(t - t_0)$ whenever this limit exists. In fact, all the geometry we are developing here can be done as well in V as in $\mathbb{R}^{n+1}$.

The converse of the above lemma is also true: if v_0 is an eigenvector of L then $f(v) = L(v) \cdot v$ is stationary at $v_0 \in S$. For if $\alpha\colon I \to S$ then $\alpha(t) \cdot \alpha(t) \equiv 1$ so $\alpha(t) \cdot (d\alpha/dt)(t) = 0$ for all $t \in I$ and if $\alpha(t_0) = v_0$ then

$$\begin{aligned}
(f \circ \alpha)'(t_0) &= \frac{d}{dt}\bigg|_{t_0} [L(\alpha(t)) \cdot \alpha(t)] \\
&= L\left(\frac{d\alpha}{dt}(t_0)\right) \cdot \alpha(t_0) + L(\alpha(t_0)) \cdot \frac{d\alpha}{dt}(t_0) \\
&= 2L(\alpha(t_0)) \cdot \frac{d\alpha}{dt}(t_0) = 2\lambda\alpha(t_0) \cdot \frac{d\alpha}{dt}(t_0) = 0.
\end{aligned}$$

Theorem 2. *Let V be a finite dimensional vector space with dot product and let $L\colon V \to V$ be a self-adjoint linear transformation on V. Then there exists an orthonormal basis for V consisting of eigenvectors of L.*

PROOF. By induction on the dimension n of V. For $n = 1$, the theorem is trivially true. Assume then that it is true for $n = k$. Suppose $n = k + 1$. By the lemma, there exists a unit vector v_1 in V which is an eigenvector of L (e.g., choose v_1 such that $L(v_1) \cdot v_1 \geq L(v) \cdot v$ for all unit vectors $v \in V$). Let $W = v_1^{\perp}$. Then

$$L(w) \cdot v_1 = w \cdot L(v_1) = w \cdot \lambda_1 v_1 = \lambda_1 (w \cdot v_1) = 0$$

for all $w \in W$, where λ_1 is the eigenvalue belonging to v_1. Thus the restriction $L|_W$ of L to W maps W into W. Clearly $L|_W$ is self adjoint. Since $\dim(W) = \dim V - 1 = k$, the induction assumption implies that there exist $\{v_2, \ldots, v_{k+1}\}$, an orthonormal basis for W consisting of eigenvectors of $L|_W$. But each eigenvector of $L|_W$ is also an eigenvector of L, so $\{v_1, \ldots, v_{k+1}\}$ is an orthonormal basis for V consisting of eigenvectors of L. □

Note that there exist at most n eigenvalues of a self-adjoint linear transformation L on an n-dimensional vector space because each eigenvalue is a root of the characteristic polynomial $\det(L - \lambda I)$ which is a polynomial in λ of degree n. Here, I is the identity transformation on V. That λ is a root of this polynomial follows from the fact that $L(v) = \lambda v$ if and only if $(L - \lambda I)(v) = 0$ so $L - \lambda I$ must be singular. Counting multiplicities then, there are exactly n eigenvalues of L. Note further that the eigen directions v_i of L are determined uniquely (up to sign) if and only if the n eigenvalues of L are distinct.

For S an oriented n-surface in $\mathbb{R}^{n+1}$ and $p \in S$, the eigenvalues $k_1(p), \ldots, k_n(p)$ of the Weingarten map $L_p: S_p \to S_p$ are called *principal curvatures* of S at p and the unit eigenvectors of L_p are called *principal curvature directions.* If the principal curvatures are ordered so that $k_1(p) \leq k_2(p) \leq \cdots \leq k_n(p)$, the discussion above shows that $k_n(p)$ is the maximum value of normal curvature $k(\mathbf{v})$ for $\mathbf{v} \in S_p$, $\|\mathbf{v}\| = 1$; $k_{n-1}(p)$ is the maximum value of $k(\mathbf{v})$ for $\mathbf{v} \in S_p$, $\|\mathbf{v}\| = 1$, and $\mathbf{v} \perp \mathbf{v}_n$ where $\mathbf{v}_n$ is a principal curvature direction corresponding to $k_n(p)$; $k_{n-2}(p) = \max\{k(\mathbf{v}): \mathbf{v} \in S_p, \|\mathbf{v}\| = 1, \mathbf{v} \perp \{\mathbf{v}_n, \mathbf{v}_{n-1}\}\}$ etc. Furthermore, all the principal curvatures $k_i(p)$ are stationary values of normal curvature, and $k_1(p)$ is the minimum value of $k(\mathbf{v})$ for $\mathbf{v} \in S_p$, $\|\mathbf{v}\| = 1$.

EXAMPLE. Let S be the hyperboloid $-x_1^2 + x_2^2 + x_3^2 = 1$ in $\mathbb{R}^3$, oriented by $\mathbf{N}(p) = (p, -x_1/\|p\|, x_2/\|p\|, x_3/\|p\|)$, $p = (x_1, x_2, x_3) \in S$. Then, as we saw earlier in this chapter, for $p = (0, 0, 1)$, $S_p = \{(p, v_1, v_2, 0): v_1, v_2 \in \mathbb{R}\}$ and, for $\|\mathbf{v}\| = 1$, $k(\mathbf{v}) = v_1^2 - v_2^2$. Thus $k(\mathbf{v})$ attains its maximum value (for $\|\mathbf{v}\|^2 = v_1^2 + v_2^2 = 1$) when $\mathbf{v} = (p, \pm 1, 0, 0)$ and its minimum value when $\mathbf{v} = (p, 0, \pm 1, 0)$, so the principal curvatures at p are $k_1(p) = -1$ and $k_2(p) = 1$.

Theorem 3. *Let S be an oriented n-surface in $\mathbb{R}^{n+1}$, let $p \in S$, and let $\{k_1(p), \ldots, k_n(p)\}$ be the principal curvatures of S at p with corresponding orthogonal* *principal curvature directions* $\{\mathbf{v}_1, \ldots, \mathbf{v}_n\}$. *Then the normal curvature* $k(\mathbf{v})$ *in the direction* $\mathbf{v} \in S_p(\|\mathbf{v}\| = 1)$ *is given by*

$$k(\mathbf{v}) = \sum_{i=1}^{n} k_i(p)(\mathbf{v} \cdot \mathbf{v}_i)^2 = \sum_{i=1}^{n} k_i(p) \cos^2 \theta_i$$

where $\theta_i = \cos^{-1}(\mathbf{v} \cdot \mathbf{v}_i)$ *is the angle between* $\mathbf{v}$ *and* $\mathbf{v}_i$.

PROOF. Since $\mathbf{v}$ can be expressed as a linear combination of the orthonormal basis vectors $\{\mathbf{v}_1, \ldots, \mathbf{v}_n\}$ by $\mathbf{v} = \sum_{i=1}^n (\mathbf{v} \cdot \mathbf{v}_i)\mathbf{v}_i = \sum_{i=1}^n (\cos \theta_i)\mathbf{v}_i$ we have

$$\begin{aligned} k(\mathbf{v}) = L_p(\mathbf{v}) \cdot \mathbf{v} &= \sum_{i=1}^{n} (\cos \theta_i) L_p(\mathbf{v}_i) \cdot \mathbf{v} \\ &= \sum_{i=1}^{n} (\cos \theta_i) k_i(p) \mathbf{v}_i \cdot \mathbf{v} = \sum_{i=1}^{n} k_i(p) \cos^2 \theta_i. \end{aligned}$$ □

The numbers $\cos \theta_i = \mathbf{v} \cdot \mathbf{v}_i$ such that $\mathbf{v} = \sum_{i=1}^n (\cos \theta_i)\mathbf{v}_i$ are called the *direction cosines* of $\mathbf{v}$ with respect to the orthonormal basis $\{\mathbf{v}_1, \ldots, \mathbf{v}_n\}$.

Associated with any self-adjoint linear transformation $L: V \to V$, where V is a vector space with a dot product, is a real valued function $\mathscr{Q}: V \to \mathbb{R}$ defined by

$$\mathscr{Q}(v) = L(v) \cdot v.$$

This function $\mathscr{Q}$ is the *quadratic form* associated with L. The quadratic form associated with the Weingarten map L_p at a point p of an oriented n-surface $S \subset \mathbb{R}^{n+1}$ is called the *second fundamental form* of S at p and is denoted by $\mathscr{S}_p$. Thus, $\mathscr{S}_p(\mathbf{v}) = L_p(\mathbf{v}) \cdot \mathbf{v} = \ddot{\alpha}(t_0) \cdot \mathbf{N}(p)$ where $\alpha: I \to S$ is any parametrized curve in S with $\alpha(t_0) = p$ and $\dot{\alpha}(t_0) = \mathbf{v}$. In particular, when $\|\mathbf{v}\| = 1$, $\mathscr{S}_p(\mathbf{v})$ is equal to the normal curvature of S at p in the direction $\mathbf{v}$.

The *first fundamental form* of S at p is the quadratic form $\mathscr{I}_p$ associated with the identity transformation on S_p. Thus $\mathscr{I}_p(\mathbf{v}) = \mathbf{v} \cdot \mathbf{v} = \|\mathbf{v}\|^2$, for all $\mathbf{v} \in S_p$.

Note that the quadratic form associated with a self-adjoint linear transformation L contains exactly the same information as L, since L can be recovered from $\mathscr{Q}$ by use of the formula

$$L(v) \cdot w = \tfrac{1}{2}[\mathscr{Q}(v + w) - \mathscr{Q}(v) - \mathscr{Q}(w)],$$

which is valid for all v and w in V.

A quadratic form $\mathscr{Q}$ is said to be

positive definite if $\mathscr{Q}(v) > 0$ for all $v \neq 0$,
negative definite if $\mathscr{Q}(v) < 0$ for all $v \neq 0$,
definite if it is either positive or negative definite,
indefinite if it is neither positive or negative definite,
positive semi-definite if $\mathscr{Q}(v) \geq 0$ for all v,
negative semi-definite if $\mathscr{Q}(v) \leq 0$ for all v, and
semi-definite if it is either positive or negative semi-definite.

Thus the first fundamental form $\mathscr{I}_p$ of an oriented n-surface $S \subset \mathbb{R}^{n+1}$ is always positive definite. The second fundamental form $\mathscr{S}_p$ is positive definite if and only if the normal curvature $k(\mathbf{v}) = \mathscr{S}_p(\mathbf{v})$ is positive for every direction $\mathbf{v}$ at p. By the previous theorem, this is the case if and only if all the principal curvatures $k_i(p)$ of S at p are positive. Similarly, $\mathscr{S}_p$ is negative definite if and only if all the principal curvatures of S at p are negative. When $\mathscr{S}_p$ is positive definite, the surface S bends toward the unit normal $\mathbf{N}(p)$ in every tangent direction $\mathbf{v}$ at p, whereas if $\mathscr{S}_p$ is negative definite S bends away from $\mathbf{N}(p)$ in all directions (see Figure 12.3).

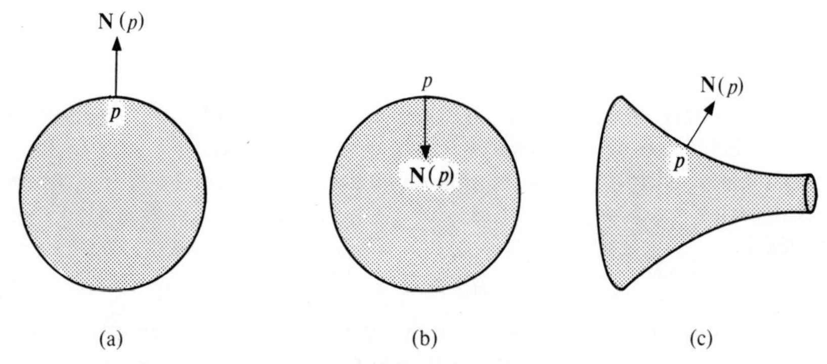

Figure 12.3 The Weingarten map at p is negative definite in (a), positive definite in (b), and indefinite in (c).

Theorem 4. *On each compact oriented n-surface S in* $\mathbb{R}^{n+1}$ *there exists a point p such that the second fundamental form at p is definite.*

PROOF. The idea of the proof is to enclose S in a large sphere and then shrink the sphere until it touches S (see Figure 12.4.) At the point of contact, the normal curvature of S will be bounded away from zero by the normal curvature of the sphere.

More precisely, define g: $\mathbb{R}^{n+1} \to \mathbb{R}$ by $g(x_1, \ldots, x_{n+1}) = x_1^2 + \cdots + x_{n+1}^2$. Since S is compact, there exists $p \in S$ such that $g(p) \geq g(q)$ for all $q \in S$. By the Lagrange multiplier theorem, there exists $\lambda \in \mathbb{R}$ such that $\nabla g(p) = \lambda \nabla f(p) = \mu \mathbf{N}(p)$ where $S = f^{-1}(c)$ and $\mu = \pm\lambda\|\nabla f(p)\|$. The sign of μ depends on the orientation of S; assume for the moment that $\mu < 0$ (i.e., S in oriented by its "inward" normal). Then $\mu = -|\mu| = -\|\mu\mathbf{N}(p)\| = -\|\nabla g(p)\| = -2\|p\|$, so that

$$\mathbf{N}(p) = \frac{1}{\mu}\nabla g(p) = -\frac{1}{\|p\|}(p, p).$$

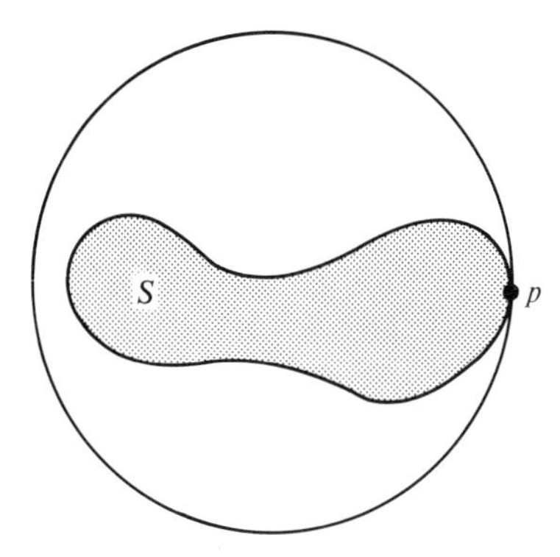

Figure 12.4 $|k(v)| \geq 1/r$ for all directions $\mathbf{v}$ at p, where r is the radius of the enveloping sphere.

Now, for $\mathbf{v} \in S_p$, $\|\mathbf{v}\| = 1$, let $\alpha\colon I \to S$ be such that $\dot{\alpha}(t_0) = \mathbf{v}$. Then $g \circ \alpha(t_0) \geq g \circ \alpha(t)$ for all $t \in I$, so

$$\begin{aligned}
0 &\geq \frac{d^2}{dt^2}\bigg|_{t_0} (g \circ \alpha) = \frac{d}{dt}\bigg|_{t_0} \nabla g(\alpha(t)) \cdot \dot{\alpha}(t) \\
&= \frac{d}{dt}\bigg|_{t_0} 2\alpha(t) \cdot \frac{d\alpha}{dt}(t) \\
&= 2[\|\dot{\alpha}(t_0)\|^2 + (\alpha(t_0), \alpha(t_0)) \cdot \ddot{\alpha}(t_0)] \\
&= 2[1 - \|p\|\mathbf{N}(p) \cdot \ddot{\alpha}(t_0)] \\
&= 2[1 - \|p\|k(\mathbf{v})].
\end{aligned}$$

Thus $k(\mathbf{v}) \geq 1/\|p\|$ for all directions $\mathbf{v} \in S_p$.

If S were oriented so that $\mu = \nabla g(p) \cdot \mathbf{N}(p) > 0$, then the normal curvature would change sign so that $k(\mathbf{v}) \leq -1/\|p\|$ for all directions $\mathbf{v}$ at p. □

The determinant and trace of the Weingarten map are of particular importance in differential geometry. The determinant $K(p) = \det L_p$ is called the *Gauss-Kronecker curvature* of S at p. It is equal to the product of the principal curvatures at p. When $n = 2$, $K(p) = k_1(p)k_2(p)$ is called simply the *Gaussian curvature* at p. $1/n$ times the trace of L_p is called the *mean curvature* $H(p)$ of S at p. Thus $H(p) = (1/n)\sum_{i=1}^{n} k_i(p)$ is the average value of the principal curvatures at p.

The following theorem is useful in computing Gauss-Kronecker curvature.

Theorem 5. *Let S be an oriented n-surface in $\mathbb{R}^{n+1}$ and let $p \in S$. Let $\mathbf{Z}$ be any non-zero normal vector field on S such that $\mathbf{N} = \mathbf{Z}/\|\mathbf{Z}\|$ and let $\{\mathbf{v}_1, \ldots, \mathbf{v}_n\}$ be* *any basis for S_p. Then*

$$K(p) = (-1)^n \det\begin{pmatrix}\nabla_{\mathbf{v}_1}\mathbf{Z}\\ \vdots\\ \nabla_{\mathbf{v}_n}\mathbf{Z}\\ \mathbf{Z}(p)\end{pmatrix} \Bigg/ \|\mathbf{Z}(p)\|^n \det\begin{pmatrix}\mathbf{v}_1\\ \vdots\\ \mathbf{v}_n\\ \mathbf{Z}(p)\end{pmatrix},$$

where, for $\mathbf{w}_1, \ldots, \mathbf{w}_{n+1} \in \mathbb{R}_p^{n+1}$, $\mathbf{w}_i = (p, w_{i,1}, \ldots, w_{i,n+1})$,

$$\det\begin{pmatrix}\mathbf{w}_1\\ \vdots\\ \mathbf{w}_{n+1}\end{pmatrix} = \begin{vmatrix} w_{1,1} & \cdots & w_{1,n+1}\\ \vdots & & \vdots\\ w_{n+1,1} & \cdots & w_{n+1,n+1}\end{vmatrix}.$$

PROOF. Since $\mathbf{Z} = \|\mathbf{Z}\|\mathbf{N}$,

$$\begin{aligned}\det\begin{pmatrix}\nabla_{\mathbf{v}_1}\mathbf{Z}\\ \vdots\\ \nabla_{\mathbf{v}_n}\mathbf{Z}\\ \mathbf{Z}(p)\end{pmatrix} &= \det\begin{pmatrix}(\nabla_{\mathbf{v}_1}\|\mathbf{Z}\|)\mathbf{N}(p) + \|\mathbf{Z}(p)\|\nabla_{\mathbf{v}_1}\mathbf{N}\\ \vdots\\ (\nabla_{\mathbf{v}_n}\|\mathbf{Z}\|)\mathbf{N}(p) + \|\mathbf{Z}(p)\|\nabla_{\mathbf{v}_n}\mathbf{N}\\ \|\mathbf{Z}(p)\|\mathbf{N}(p)\end{pmatrix}\\
&= \|\mathbf{Z}(p)\|^n \det\begin{pmatrix}\nabla_{\mathbf{v}_1}\mathbf{N}\\ \vdots\\ \nabla_{\mathbf{v}_n}\mathbf{N}\\ \|\mathbf{Z}(p)\|\mathbf{N}(p)\end{pmatrix}\\
&= (-1)^n\|\mathbf{Z}(p)\|^n \det\begin{pmatrix}L_p(\mathbf{v}_1)\\ \vdots\\ L_p(\mathbf{v}_n)\\ \mathbf{Z}(p)\end{pmatrix}\\
&= (-1)^n\|\mathbf{Z}(p)\|^n \det\left[\begin{pmatrix} & & & 0\\ & A^t & & \vdots\\ & & & 0\\ 0 & \cdots & 0 & 1\end{pmatrix}\begin{pmatrix}\mathbf{v}_1\\ \vdots\\ \mathbf{v}_n\\ \mathbf{Z}(p)\end{pmatrix}\right]\\
&= (-1)^n\|\mathbf{Z}(p)\|^n(\det A) \det\begin{pmatrix}\mathbf{v}_1\\ \vdots\\ \mathbf{v}_n\\ \mathbf{Z}(p)\end{pmatrix}\\
&= (-1)^n\|\mathbf{Z}(p)\|^n K(p) \det\begin{pmatrix}\mathbf{v}_1\\ \vdots\\ \mathbf{v}_n\\ \mathbf{Z}(p)\end{pmatrix},\end{aligned}$$

where A is the matrix for L_p with respect to the basis $\{\mathbf{v}_1, \ldots, \mathbf{v}_n\}$ for S_p, and A^t denotes the transpose of A. Solving for $K(p)$ completes the proof. □

EXAMPLE. Let S be the ellipsoid $(x_1^2/a^2) + (x_2^2/b^2) + (x_3^2/c^2) = 1$ (a, b, and c all $\neq 0$), oriented by its outward normal. Let $\mathbf{Z}(p) = \frac{1}{2}\nabla f(p) =$

$(p, x_1/a^2, x_2/b^2, x_3/c^2)$ for $p = (x_1, x_2, x_3) \in S$. A basis for S_p will consist of any independent pair of vectors orthogonal to $\mathbf{Z}(p)$. For $x_1 \neq 0$, we may take $\mathbf{v}_1 = (p, x_2/b^2, -x_1/a^2, 0)$ and $\mathbf{v}_2 = (p, x_3/c^2, 0, -x_1/a^2)$. Then

$$\det\begin{pmatrix} \nabla_{\mathbf{v}_1}\mathbf{Z} \\ \nabla_{\mathbf{v}_2}\mathbf{Z} \\ \mathbf{Z}(p) \end{pmatrix} = \begin{vmatrix} x_2/a^2b^2 & -x_1/a^2b^2 & 0 \\ x_3/a^2c^2 & 0 & -x_1/a^2c^2 \\ x_1/a^2 & x_2/b^2 & x_3/c^2 \end{vmatrix}$$

$$= \frac{x_1}{a^4b^2c^2}\left(\frac{x_1^2}{a^2} + \frac{x_2^2}{b^2} + \frac{x_3^2}{c^2}\right) = \frac{x_1}{a^4b^2c^2},$$

$$\det\begin{pmatrix} \mathbf{v}_1 \\ \mathbf{v}_2 \\ \mathbf{Z}(p) \end{pmatrix} = \begin{vmatrix} x_2/b^2 & -x_1/a^2 & 0 \\ x_3/c^2 & 0 & -x_1/a^2 \\ x_1/a^2 & x_2/b^2 & x_3/c^2 \end{vmatrix} = \frac{x_1}{a^2}\left(\frac{x_1^2}{a^4} + \frac{x_2^2}{b^4} + \frac{x_3^2}{c^4}\right),$$

and

$$\|\mathbf{Z}(p)\| = \left(\frac{x_1^2}{a^4} + \frac{x_2^2}{b^4} + \frac{x_3^2}{c^4}\right)^{1/2}$$

so the Gaussian curvature of the ellipsoid is

$$K(p) = \frac{1}{a^2b^2c^2\left(\dfrac{x_1^2}{a^4} + \dfrac{x_2^2}{b^4} + \dfrac{x_3^2}{c^4}\right)^2}.$$

Note that, although this formula for K was derived under the assumption that $x_1 \neq 0$, it is valid for all $p \in S$ by continuity.

Theorem 4 of this chapter is an example of a global theorem in differential geometry. A property of an n-surface S is said to be a *global property* if it expresses a fact about the surface S as a whole (such as, the n-surface S is compact, or the n-surface S is connected, or the 1-surface C has finite length). On the other hand, a property of S is said to be a *local property* if it expresses a fact about the surface S at or near a particular point of S, a fact which can be verified by computations in an arbitrarily small open set containing the point (for example, the second fundamental form $\mathscr{S}_p$ of S at p is definite, or the Gauss-Kronecker curvature of S at p is positive). A *global theorem* is a theorem in which a global property is among the essential hypotheses or among the conclusions. A theorem in which all the essential hypotheses and conclusions are local properties is called a *local theorem*. Thus, for example, Theorems 3 and 5 of this chapter are local theorems (the hypothesis that S be oriented, although a global hypothesis, is in fact inessential; these theorems are independent of the orientation chosen and in fact all that is required for the validity of these theorems is a choice of smooth unit normal vector field $\mathbf{N}$ defined at and near the point p). In contrast, Theorem 4 of this chapter is a global theorem; its validity depends crucially on the hypothesis that S be compact. The theorem of Chapter 6 is another example of a global theorem.

Our next theorem is an especially interesting type of global theorem. It asserts the equivalence of two local properties, at each point of an n-surface S, but only in the presence of a global hypothesis. Note that the theorem fails to be true if the compactness hypothesis is removed. (Consider a one-sheeted hyperboloid in $\mathbb{R}^3$.)

Theorem 6. *Let S be a compact connected oriented n-surface in $\mathbb{R}^{n+1}$. Then the Gauss-Kronecker curvature $K(p)$ of S at p is non-zero for all $p \in S$ if and only if the second fundamental form $\mathscr{S}_p$ of S at p is definite for all $p \in S$.*

PROOF. If $\mathscr{S}_p$ is definite for all $p \in S$ then the normal curvature $k(\mathbf{v}) = \mathscr{S}_p(\mathbf{v})$ is non-zero for each direction $\mathbf{v} \in S_p$ so in particular all principal curvatures at p are non-zero and hence so is their product $K(p)$.

Conversely, by Theorem 4 there exists a point $p_0 \in S$ such that $\mathscr{S}_{p_0}$ is definite. Suppose $\mathscr{S}_{p_0}$ is in fact positive definite. Then the minimum principal curvature k_1 of S is positive at p_0. Since $k_1 \colon S \to \mathbb{R}$ is continuous, S is connected, and k_1 is nowhere zero (since, by hypothesis, K is nowhere zero), k_1 must be everywhere positive. Hence all the principal curvatures are everywhere positive and $\mathscr{S}_p$ is positive definite for all $p \in S$. If $\mathscr{S}_{p_0}$ were negative definite, a similar argument, with k_1 replaced by the maximum principal curvature k_n, would show that $\mathscr{S}_p$ is negative definite for all $p \in S$. □

Remark. The connectedness hypothesis in Theorem 6 is actually inessential since it can be shown that every compact n-surface in $\mathbb{R}^{n+1}$ is a finite union of connected ones, and Theorem 6 can then be applied to each of these.

EXERCISES

12.1. Let $S = f^{-1}(c)$ be an n-surface in $\mathbb{R}^{n+1}$, oriented by $\nabla f/\|\nabla f\|$. Show that, for $\mathbf{v} = (p, v_1, \ldots, v_{n+1})$ a vector tangent to S at $p \in S$, the value of the second fundamental form of S at p on $\mathbf{v}$ is given by

$$\mathscr{S}_p(\mathbf{v}) = -(1/\|\nabla f(p)\|) \sum_{i,j=1}^{n+1} \frac{\partial^2 f}{\partial x_i\, \partial x_j}(p) v_i v_j.$$

(When $\|\mathbf{v}\| = 1$, this formula provides a straightforward way to compute the normal curvature $k(\mathbf{v}) = \mathscr{S}_p(\mathbf{v})$ in the direction $\mathbf{v}$.)

In Exercises 12.2–12.6, find the normal curvature $k(\mathbf{v})$ for each tangent direction $\mathbf{v}$, the principal curvatures and principal curvature directions, and the Gauss-Kronecker and mean curvatures, at the given point p of the given n-surface $f(x_1, \ldots, x_{n+1}) = c$ oriented by $\nabla f/\|\nabla f\|$.

12.2. $x_1 + x_2 + \cdots + x_{n+1} = 1$, $p = (1, 0, \ldots, 0)$.

12.3. $x_2^2 + x_3^2 + \cdots + x_{n+1}^2 = r^2$, $r > 0$, $p = (0, \ldots, 0, r)$.

12.4. $(x_1^2/a^2) + (x_2^2/b^2) + (x_3^2/c^2) = 1$, $p = (a, 0, 0)$ (in $\mathbb{R}^3$)

12.5. $(x_1^2/a^2) + (x_2^2/b^2) - (x_3^2/c^2) = 1$, $p = (a, 0, 0)$ (in $\mathbb{R}^3$)

12.6. $x_1^2 + (\sqrt{x_2^2 + x_3^2} - 2)^2 = 1$ (torus in $\mathbb{R}^3$)

(a) $p = (0, 3, 0)$
(b) $p = (0, 1, 0)$

12.7. Show that if S and $\tilde{S}$ denote the same n-surface in $\mathbb{R}^{n+1}$ but with opposite orientations, then $\tilde{K} = (-1)^n K$ where K and $\tilde{K}$ are the Gauss-Kronecker curvatures of S and $\tilde{S}$ respectively. (In particular, Gauss-Kronecker curvature is independent of the choice of orientation if n is even.)

In Exercises 12.8–12.11, find the Gaussian curvature $K: S \to \mathbb{R}$ where S is the given surface.

12.8. $x_1^2 + x_2^2 - x_3^2 = 0$, $x_3 > 0$ (cone)

12.9. $(x_1^2/a^2) + (x_2^2/b^2) - (x_3^2/c^2) = 1$ (hyperboloid)

12.10. $(x_1^2/a^2) + (x_2^2/b^2) - x_3 = 0$ (elliptic paraboloid)

12.11. $(x_1^2/a^2) - (x_2^2/b^2) - x_3 = 0$ (hyperbolic paraboloid)

12.12. (a) Find the Gaussian curvature of a cylinder over a plane curve.
(b) Find the Gauss-Kronecker curvature of a cylinder over an n-surface.

12.13. Let $g: \mathbb{R}^n \to \mathbb{R}$ be a smooth function. Show that the Gauss-Kronecker curvature K of the graph of g is given by the formula

$$K = \det\left(\frac{\partial^2 g}{\partial x_i\, \partial x_j}\right) \Bigg/ \left(1 + \sum_{i=1}^{n} \left(\frac{\partial g}{\partial x_i}\right)^2\right)^{(n/2)+1}$$

where the orientation $\mathbf{N}$ is chosen so that $\mathbf{N}(p) \cdot (p, 0, \ldots, 0, 1) > 0$ for all p in the graph.

12.14. Let S be an oriented 2-surface in $\mathbb{R}^3$ and let $p \in S$. Show that, for each $\mathbf{v}$, $\mathbf{w} \in S_p$, $L_p(\mathbf{v}) \times L_p(\mathbf{w}) = K(p)\mathbf{v} \times \mathbf{w}$.

12.15. Show that, for S an oriented 2-surface in $\mathbb{R}^3$,

$$K(p) = \mathbf{Z}(p) \cdot \nabla_{\mathbf{v}} \mathbf{Z} \times \nabla_{\mathbf{w}} \mathbf{Z} / \|\mathbf{Z}(p)\|^4$$

where $\mathbf{Z}$ is any nowhere zero normal vector field on S and $\mathbf{v}$ and $\mathbf{w}$ are any two vectors in S_p such that $\mathbf{v} \times \mathbf{w} = \mathbf{Z}(p)$.

12.16. Show that the mean curvature at a point p of an oriented n-surface S can be computed from the values of normal curvature on any orthonormal basis $\{\mathbf{v}_1, \ldots, \mathbf{v}_n\}$ for S_p by the formula

$$H(p) = \frac{1}{n} \sum_{i=1}^{n} k(\mathbf{v}_i).$$

12.17. Let S be an oriented 2-surface in $\mathbb{R}^3$ and let $\{\mathbf{v}_1, \mathbf{v}_2\}$ be an orthonormal basis for S_p consisting of eigenvectors of L_p. Let $k_i = k(\mathbf{v}_i)$.

(a) Show that for $\mathbf{v}(\theta) = (\cos\theta)\mathbf{v}_1 + (\sin\theta)\mathbf{v}_2 \in S_p$,

$$k(\mathbf{v}(\theta)) = k_1 \cos^2\theta + k_2 \sin^2\theta.$$

(b) Show that the mean curvature at p is given by the formula

$$H(p) = \frac{1}{2\pi}\int_0^{2\pi} k(\mathbf{v}(\theta))\,d\theta.$$

12.18. Let $S = f^{-1}(c)$ be an oriented n-surface, oriented by $\mathbf{N} = \nabla f/\|\nabla f\|$. Show that $H(p) = -(1/n)\operatorname{div}\mathbf{N}$. [*Hint*: First note that $\operatorname{div}\mathbf{N} = \operatorname{trace}\{\mathbf{v} \mapsto \nabla_{\mathbf{v}}\mathbf{N}\}$ and then evaluate this trace using the basis $\{\mathbf{v}_1, \ldots, \mathbf{v}_n, \mathbf{N}(p)\}$ where $\{\mathbf{v}_1, \ldots, \mathbf{v}_n\}$ is an orthonormal basis of S_p consisting of eigenvectors of L_p.]

12.19. Let $S = f^{-1}(c)$ be an n-surface in $\mathbb{R}^{n+1}$. Given $a > 0$, let $\tilde{S} = g^{-1}(c)$ where $g(p) = f(p/a)$ for all p such that p/a is in the domain of f.

(a) Show that $\tilde{S}$ is an n-surface in $\mathbb{R}^{n+1}$ and that $p \in S$ if and only if $ap \in \tilde{S}$.
(b) Letting S be oriented by $\nabla f/\|\nabla f\|$ and $\tilde{S}$ by $\nabla g/\|\nabla g\|$, show that the spherical images of S and of $\tilde{S}$ are the same.
(c) Show that the mean curvatures H and $\tilde{H}$ of S and $\tilde{S}$ are related by $\tilde{H}(ap) = (1/a)H(p)$.
(d) Show that the Gauss-Kronecker curvatures K and $\tilde{K}$ of S and $\tilde{S}$ are related by $\tilde{K}(ap) = (1/a^n)K(p)$.

Convex Surfaces 13

An oriented n-surface S in $\mathbb{R}^{n+1}$ is *convex* (or *globally convex*) if, for each $p \in S$, S is contained in one of the closed half-spaces

$$H_p^+ = \{q \in \mathbb{R}^{n+1}: (q - p) \cdot N(p) \geq 0\}$$

or

$$H_p^- = \{q \in \mathbb{R}^{n+1}: (q - p) \cdot N(p) \leq 0\},$$

where N is the Gauss map of S (see Figure 13.1). An oriented n-surface S is *convex at* $p \in S$ if there exists an open set $V \subset \mathbb{R}^{n+1}$ containing p such that $S \cap V$ is contained either in H_p^+ or in H_p^-. Thus a convex n-surface is necessarily convex at each of its points, but an n-surface convex at each point need not be a convex n-surface (see Figure 13.2).

If S is convex and $S \cap H_p = \{p\}$ for each $p \in S$, where

$$H_p = \{q \in \mathbb{R}^{n+1}: (q - p) \cdot N(p) = 0\},$$

then S is said to be *strictly convex*. Similarly, if S is convex at p for some $p \in S$ and $S \cap V \cap H_p = \{p\}$ for some open set V containing p, then S is *strictly convex at p*.

The goal of this chapter is to relate curvature to convexity. The first result is an easy one.

Theorem 1. *Let S be an oriented n-surface in $\mathbb{R}^{n+1}$ which is convex at $p \in S$. Then the second fundamental form $\mathscr{S}_p$ of S at p is semi-definite.*

PROOF. Suppose $S \cap V \subset H_p^+$ for some open set V in $\mathbb{R}^{n+1}$ containing p. For $\mathbf{v} \in S_p$, let $\alpha: I \to S \cap V$ be such that $\alpha(t_0) = p$ and $\dot{\alpha}(t_0) = \mathbf{v}$. Define $h: I \to \mathbb{R}$ by $h(t) = (\alpha(t) - p) \cdot N(p)$. Then $h(t) \geq 0$ for all t, since $\alpha(t) \in H_p^+$

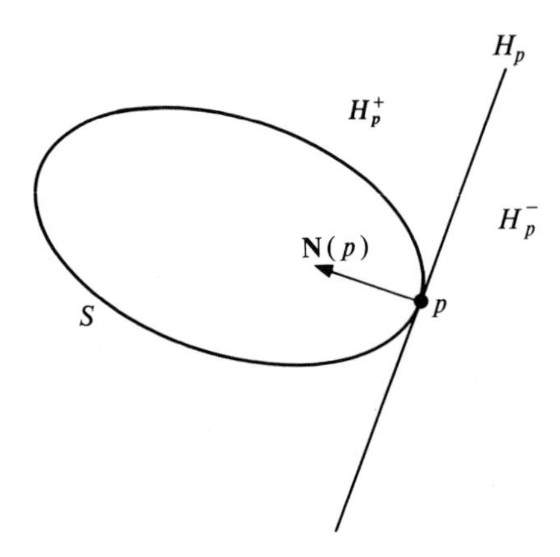

Figure 13.1 S is convex if, for each $p \in S$, S is contained in one of the half-spaces $H_p^{\pm}$. Note that the common boundary of these two half-spaces is the n-plane $H_p = \{q \in \mathbb{R}^{n+1}: q \cdot N(p) = 0\}$ tangent to S at p.

for all t, and $h(t_0) = 0$, so h attains an absolute minimum at t_0. Hence

$$\mathscr{S}_p(\mathbf{v}) = \ddot{\alpha}(t_0) \cdot \mathbf{N}(\alpha(t_0)) = h''(t_0) \geq 0.$$

If $S \subset H_p^-$, the inequality is reversed, for all $\mathbf{v} \in S_p$. □

The converse of Theorem 1 is not true; for example, the 2-surface $x_3 = x_1^2 - x_2^4$ in $\mathbb{R}^3$ has $\mathscr{S}_0$ semi-definite yet is not convex at 0. We can prove, however, (Theorem 3) that if $\mathscr{S}_p$ is definite, then S is convex (in fact, strictly convex) at p.

The key idea in studying convexity is the observation that S is convex at $p \in S$ if and only if the "height function" $h: S \to \mathbb{R}$, defined by $h(q) = q \cdot N(p)$, attains either a local minimum or a local maximum at p. In order to capitalize on this observation, we shall need to develop further the calculus of smooth functions on n-surfaces.

Let $h: S \to \mathbb{R}$ be any smooth function on the n-surface $S \subset \mathbb{R}^{n+1}$. The *gradient vector field* of h is the smooth tangent vector field grad h on S defined by

$$(\text{grad } h)(p) = \nabla\tilde{h}(p) - (\nabla\tilde{h}(p) \cdot \mathbf{N}(p))\mathbf{N}(p)$$

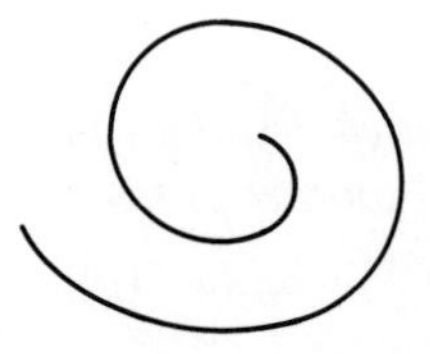

Figure 13.2 A non-convex plane curve which is convex at each point.

where $\tilde{h}$ is any extension of h to a smooth function on an open set containing S, and $\mathbf{N}$ is any orientation on S. Thus grad h is the tangential component of $\nabla\tilde{h}$. The gradient of $h: S \to \mathbb{R}$ has the following properties:

(i) $\nabla_{\mathbf{v}} h = (\text{grad } h)(p) \cdot \mathbf{v}$ for all $\mathbf{v} \in S_p$, $p \in S$.
(ii) $(h \circ \alpha)'(t) = (\text{grad } h)(\alpha(t)) \cdot \dot{\alpha}(t)$ for all $t \in I$, where $\alpha: I \to S$ is any parametrized curve in S.
(iii) $(\text{grad } h)(p) = \sum_{i=1}^{n} (\nabla_{\mathbf{v}_i} h)\mathbf{v}_i$, where $\{\mathbf{v}_1, \ldots, \mathbf{v}_n\}$ is any orthonormal basis for S_p, $p \in S$.
(iv) $(\text{grad } h)(p) = \mathbf{0}$ if and only if h is stationary at p, $p \in S$.

In particular, grad h is independent of the choice of extension $\tilde{h}$, by (iii). Property (ii) is a form of the chain rule.

Property (i) is true because

$$\nabla_{\mathbf{v}} h = \nabla_{\mathbf{v}} \tilde{h} = \nabla\tilde{h}(p) \cdot \mathbf{v} = (\text{grad } h)(p) \cdot \mathbf{v}.$$

(ii) follows from (i), since $(h \circ \alpha)'(t) = \nabla_{\dot{\alpha}(t)} h$. To check (iii), dot both sides with $\mathbf{v}_j (j \in \{1, \ldots, n\})$ and use (i). Finally, (i) and (iii) together imply that $(\text{grad } h)(p) = \mathbf{0}$ if and only if $\nabla_{\mathbf{v}} h = 0$ for all $\mathbf{v} \in S_p$, which establishes (iv). (Recall that $h: S \to \mathbb{R}$ is *stationary* at $p \in S$ if $\nabla_{\mathbf{v}} h = 0$ for all $\mathbf{v} \in S_p$; that is, if $(h \circ \alpha)'(t_0) = 0$ for all parametrized curves α in S with $\alpha(t_0) = p$.)

A point $p \in S$ at which $h: S \to \mathbb{R}$ is stationary is called a *critical point* of h. Critical points of smooth functions $h: S \to \mathbb{R}$ come in three varieties: local minima, local maxima, and saddle points (see Figure 13.3):

$h: S \to \mathbb{R}$ attains a *local minimum* at $p \in S$ if there is an open set V in S, containing p, such that $h(q) \geq h(p)$ for all $q \in V$ ($V \subset S$ is an *open set* in S if $V = W \cap S$ for some open set W in $\mathbf{R}^{n+1}$.)

$h: S \to \mathbb{R}$ attains a *local maximum* at $p \in S$ if there is an open set V in S, containing p, such that $h(q) \leq h(p)$ for all $q \in V$.

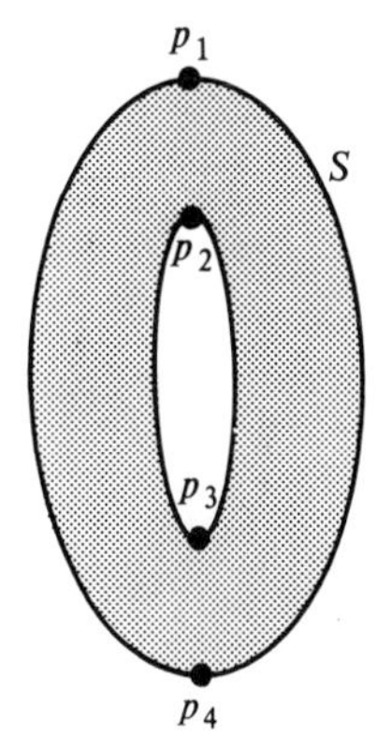

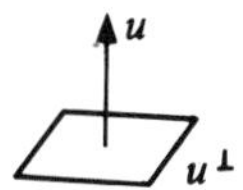

Figure 13.3 Critical points of the height function $h: S \to \mathbb{R}$, $h(q) = q \cdot u$ where u is a unit vector in $\mathbb{R}^{n+1}$. h measures height above the n-plane $u^\perp$. h attains a local maximum at p_1 and a local minimum at p_4; p_2 and p_3 are saddle points.

$p \in S$ is a *saddle point* of $h: S \to \mathbb{R}$ if h is stationary at p but h attains neither a local minimum nor a local maximum at p.

If, in the definitions of local minimum and local maximum, the inequalities are strict then h is said to attain a *strict local minimum* ($h(q) > h(p)$ for all $q \in V$, $q \neq p$) or a *strict local maximum* ($h(q) < h(p)$ for all $q \in V$, $q \neq p$) at p.

The condition $(\text{grad } h)(p) = \mathbf{0}$ is the "first derivative test" for a critical point of $h: S \to \mathbb{R}$. We shall need a "second derivative test" to help distinguish between the types of critical points.

Let $p \in S$ be a critical point of $h: S \to \mathbb{R}$. The *Hessian* of h at p is the quadratic form $\mathscr{H}_p: S_p \to \mathbb{R}$ defined by

$$\mathscr{H}_p(\mathbf{v}) = \nabla_{\mathbf{v}}(\text{grad } h) \cdot \mathbf{v}.$$

Thus $\mathscr{H}_p$ is the quadratic form associated with the self-adjoint linear transformation on S_p which sends $\mathbf{v}$ to $\nabla_{\mathbf{v}}(\text{grad } h)$. Note that $\nabla_{\mathbf{v}}(\text{grad } h)$ does lie in S_p for each $\mathbf{v} \in S_p$ since

$$\begin{aligned}\nabla_{\mathbf{v}}(\text{grad } h) \cdot \mathbf{N}(p) &= \nabla_{\mathbf{v}}((\text{grad } h) \cdot \mathbf{N}) - (\text{grad } h)(p) \cdot \nabla_{\mathbf{v}} \mathbf{N} \\ &= \nabla_{\mathbf{v}}(0) - \mathbf{0} \cdot \nabla_{\mathbf{v}} \mathbf{N} = 0.\end{aligned}$$

Verification that this linear transformation is in fact self-adjoint is left as an exercise (Exercise 13.2).

Theorem 2. (*Second derivative test for local minima and maxima*). *Let S be an n-surface in $\mathbb{R}^{n+1}$, let $h: S \to \mathbb{R}$ be a smooth function which is stationary at $p \in S$ and let $\mathscr{H}_p$ denote the Hessian of h at p.*

(i) *If h attains a local minimum at p then $\mathscr{H}_p$ is positive semi-definite. If h attains a local maximum at p then $\mathscr{H}_p$ is negative semi-definite.*

(ii) *If $\mathscr{H}_p$ is positive definite then h attains a strict local minimum at p. If $\mathscr{H}_p$ is negative definite then h attains a strict local maximum at p.*

PROOF. (i) Suppose h attains a local minimum at p. For $\mathbf{v} \in S_p$, let $\alpha: I \to S$ be such that $\dot{\alpha}(t_0) = \mathbf{v}$. Then

$$\begin{aligned}(h \circ \alpha)'(t) &= (\text{grad } h)(\alpha(t)) \cdot \dot{\alpha}(t), \quad \text{for all } t \in I, \quad \text{and} \\ 0 \leq (h \circ \alpha)''(t_0) &= \nabla_{\mathbf{v}}(\text{grad } h) \cdot \dot{\alpha}(t_0) + (\text{grad } h)(\alpha(t_0)) \cdot \ddot{\alpha}(t_0) \\ &= \mathscr{H}_p(\mathbf{v}),\end{aligned}$$

since $(\text{grad } h)(p) = \mathbf{0}$. Thus $\mathscr{H}_p$ is positive semi-definite. The proof for local maxima is similar.

(ii) To prove the first statement of (ii) it suffices to show that if h does not attain a strict local minimum at p then $\mathscr{H}_p$ cannot be positive definite. So suppose h does not attain a strict local minimum at p. Then there must be a sequence $\{p_k\}$ in $S - \{p\}$, with $\lim_{k \to \infty} p_k = p$, such that $h(p_k) \leq h(p)$ for all k. For each k, set $v_k = (p_k - p)/\|p_k - p\|$. Then $\{v_k\}$ is a sequence in the unit sphere S^n. Since S^n is compact, the sequence $\{v_k\}$ must have a convergent subsequence, which we may as well assume is $\{v_k\}$ itself; let $v = \lim_{k\to\infty} v_k$. We shall show that $\mathbf{v} = (p, v) \in S_p$ and that $\mathscr{H}_p(\mathbf{v}) \leq 0$.

Let W be an open ball in $\mathbb{R}^{n+1}$, containing p, such that both $\tilde{h}$, a smooth extension of h, and f, a smooth function defining S as $f^{-1}(c)$, are defined on W. Then $p_k \in W$ for sufficiently large k. Applying the mean value theorem to $g(t) = f(p + tv_k)$ we find

$$\begin{aligned} 0 = \frac{f(p_k) - f(p)}{\|p_k - p\|} &= \frac{g(\|p_k - p\|) - g(0)}{\|p_k - p\| - 0} \\ &= g'(t_k) = \nabla f(p + t_k v_k) \cdot \mathbf{v}_k \end{aligned}$$

for some $t_k \in (0, \|p_k - p\|)$, where $\mathbf{v}_k = (p + t_k v_k, v_k)$. Taking the limit as $k \to \infty$ yields $0 = \nabla f(p) \cdot \mathbf{v}$ (since $\lim_{k\to\infty} t_k = 0$) so $\mathbf{v} \in S_p$.

To see that $\mathscr{H}_p(\mathbf{v}) \leq 0$, note that $\nabla\tilde{h}(p) = \lambda \nabla f(p)$ for some $\lambda \in \mathbb{R}$, since $(\text{grad } h)(p) = \mathbf{0}$, and $\lambda = \nabla\tilde{h}(p) \cdot \nabla f(p)/\|\nabla f(p)\|^2$. We shall apply Taylor's theorem to $g_k(t) = (\tilde{h} - \lambda f)(\alpha_k(t))$, where $\alpha_k(t) = p + tv_k$. Since

$$g_k'(t) = \nabla(\tilde{h} - \lambda f)(\alpha_k(t)) \cdot \dot{\alpha}_k(t) = (\nabla\tilde{h} - \lambda\nabla f)(\alpha_k(t)) \cdot (\alpha_k(t), v_k)$$

and

$$g_k''(t) = (\nabla_{\dot{\alpha}_k(t)}(\nabla\tilde{h} - \lambda\nabla f)) \cdot (\alpha_k(t), v_k)$$

we find, for some t_k between 0 and $\|p_k - p\|$,

$$\begin{aligned} g_k(\|p_k - p\|) &= g_k(0) + g_k'(0)\|p_k - p\| + \tfrac{1}{2}g_k''(t_k)\|p_k - p\|^2 \\ &= g_k(0) + ((\nabla\tilde{h} - \lambda\nabla f)(p) \cdot (p, v_k))\|p_k - p\| \\ &\quad + \tfrac{1}{2}(\nabla_{\dot{\alpha}_k(t_k)}(\nabla\tilde{h} - \lambda\nabla f)) \cdot (\alpha_k(t_k), v_k)\|p_k - p\|^2. \end{aligned}$$

The middle term is zero since $\nabla\tilde{h}(p) = \lambda\nabla f(p)$. Hence

$$\begin{aligned} 0 \geq \frac{h(p_k) - h(p)}{\|p_k - p\|^2} &= \frac{\tilde{h}(p_k) - \tilde{h}(p)}{\|p_k - p\|^2} \\ &= \frac{(\tilde{h} - \lambda f)(p_k) - (\tilde{h} - \lambda f)(p)}{\|p_k - p\|^2} \quad (\text{since } f(p_k) = f(p) = c) \\ &= \frac{g_k(\|p_k - p\|) - g_k(0)}{\|p_k - p\|^2} \\ &= \tfrac{1}{2}(\nabla_{\dot{\alpha}_k(t_k)}(\nabla\tilde{h} - \lambda\nabla f)) \cdot (\alpha_k(t_k), v_k). \end{aligned}$$

Taking the limit as $k \to \infty$ yields

$$0 \geq \tfrac{1}{2}\nabla_{\mathbf{v}}(\nabla\tilde{h} - \lambda\nabla f) \cdot \mathbf{v}.$$

But this last expression is just $\frac{1}{2}\mathscr{H}_p(\mathbf{v})$, since

$$\begin{aligned}\mathscr{H}_p(\mathbf{v}) &= \nabla_{\mathbf{v}}(\text{grad } h) \cdot \mathbf{v} \\ &= \nabla_{\mathbf{v}}(\nabla \tilde{h} - (\nabla \tilde{h} \cdot \mathbf{N})\mathbf{N}) \cdot \mathbf{v} \\ &= \nabla_{\mathbf{v}}\left(\nabla \tilde{h} - \frac{\nabla \tilde{h} \cdot \nabla f}{\|\nabla f\|^2} \nabla f\right) \cdot \mathbf{v} \\ &= \nabla_{\mathbf{v}}(\nabla \tilde{h}) \cdot \mathbf{v} - \nabla_{\mathbf{v}}\left(\frac{\nabla \tilde{h} \cdot \nabla f}{\|\nabla f\|^2}\right) \nabla f(p) \cdot \mathbf{v} - \left(\frac{\nabla \tilde{h} \cdot \nabla f}{\|\nabla f\|^2}\right)(p) \nabla_{\mathbf{v}}(\nabla f) \cdot \mathbf{v} \\ &= \nabla_{\mathbf{v}}(\nabla \tilde{h}) \cdot \mathbf{v} - \lambda \nabla_{\mathbf{v}}(\nabla f) \cdot \mathbf{v} \\ &= \nabla_{\mathbf{v}}(\nabla \tilde{h} - \lambda \nabla f) \cdot \mathbf{v},\end{aligned}$$

so $0 \geq \mathscr{H}_p(\mathbf{v})$, as claimed.

Thus we have shown that if h does not attain a strict local minimum at the critical point $p \in S$ then its Hessian $\mathscr{H}_p$ cannot be positive definite. The proof that if h does not attain a strict local maximum at p then $\mathscr{H}$ cannot be negative definite is similar. □

When S is displayed as a level set $S = f^{-1}(c)$ ($f\colon U \to \mathbb{R}^{n+1}$ such that $\nabla f(p) \neq 0$ for all $p \in S$) and h is described as the restriction to S of a smooth function $\tilde{h}\colon U \to \mathbb{R}$, the local minima and local maxima are most easily found using the following facts, which are evident from the above proof. The critical points of $h = \tilde{h}|_S$ are those points $p \in S$ such that $\nabla \tilde{h}(p) = \lambda \nabla f(p)$ for some $\lambda \in \mathbb{R}$ (λ is the *Lagrange multiplier* at p of the pair of functions $\tilde{h}, f$). Then a sufficient condition that $h = \tilde{h}|_S$ attain a local minimum at the critical point p is that the quadratic form

$$\mathscr{H}_p(\mathbf{v}) = \nabla_{\mathbf{v}}(\nabla \tilde{h} - \lambda \nabla f) \cdot \mathbf{v}, \qquad \mathbf{v} \in S_p,$$

be positive definite; a sufficient condition that h attain a local maximum at p is that this quadratic form be negative definite.

The critical point theory becomes especially simple when $h\colon S \to \mathbb{R}$ is a height function $h_u(q) = q \cdot u$, where u is a unit vector in $\mathbb{R}^{n+1}$ (Figure 13.3). Then $h_u = \tilde{h}_u|_S$ where

$$\begin{aligned}\tilde{h}_u(q) &= q \cdot u, \\ \nabla \tilde{h}_u(q) &= (q, u), \quad \text{and} \\ \nabla_{\mathbf{v}}(\nabla \tilde{h}_u) &= \mathbf{0}\end{aligned}$$

for all $q \in \mathbb{R}^{n+1}$ and all $\mathbf{v} \in \mathbb{R}^{n+1}_q$. It follows that $p \in S$ is a critical point of $h_u = \tilde{h}_u|_S$ if and only if $(p, u) = \lambda \nabla f(p)$ for some $\lambda \in \mathbb{R}$. Since u is a unit vector, $|\lambda|$ must equal $1/\|\nabla f(p)\|$ so p is a critical point if and only if

$(p, u) = \pm\mathbf{N}(p)$. Moreover, if p is a critical point of h_u and $\mathbf{v} \in S_p$, then

$$\begin{aligned}\mathscr{H}_p(\mathbf{v}) &= \nabla_{\mathbf{v}}(\nabla\tilde{h}_u - (\nabla\tilde{h}_u \cdot \mathbf{N})\mathbf{N}) \cdot \mathbf{v} \\ &= (\nabla\tilde{h}_u \cdot \mathbf{N})(p)(-\nabla_{\mathbf{v}}\mathbf{N} \cdot \mathbf{v}) \\ &= (u \cdot N(p))\mathscr{S}_p(\mathbf{v}) \\ &= \pm\mathscr{S}_p(\mathbf{v}),\end{aligned}$$

where $\mathscr{S}_p$ is the second fundamental form of S at p. We conclude, then, that *$p \in S$ is a critical point of the height function $h_u\colon S \to \mathbb{R}$, $h_u(q) = q \cdot u$ where u is a unit vector in $\mathbb{R}^{n+1}$, if and only if $N(p) = \pm u$, and at a critical point p of h_u the Hessian is equal to $\pm$ the second fundamental form of S at p, the sign being the same as that of $N(p) \cdot u$.*

An immediate consequence of these facts is the following partial converse to Theorem 1.

Theorem 3. *Let S be an oriented n-surface in $\mathbb{R}^{n+1}$. Suppose $p \in S$ is such that the second fundamental form $\mathscr{S}_p$ of S at p is definite. Then S is strictly convex at p.*

PROOF. S is strictly convex at p if and only if the height function $h_{N(p)}\colon S \to \mathbb{R}$ attains either a strict local minimum or a strict local maximum at p. But this is the case since $h_{N(p)}$ is stationary at p and $\mathscr{H}_p = \pm\mathscr{S}_p$ is definite. □

By combining Theorem 3 with Theorem 6 of Chapter 12 we see that if S is a compact connected oriented n-surface in $\mathbb{R}^{n+1}$ whose Gauss-Kronecker curvature is nowhere zero then S is strictly convex at each point. The remainder of this chapter will be devoted to proving (Theorem 5) that such an S is globally convex. For this, we must show that if $u = N(p)$ for some $p \in S$ then the height function h_u attains not just a local maximum or minimum at p but in fact attains a global maximum or minimum at p. The idea of the proof is to show that each h_u can have only two critical points, namely the point at which h_u attains its global maximum and the point at which h_u attains its global minimum. The proof requires some further facts about differential equations.

Recall that, given a smooth vector field $\mathbf{X}$ on an open set $U \subset \mathbb{R}^{n+1}$ and a point $q \in U$, there exists an open interval I_q containing 0 and a unique integral curve $\alpha_q\colon I_q \to U$ of $\mathbf{X}$ with $\alpha_q(0) = q$. Theorem 4 says that, at least for small t, $\alpha_q(t)$ is a smooth function of q.

Theorem 4. *Let $\mathbf{X}$ be a smooth vector field on an open set $U \subset \mathbb{R}^{n+1}$ and let $p \in U$. Then there exists an open set V in $\mathbb{R}^{n+1}$ with $p \in V \subset U$ and an $\varepsilon > 0$ such that for each $q \in V$ there is an integral curve $\alpha_q\colon (-\varepsilon, \varepsilon) \to U$ of $\mathbf{X}$ with $\alpha_q(0) = q$. Moreover, for each such V and ε, the map $\psi\colon V \times (-\varepsilon, \varepsilon) \to U$ defined by $\psi(q, t) = \alpha_q(t)$ is smooth.*

For a proof of Theorem 4, see, e.g., W. Hurewicz, *Lectures on Ordinary Differential Equations*, M.I.T. Press, Cambridge, Mass. 1958, pp. 28–29.

Corollary. *Let* $\mathbf{X}$ *be a smooth tangent vector field on a compact n-surface* $S \subset \mathbb{R}^{n+1}$. *Then* $\mathbf{X}$ *is complete; that is, every maximal integral curve of* $\mathbf{X}$ *has domain the whole real line. Moreover, for each* $t \in \mathbb{R}$, *the map* $\psi_t\colon S \to S$ *defined by* $\psi_t(q) = \alpha_q(t)$, *where* α_q *is the maximal integral curve of* $\mathbf{X}$ *through q, is smooth.*

PROOF. Extend $\mathbf{X}$ to a smooth vector field $\tilde{\mathbf{X}}$ on an open set $U \subset \mathbb{R}^{n+1}$ containing S. We shall show first that there exists an $\bar{\varepsilon} > 0$ (a uniform $\bar{\varepsilon}$) such that for every $p \in S$ there is an open set V_p in $\mathbb{R}^{n+1}$, with $p \in V_p \subset U$, such that the conclusions of Theorem 4 hold for $\tilde{\mathbf{X}}$, for this open set V_p, and for this $\bar{\varepsilon}$. Suppose there were no such $\bar{\varepsilon}$. Then for each positive integer k there would be a point p_k in S such that for every open set V containing p_k the maximal integral curve α_q of $\tilde{\mathbf{X}}$ through q does not contain the interval $(-1/k, 1/k)$, for some $q \in V$. Since S is compact, the sequence $\{p_k\}$ has a subsequence converging to some point $p \in S$. Now let V and ε be as in Theorem 4, for this p and for $\mathbf{X} = \tilde{\mathbf{X}}$. Then for every $q \in V$ the maximal integral curve of $\tilde{\mathbf{X}}$ through q will have domain containing the interval $(-\varepsilon, \varepsilon)$. But $p_k \in V$ for arbitrarily large k, and in particular for some k with $1/k < \varepsilon$. This contradicts the existence of $\{p_k\}$ and thereby establishes the existence of a uniform $\bar{\varepsilon}$.

It follows that the domain I of each maximal integral curve α of $\mathbf{X}$ is the whole real line. For suppose there were an end point $b \in \mathbb{R}$ of the open interval I. Choosing $t_0 \in I$ so that $|t_0 - b| < \bar{\varepsilon}$, let β be the maximal integral curve of $\mathbf{X}$ through $\alpha(t_0)$. Then $\alpha(t)$ and $\beta(t - t_0)$ are integral curves of $\mathbf{X}$ which agree at t_0 and so agree for all t in their common domain. Hence the parametrized curve which sends t to $\alpha(t)$ for $t \subset I$ and sends t to $\beta(t - t_0)$ for $|t - t_0| < \bar{\varepsilon}$ is an integral curve of $\mathbf{X}$ which extends α beyond b, contradicting the fact that α is maximal. Hence $I = \mathbb{R}$, as claimed.

Thus for each $t \in \mathbb{R}$ there is a map $\psi_t\colon S \to S$ defined by $\psi_t(q) = \alpha_q(t)$. Note that, by uniqueness of integral curves, $\psi_t \circ \psi_s = \psi_{s+t}$ for all $s, t \in \mathbb{R}$. Indeed, for each $s \in \mathbb{R}$ and $q \in S$, $\psi_t(\psi_s(q))$ and $\psi_{s+t}(q)$ both describe the unique maximal integral curve of $\mathbf{X}$ through $\psi_s(q)$. It follows that $\psi_t = \psi_{t/k} \circ \psi_{t/k} \circ \cdots \circ \psi_{t/k}$ (composition k times), where the positive integer k is chosen so that $|t/k| < \bar{\varepsilon}$. But $\psi_{t/k}$ is smooth by Theorem 4; hence, so is ψ_t. □

Remark. It can be shown, further, that the map $\psi\colon S \times \mathbb{R} \to \mathbb{R}$ defined by $\psi(q, t) = \psi_t(q) = \alpha_q(t)$ is also smooth, but we shall not need this fact.

Consider now the smooth tangent vector field grad h, where $h\colon S \to \mathbb{R}$ is a smooth function on the n-surface $S \subset \mathbb{R}^{n+1}$. The integral curves of grad h are called *gradient lines* of h (see Figure 13.4). If $p \in S$ is a critical point of h, then $(\text{grad } h)(p) = \mathbf{0}$ so any gradient line $\alpha\colon I \to S$ of h passing through p will be simply a constant curve, $\alpha(t) = p$ for all $t \in I$. Along all other gradient lines, h

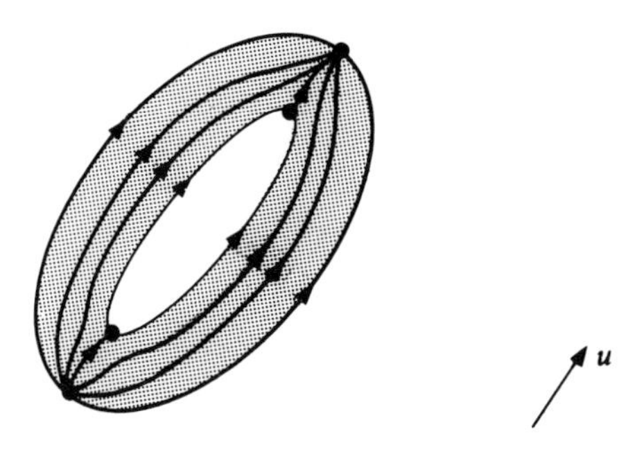

Figure 13.4 Gradient lines of the height function $h: S \to \mathbb{R}$, $h(q) = q \cdot u$.

is strictly increasing. Indeed, if $\alpha: I \to S$ is any gradient line of h not passing through a critical point of h,

$$(h \circ \alpha)'(t) = (\text{grad } h)(\alpha(t)) \cdot \dot{\alpha}(t) = \|(\text{grad } h)(\alpha(t))\|^2 > 0$$

for all $t \in I$. In fact, the gradient lines of h are the curves in S along which h increases fastest among all curves in S with comparable speed (see Exercise 13.4).

A critical point p of a smooth function $h: S \to \mathbb{R}$ is *non-degenerate* if $\nabla_v(\text{grad } h) \neq 0$ for all $\mathbf{v} \in S_p$, $\mathbf{v} \neq 0$. Note that non-degenerate critical points are *isolated* in that for each such critical point p there is an open set V in S about p such that V contains no other critical points of h. For otherwise there would be a sequence $\{p_k\}$ of critical points of h converging to p and a subsequence $\{p_{k_i}\}$ such that $(p_{k_i} - p)/\|p_{k_i} - p\|$ converges to a point v in the unit sphere S^n; setting $\mathbf{v} = (p, v)$ it would follow, as in the proof of Theorem 2, that $\mathbf{v} \in S_p$ and that $\nabla_{\mathbf{v}}(\text{grad } h) = 0$ (since $(\text{grad } h)(p_{k_i}) = 0$ for all k_i) contradicting non-degeneracy.

Lemma 1. *Let S be a compact n-surface and let $h: S \to \mathbb{R}$ be a smooth function all of whose critical points are non-degenerate. Then the gradient lines of h run from one critical point of h to another; that is, if $\alpha: \mathbb{R} \to S$ is any maximal gradient line of h then there exist critical points p and q of h such that* $\lim_{t \to -\infty} \alpha(t) = q$ *and* $\lim_{t \to \infty} \alpha(t) = p$ *(see Figure* 13.4).

PROOF. Let $\alpha: \mathbb{R} \to S$ be a maximal gradient line of h. Since S is compact, the sequence $\{\alpha(k): k = 1, 2, \ldots\}$ has a convergent subsequence $\{\alpha(t_k)\}$. Let $p = \lim_{k \to \infty} \alpha(t_k)$. Then

$$\begin{aligned} h(p) - h(\alpha(0)) &= \lim_{k \to \infty} [h(\alpha(t_k)) - h(\alpha(0))] \\ &= \lim_{k \to \infty} \int_0^{t_k} (h \circ \alpha)'(t)\, dt \\ &= \int_0^{\infty} \|(\text{grad } h)(\alpha(t))\|^2\, dt \end{aligned}$$

so the integral $\int_0^\infty \|(\text{grad } h)(\alpha(t))\|^2 \, dt$ converges, which can happen only if $\lim_{t\to\infty} \|(\text{grad } h)(\alpha(t))\|^2 = 0$. In particular,

$$\|(\text{grad } h)(p)\| = \lim_{k\to\infty} \|(\text{grad } h)(\alpha(t_k))\| = 0$$

so p is a critical point of h.

We still must check that $\lim_{t\to\infty} \alpha(t) = p$. But if not, there will be an $\varepsilon > 0$ such that, for each k, $\|\alpha(s_k) - p\| \geq \varepsilon$ for some $s_k > t_k$. Since $\|\alpha(t_k) - p\| < \varepsilon$ for k sufficiently large, this says that $\alpha(t)$ enters and leaves the ball $\{q \in \mathbb{R}^{n+1} : \|q - p\| < \varepsilon\}$ repeatedly as $t \to \infty$ (see Figure 13.5). Since

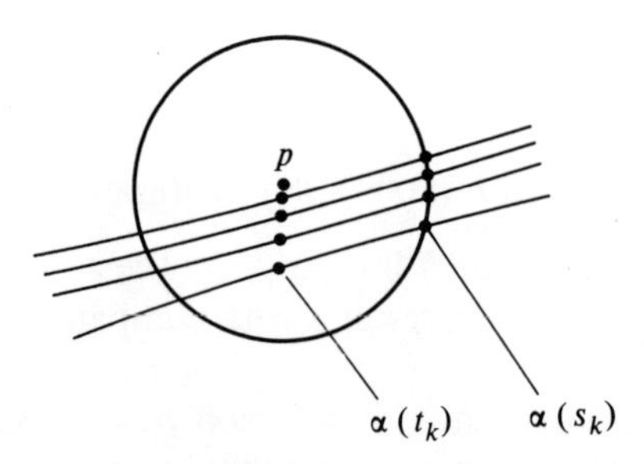

Figure 13.5 If $\lim_{t\to\infty} \alpha(t) \neq p$ then α must repeatedly enter and leave the ε-ball about $p = \lim_{k\to\infty} \alpha(t_k)$.

$\|\alpha(t) - p\|$ must equal ε for some t between t_k and s_k, we can choose s_k so that in fact $\|\alpha(s_k) - p\| = \varepsilon$. Since $\{q \in S : \|q - p\| = \varepsilon\}$ is compact, the sequence $\{\alpha(s_k)\}$ has a subsequence converging to some point $p_1 \in S$ with $\|p_1 - p\| = \varepsilon$. Since $\lim_{k\to\infty} s_k = \infty$, the same argument that showed p was a critical point of h also shows that p_1 is a critical point of h. Repeating this construction with ε replaced by ε/m leads to a critical point $p_m \in S$ of h with $\|p_m - p\| = \varepsilon/m$, for each positive integer m. But this contradicts the fact that p is a non-degenerate, hence isolated, critical point of h. So, indeed, $\lim_{t\to\infty} \alpha(t) = p$.

Consideration of the sequence $\{\alpha(-k) : k = 1, 2, \ldots\}$ will yield in the same way a critical point q of h such that $\lim_{t\to-\infty} \alpha(t) = q$. □

Theorem 5. *Let S be a compact connected oriented n-surface in $\mathbb{R}^{n+1}$ whose Gauss-Kronecker curvature is nowhere zero. Then*

(i) *The Gauss map $N: S \to S^n$ is one to one and onto, and*
(ii) *S is strictly convex.*

PROOF. (i) By Theorem 6 of Chapter 12, the second fundamental form $\mathscr{S}_p$ of S at p is definite for all $p \in S$. $\mathscr{S}_p$ is either positive definite for all p or negative definite for all p because the minimum and maximum normal curvatures k_1 and k_n, being continuous nowhere zero functions on the connected n-surface S, cannot change sign. By reversing the orientation on S if necessary, we may assume that $\mathscr{S}_p$ is negative definite for all $p \in S$.

Now let $u \in S^n$. By the discussion preceding Theorem 3, $p \in S$ is a critical point of the height function $h_u: S \to \mathbb{R}$, $h_u(q) = q \cdot u$, if and only if $N(p) = \pm u$. Furthermore, since $\mathscr{H}_p = u \cdot N(p)\mathscr{S}_p$, $N(p) = +u$ if and only if h_u attains a local maximum at p and $N(p) = -u$ if and only if h_u attains a local minimum at p. In particular, all critical points of h_u are non-degenerate, hence isolated, and h_u attains either a strict local maximum or a strict local minimum at each one.

The fact that N is onto is now evident: given $u \in S^n$ we find that $u = N(p)$ where p is any point where h_u attains its maximum.

To see that N is one to one, let p be a point in S with $N(p) = u$. Then the height function h_u must attain a strict local maximum at p. We shall show that the set

$$U_p = \{q \in S: \lim_{t \to \infty} \alpha_q(t) = p\},$$

where $\alpha_q: \mathbb{R} \to S$ is the maximal gradient line of h_u with $\alpha_q(0) = q$, is an open set in S.

First note that there exists an open set V_p in S about p such that all gradient lines of h_u which enter V_p must run to p. Indeed, choose $\varepsilon > 0$ small enough so that

(1) p is the only critical point of h_u in $A_\varepsilon = \{q \in S: \|q - p\| \leq \varepsilon\}$,
(2) $h_u(p) > h_u(q)$ for all $q \in A_\varepsilon$, $q \neq p$, and
(3) $B_\varepsilon = \{q \in S: \|q - p\| = \varepsilon\}$ is non-empty,

let M_ε denote the maximum value of h_u on the compact set B_ε, and set $V_p = \{q \in S: \|q - p\| < \varepsilon$ and $h(q) > M_\varepsilon\}$. Then a gradient line α of h with $\alpha(t_0) \in V_p$ for some $t_0 \in \mathbb{R}$ cannot have $\alpha(t) \in B_\varepsilon$ for any $t \geq t_0$ (since h increases along α) and so $\alpha(t)$ must stay in A_ε for all $t \geq t_0$. α must therefore run to a critical point in A_ε and p is the only one there.

The fact that U_p is open now follows from the Corollary to Theorem 4. Given any $q_0 \in U_p$ there will be a $t_0 \in \mathbb{R}$ such that $\alpha_{q_0}(t_0) \in V_p$. By continuity of ψ_{t_0}, $\alpha_q(t_0) = \psi_{t_0}(q) \in V_p$, and hence α_q runs to p, for all q sufficiently close to q_0. Thus U_p is open in S.

Finally, let $\{p_1, \ldots, p_k\}$ be the set of points in S where h_u attains a local maximum (i.e., where $N(p_i) = u$) and let $\{q_1, \ldots, q_l\}$ be the set of points where h_u attains a local minimum. These sets are finite because the critical points of h_u are isolated and S is compact. By the lemma, $S - \{q_1, \ldots, q_l\}$ is the union of the mutually disjoint open sets $U_{p_1}, \ldots, U_{p_k}$. But, since S is connected, $S - \{q_1, \ldots, q_l\}$ is also connected (see Figure 13.6), provided $n > 1$, so this is possible only if there is just one p_i; i.e., only if N is one to one. The last step in the argument breaks down if $n = 1$, but Exercise 11.19 takes care of this special case.

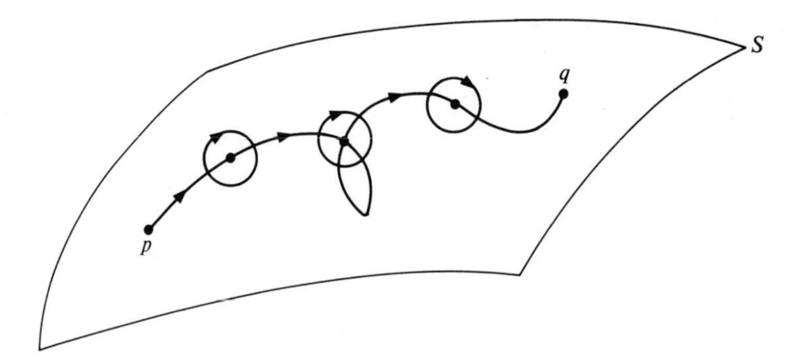

Figure 13.6 A connected n-surface S $(n > 1)$ cannot be disconnected by the removal of a finite set of points: take a small ball about each point and reroute each continuous curve from p to q around the boundaries of these balls. That the intersections of sufficiently small balls with S are connected can be established rigorously by applying the inverse function theorem (see Chapter 15).

(ii) We have just seen that for each $u \in S^n$ there is only one point $p_1 \in S$ at which the height function h_u attains a local maximum. The same arguments, applied to $-h_u$, show that there is only one point $q_1 \in S$ at which h_u attains a local minimum. Moreover, these two points are the only critical points of h_u. Hence for each $p \in S$ the height function h_u, where $u = N(p)$, must attain either its strict global maximum or its strict global minimum at the critical point p. This says that S is strictly convex. □

Exercises

13.1. Show that if n is even and if the Gauss-Kronecker curvature at a point p of an n-surface $S \subset \mathbb{R}^{n+1}$ is negative then S is not convex at p.

13.2. Let S be an n-surface in $\mathbb{R}^{n+1}$, let $h: S \to \mathbb{R}$ be smooth, and let $p \in S$ be a critical point of h. Show that the linear transformation from S_p into itself which sends $\mathbf{v}$ into $\nabla_{\mathbf{v}}(\text{grad } h)$ is self-adjoint.

13.3. Let V be a vector space with a dot product, let $L: V \to V$ be a self-adjoint linear transformation, and let $\mathscr{Q}$ be the quadratic form associated with L.
(a) Show that $\mathscr{Q}$ is positive definite if and only if all the eigenvalues of L are positive.
(b) Show that, if V has dimension 2, then $\mathscr{Q}$ is positive definite if and only if (i) $\mathscr{Q}(v) > 0$ for some $v \in V$ and (ii) $\det L > 0$.
(c) $\mathscr{Q}$ is said to be *non-degenerate* if L is non-singular. Show that $\mathscr{Q}$ is non-degenerate if and only if all the eigenvalues of L are non-zero. (*Remark.* Note that a critical point p of a smooth function $h: S \to \mathbb{R}$ is non-degenerate if and only if the Hessian $\mathscr{H}_p$ of h at p is a non-degenerate quadratic form.)

13.4. Let S be an n-surface in $\mathbb{R}^{n+1}$ and let $h: S \to \mathbb{R}$ be smooth. Show that the curves in S along which h increases fastest are the gradient lines of h by showing that if $\alpha: [a, b] \to S$ is an integral curve of grad h and $\beta: [a, b] \to S$ is any other curve with the same speed ($\|\dot{\beta}(t)\| = \|\dot{\alpha}(t)\|$ for all $t \in [a, b]$) and with $\beta(a) = \alpha(a)$, then $h(\alpha(b)) \geq h(\beta(b))$. Show further that equality holds if and only if $\beta = \alpha$.

13.5. Let S be an n-surface in $\mathbb{R}^{n+1}$ and let $h: S \to \mathbb{R}$ be smooth. Show that the gradient lines of h are everywhere orthogonal to the level sets of h; i.e., show that if α is a gradient line of h and if β is any parametrized curve in S such that $h \circ \beta$ is constant, then $\dot{\beta}(t_1) \cdot \dot{\alpha}(t_0) = 0$ whenever $\beta(t_1) = \alpha(t_0)$.

14 Parametrized Surfaces

We have seen that every connected oriented plane curve C has a global parametrization and that, using one, we can (i) find a useful formula for curvature ($\kappa \circ \alpha = \ddot{\alpha} \cdot \mathbf{N} \circ \alpha / \|\dot{\alpha}\|^2$) and (ii) define various integrals over C. We shall now carry out a similar program for n-surfaces ($n > 1$). It will turn out that oriented n-surfaces (even connected ones) in general admit only local parametrizations, but that will be adequate for our needs.

The first property that a parametrization must have is regularity. In order to define regularity, we need the differential of a map. Let U be an open set in $\mathbb{R}^n$ and let $\varphi: U \to \mathbb{R}^m$ be a smooth map. The *differential* of φ is the smooth map $d\varphi: U \times \mathbb{R}^n \to \mathbb{R}^m \times \mathbb{R}^m$ defined as follows. A point $\mathbf{v} \in U \times \mathbb{R}^n$ is a vector $\mathbf{v} = (p, v)$ at a point $p \in U$. Given $\mathbf{v}$, let $\alpha: I \to U$ be any parametrized curve in U with $\dot{\alpha}(t_0) = \mathbf{v}$. Then $d\varphi(\mathbf{v})$ is the vector at $\varphi(p)$ ($d\varphi(\mathbf{v}) \in \mathbb{R}^m_{\varphi(p)} \subset \mathbb{R}^m \times \mathbb{R}^m$) defined by

$$d\varphi(\mathbf{v}) = \varphi \mathbin{\dot{\circ}} \alpha(t_0)$$

(see Figure 14.1). Note that the value of $d\varphi(\mathbf{v})$ does not depend on the choice of parametrized curve α, because

$$\begin{aligned}\varphi \mathbin{\dot{\circ}} \alpha(t_0) &= (\varphi \circ \alpha(t_0), (\varphi_1 \circ \alpha)'(t_0), \ldots, (\varphi_m \circ \alpha)'(t_0)) \\ &= (\varphi(p), \nabla\varphi_1(\alpha(t_0)) \cdot \dot{\alpha}(t_0), \ldots, \nabla\varphi_m(\alpha(t_0)) \cdot \dot{\alpha}(t_0)) \\ &= (\varphi(p), \nabla\varphi_1(p) \cdot \mathbf{v}, \ldots, \nabla\varphi_m(p) \cdot \mathbf{v}),\end{aligned}$$

where the φ_i are the component functions of φ ($\varphi(q) = (\varphi_1(q), \ldots, \varphi_m(q))$ for all $q \in U$), so

$$d\varphi(\mathbf{v}) = (\varphi(p), \nabla_{\mathbf{v}}\varphi_1, \ldots, \nabla_{\mathbf{v}}\varphi_m).$$

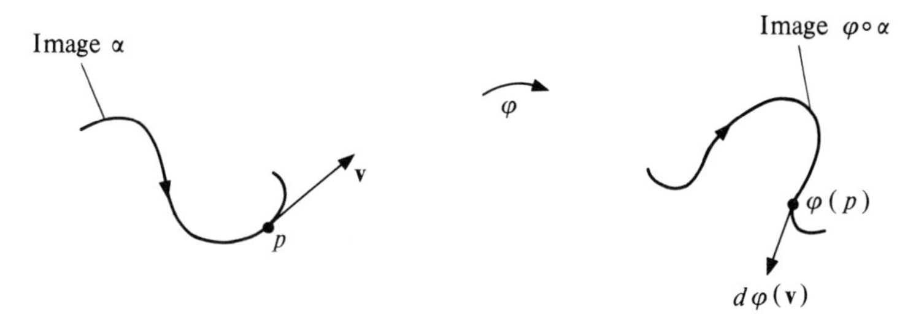

Figure 14.1 The differential of a map.

This formula for $d\varphi(\mathbf{v})$ not only shows independence of α but also provides a straightforward method of computing $d\varphi$. The smoothness of $d\varphi$ is also now evident.

It follows immediately from the above formula that the restriction $d\varphi_p$ of $d\varphi$ to $\mathbb{R}^n_p$ is a linear map $d\varphi_p\colon \mathbb{R}^n_p \to \mathbb{R}^m_{\varphi(p)}$. Its matrix relative to the standard bases for $\mathbb{R}^n_p$ and $\mathbb{R}^m_{\varphi(p)}$ is just the Jacobian matrix $((\partial\varphi_i/\partial x_j)(p))$ of φ at p. Indeed, if $\mathbf{e}_j = (p, 0, \ldots, 1, \ldots, 0)$ and $\mathbf{e}'_i = (\varphi(p), 0, \ldots, 1, \ldots, 0)$ with the 1's in the $(j+1)$th and $(i+1)$th spots respectively, then the matrix (a_{ij}) for $d\varphi_p$ is defined by

$$d\varphi_p(\mathbf{e}_j) = \sum a_{ij}\mathbf{e}'_i \qquad (j \in \{1, \ldots, n\})$$

and

$$\begin{aligned} a_{ij} &= d\varphi_p(\mathbf{e}_j) \cdot \mathbf{e}'_i = (\varphi(p), \nabla_{\mathbf{e}_j}\varphi_1, \ldots, \nabla_{\mathbf{e}_j}\varphi_m) \cdot \mathbf{e}'_i \\ &= \left(\varphi(p), \frac{\partial\varphi_1}{\partial x_j}(p), \ldots, \frac{\partial\varphi_m}{\partial x_j}(p)\right) \cdot \mathbf{e}'_i = \frac{\partial\varphi_i}{\partial x_j}(p). \end{aligned}$$

The set $U \times \mathbb{R}^n = \bigcup_{p \in U} \mathbb{R}^n_p$ is called the *tangent bundle* of the open set U in $\mathbb{R}^n$, and is denoted by $T(U)$. Thus the differential of the smooth map $\varphi\colon U \to \mathbb{R}^m$ maps $T(U)$ into $T(\mathbb{R}^m)$. Similarly, if S is an n-surface in $\mathbb{R}^{n+1}$ *its tangent bundle* is the set $T(S) = \bigcup_{p \in S} S_p \subset S \times \mathbb{R}^{n+1}$. Given a smooth map $\varphi\colon S \to \mathbb{R}^m$, *its differential* is the map $d\varphi\colon T(S) \to T(\mathbb{R}^m)$ defined by

$$d\varphi(\mathbf{v}) = (\varphi \mathbin{\dot\circ} \alpha)(t_0)$$

where $\alpha\colon I \to S$ is any parametrized curve in S with $\dot\alpha(t_0) = \mathbf{v}$. Note that $d\varphi$ is just the restriction to $T(S)$ of the differential $d\tilde\varphi$ of any smooth extension of φ to an open set in $\mathbb{R}^{n+1}$ and hence in particular $d\varphi(\mathbf{v})$ is independent of the choice of α. It follows also that the restriction $d\varphi_p$ of $d\varphi$ to $S_p (p \in S)$ is a linear map $d\varphi_p\colon S_p \to \mathbb{R}^m_{\varphi(p)}$.

Remark. For $\varphi\colon I \to \mathbb{R}$, I an open interval in $\mathbb{R}$, the symbol $d\varphi$ now has two meanings. On the one hand, in Chapter 11 we defined $d\varphi$, call it now $(d\varphi)^{(1)}$, to be a 1-form on I, so $(d\varphi)^{(1)}$ is a smooth map from $I \times \mathbb{R}$ into $\mathbb{R}$. Now we have defined $d\varphi$, call it $(d\varphi)^{(2)}$, to be a map from $I \times \mathbb{R}$ into $\mathbb{R} \times \mathbb{R}$.

These two maps are related by the formula

$$(d\varphi)^{(2)}(t, u) = (\varphi(t), (d\varphi)^{(1)}(t, u))$$

and hence either can be recovered directly from the other. We shall continue to use the notation $d\varphi$ for both; which of the two we mean will be clear in context.

A *parametrized n-surface* in $\mathbb{R}^{n+k}(k \geq 0)$ is a smooth map $\varphi\colon U \to \mathbb{R}^{n+k}$, where U is a connected open set in $\mathbb{R}^n$, which is *regular*; i.e., which is such that $d\varphi_p$ is non-singular (has rank n) for each $p \in U$. The regularity condition guarantees that the image of $d\varphi_p$ is an n-dimensional subspace of $\mathbb{R}^{n+k}_{\varphi(p)}$ for each $p \in U$. Image $d\varphi_p$ is the *tangent space* to φ corresponding to the point $p \in U$. Note that φ need not be one to one, and that $\varphi(q) = \varphi(p)$ for $q \neq p$ does not necessarily imply that Image $d\varphi_p =$ Image $d\varphi_q$.

EXAMPLE 1. A parametrized 1-surface is simply a regular parametrized curve.

EXAMPLE 2. A parametrized n-surface in $\mathbb{R}^n$ is simply a regular smooth map from one open set U in $\mathbb{R}^n$ onto another.

EXAMPLE 3. Let $f\colon U \to \mathbb{R}$ (U open in $\mathbb{R}^n$) be a smooth function. Define $\varphi\colon U \to \mathbb{R}^{n+1}$ by $\varphi(p) = (p, f(p))$. Then φ is a parametrized n-surface in $\mathbb{R}^{n+1}$ whose image is the graph of f.

EXAMPLE 4. Let $\varphi\colon U \to \mathbb{R}^3$ be given by

$$\varphi(\theta, \phi) = (r \cos \theta \sin \phi, r \sin \theta \sin \phi, r \cos \phi)$$

where $U = \{(\theta, \phi) \in \mathbb{R}^2 \colon 0 < \phi < \pi\}$ and $r > 0$. Then φ is a parametrized 2-surface whose image is the 2-sphere of radius r in $\mathbb{R}^3$, with the north and south poles missing (see Figure 14.2). Note that φ is not one to one; in fact, φ wraps the strip U in $\mathbb{R}^2$ around the sphere infinitely many times. The north and south poles are excluded from Image φ because $d\varphi_p$ is singular

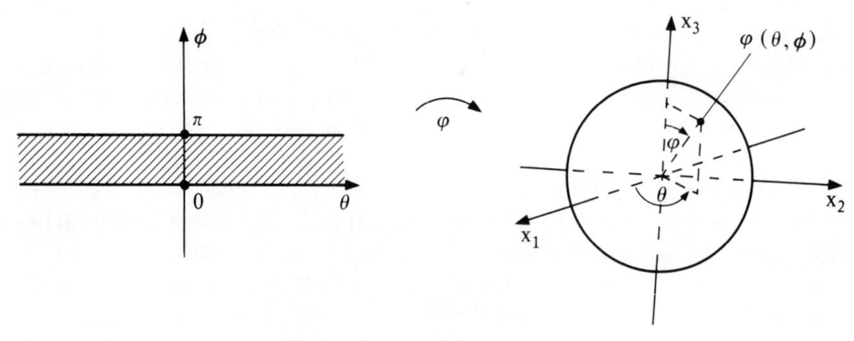

Figure 14.2 Spherical coordinates.

along the edges $\phi = 0$ and $\phi = \pi$ of the strip U. When $-\pi < \theta \leq \pi$, the numbers θ and ϕ are called the *spherical coordinates* of the point $\varphi(\theta, \phi)$ on the sphere.

EXAMPLE 5. Let $L: \mathbb{R}^n \to \mathbb{R}^{n+k}$ $(k \geq 1)$ be a non-singular linear map and let $w \in \mathbb{R}^{n+k}$. The map $\varphi: \mathbb{R}^n \to \mathbb{R}^{n+k}$ defined by

$$\varphi(p) = L(p) + w$$

is a *parametrized n-plane* through w in $\mathbb{R}^{n+k}$. Note that

$$d\varphi_p(p, v) = (\varphi(p), L(v))$$

for all $(p, v) \in \mathbb{R}^n_p$, $p \in \mathbb{R}^n$, since if $\alpha(t) = p + tv$ then $\dot{\alpha}(0) = (p, v)$ and

$$\begin{aligned}(\varphi \mathbin{\dot{\circ}} \alpha)(0) &= \left(\varphi \circ \alpha(0), \left.\frac{d}{dt}\right|_0 (L(p + tv) + w)\right) \\ &= \left(\varphi(p), \left.\frac{d}{dt}\right|_0 (L(p) + tL(v) + w)\right) \\ &= (\varphi(p), L(v)).\end{aligned}$$

EXAMPLE 6. Let $\varphi: U \to \mathbb{R}^{n+k}(U \subset \mathbb{R}^n)$ be a parametrized n-surface in $\mathbb{R}^{n+k}$. The *cylinder* over φ is the parametrized $(n + 1)$-surface $\tilde{\varphi}: U \times \mathbb{R} \to \mathbb{R}^{n+k+1}$ defined by

$$\tilde{\varphi}(u_1, \ldots, u_{n+1}) = (\varphi(u_1, \ldots, u_n), u_{n+1}), \qquad (u_1, \ldots, u_n) \in U, u_{n+1} \in \mathbb{R}$$

(see Figure 14.3).

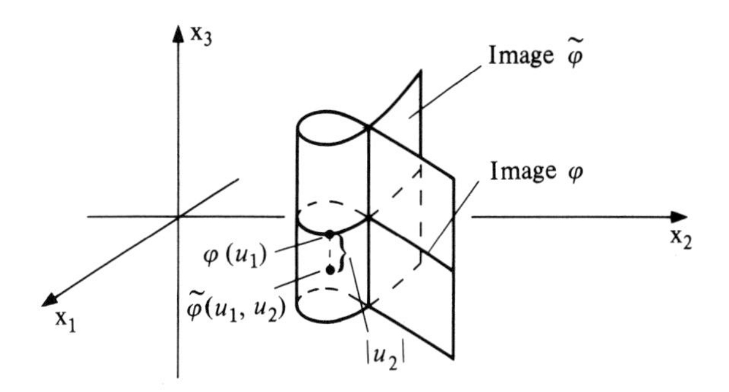

Figure 14.3 The cylinder over a parametrized curve φ.

EXAMPLE 7. Let $\alpha: I \to \mathbb{R}^2$ be a regular parametrized curve in $\mathbb{R}^2$, I open in $\mathbb{R}$, whose image lies above the x_1-axis; i.e., $y(t) > 0$ for all $t \in I$ where $\alpha(t) = (x(t), y(t))$. Define $\varphi: I \times \mathbb{R} \to \mathbb{R}^3$ by

$$\varphi(t, \theta) = (x(t), y(t) \cos \theta, y(t) \sin \theta).$$

φ is the *parametrized surface of revolution* obtained by rotating α about the

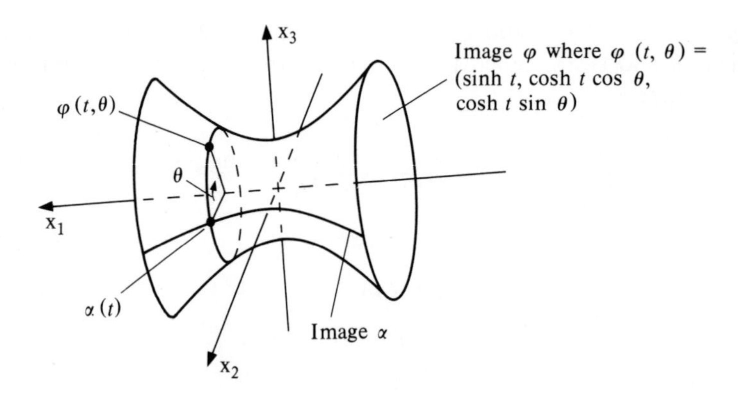

Figure 14.4 The parametrized hyperboloid of revolution obtained by rotating the parametrized curve $\alpha(t) = (\sinh t, \cosh t)$ about the x_1-axis.

x_1-axis (see Figure 14.4). Note that φ wraps the strip $I \times \mathbb{R}$ around Image φ infinitely many times.

EXAMPLE 8. Let $a > b > 0$ and define $\varphi: \mathbb{R}^2 \to \mathbb{R}^3$ by

$$\varphi(\theta, \phi) = ((a + b \cos \phi)\cos \theta, (a + b \cos \phi)\sin \theta, b \sin \phi).$$

A comparison with Example 7 with the axes interchanged ($x_1 \to x_3, x_2 \to x_1, x_3 \to x_2$) shows that φ is the parametrized surface of revolution obtained by rotating the parametrized circle

$$\alpha(\phi) = (a + b \cos \phi, b \sin \phi)$$

in the (x_1, x_3)-plane about the x_2-axis. φ is a *parametrized torus* in $\mathbb{R}^3$ (see Figure 14.5). Note that the parametrized torus is doubly periodic. In fact, $\varphi(\theta + 2k\pi, \phi) = \varphi(\theta, \phi + 2k\pi) = \varphi(\theta, \phi)$ for all $(\theta, \phi) \in \mathbb{R}^2$, $k \subset \mathbb{Z}$. Hence

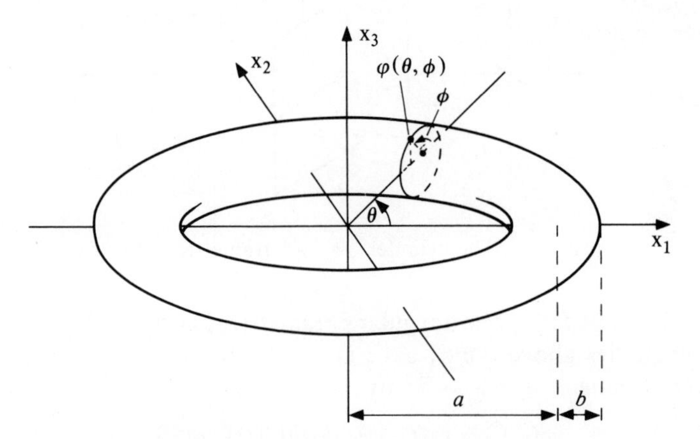

Figure 14.5 A parametrized torus in $\mathbb{R}^3$.

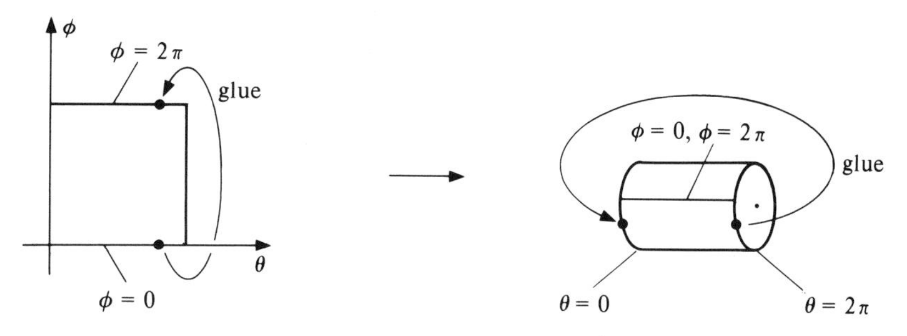

Figure 14.6 Glueing a torus from a square.

each point of Image φ corresponds to a unique point $(\theta, \phi) \in \mathbb{R}^2$ with $0 \leq \theta < 2\pi$ and $0 \leq \phi < 2\pi$. Image φ can be viewed as obtained by taking the square $\{(\theta, \phi) \in \mathbb{R}^2: 0 \leq \theta, \phi \leq 2\pi\}$, glueing the point $(\theta, 0)$ to the point $(\theta, 2\pi)$ for each $\theta \in [0, 2\pi]$ to get a cylinder and then glueing the ends of the cylinder by glueing the point $(0, \phi)$ to the point $(2\pi, \phi)$ for each $\phi \in [0, 2\pi]$ (see Figure 14.6).

EXAMPLE 9. Let $\varphi: \mathbb{R}^2 \to \mathbb{R}^4$ be defined by

$$\varphi(\theta, \phi) = (\cos\theta, \sin\theta, \cos\phi, \sin\phi).$$

This example is similar to Example 8 in that φ is doubly periodic and Image φ can be visualized as the square with opposite edges identified. Another way of visualizing Image φ is to observe that

$$\text{Image } \varphi = \{(p, q) \in \mathbb{R}^2 \times \mathbb{R}^2: p = (\cos\theta, \sin\theta) \text{ for some } \theta,$$
$$q = (\cos\phi, \sin\phi) \text{ for some } \phi\}.$$

Thus Image φ is the Cartesian product of two circles, the unit circle in the (x_1, x_2)-plane and the unit cycle in the (x_3, x_4)-plane, in $\mathbb{R}^4$. φ is called a *parametrized torus in* $\mathbb{R}^4$.

EXAMPLE 10. Let $\varphi: I \times \mathbb{R} \to \mathbb{R}^3$ be defined by

$$\varphi(t, \theta) = \left(\left(1 + t\cos\frac{\theta}{2}\right)\cos\theta, \left(1 + t\cos\frac{\theta}{2}\right)\sin\theta, t\sin\frac{\theta}{2}\right)$$

where $I = \{t \in \mathbb{R}: -\frac{1}{4} < t < \frac{1}{4}\}$. Then Image φ is the Möbius band (Figure 5.3). Note that the curves $t \mapsto \varphi(t, \theta)$ (θ fixed) are straight line segments centered on the unit circle in $\mathbb{R}^2$ and making an angle $\theta/2$ with the (x_1, x_2)-plane. The curves $\theta \mapsto \varphi(t, \theta)$ (t fixed) are periodic, with period 2π if $t = 0$ and period 4π if $t \neq 0$.

Now let $\varphi: U \to \mathbb{R}^{n+k}$ be any smooth map, U open in $\mathbb{R}^n$. A *vector field along* φ is a map $\mathbf{X}$ which assigns to each point $p \in U$ a vector $\mathbf{X}(p) \in \mathbb{R}^{n+k}_{\varphi(p)}$.

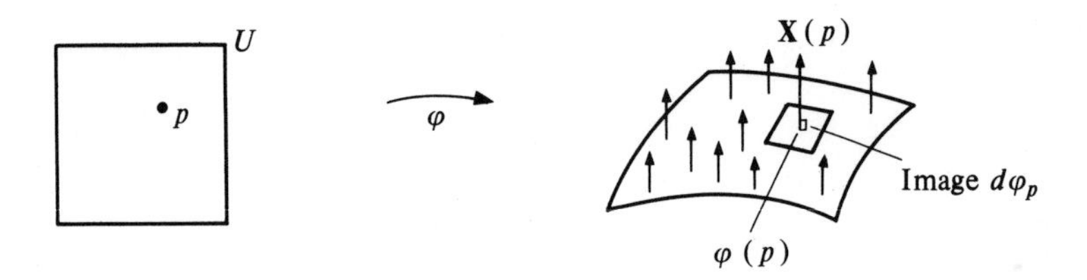

Figure 14.7 A normal vector field along a parametrized 2-surface in $\mathbb{R}^3$.

$\mathbf{X}$ is *smooth* if it is smooth as a map $\mathbf{X}: U \to \mathbb{R}^{2(n+k)}$; that is, if each $X_i: U \to \mathbb{R}$ is smooth on U, where $\mathbf{X}(p) = (\varphi(p), X_1(p), \ldots, X_{n+k}(p))$ for $p \in U$. The vector field $\mathbf{X}$ is *tangent* to φ if it is of the form $\mathbf{X}(p) = d\varphi_p(\mathbf{Y}(p))$ $(p \in U)$ for some vector field $\mathbf{Y}$ on U; $\mathbf{X}$ is *normal* to φ if $\mathbf{X}(p) \perp$ Image $d\varphi_p$ for all $p \in U$ (see Figure 14.7).

For φ a parametrized curve, the velocity field $\dot{\varphi}$ is a tangent vector field along φ since $\dot{\varphi}(t) = d\varphi_t(t, 1)$ for all t. The velocity field generalizes as follows. For $\varphi: U \to \mathbb{R}^{n+k}$ a smooth map, U open in $\mathbb{R}^n$, let $\mathbf{E}_i$ $(i \in \{1, \ldots, n\})$ denote the tangent vector fields along φ defined by

$$\mathbf{E}_i(p) = d\varphi_p(p, 0, \ldots, 1, \ldots, 0)$$

where the 1 is in the $(i+1)$th spot (i spots after the p). Note that the components of $\mathbf{E}_i$ are just the entries in the ith column of the Jacobian matrix for φ at p:

$$\mathbf{E}_i(p) = \left(\varphi(p), \frac{\partial \varphi}{\partial u_i}(p)\right) = \left(\varphi(p), \frac{\partial \varphi_1}{\partial u_i}(p), \ldots, \frac{\partial \varphi_{n+k}}{\partial u_i}(p)\right),$$

where $\varphi(p) = (\varphi_1(p), \ldots, \varphi_{n+k}(p))$ for $p \in U$. The $\mathbf{E}_i$ are called the *coordinate vector fields* along φ. Note that $\mathbf{E}_i(p)$ is simply the velocity at p of the *coordinate curve* $u_i \mapsto \varphi(u_1, \ldots, u_n)$ (all u_j held constant except u_i) passing through $\varphi(p)$ (see Figure 14.8). When φ is a parametrized n-surface (i.e., when φ is regular) these vector fields are linearly independent at each point $p \in U$, since $d\varphi_p$ is non-singular, and so they form a basis for the tangent Image $d\varphi_p$ for each $p \in U$.

For $\varphi: U \to \mathbb{R}^{n+k}$ a smooth map, U open in $\mathbb{R}^n$, and $\mathbf{X}$ a smooth vector field along φ, the *derivative* $\nabla_{\mathbf{v}} \mathbf{X} \in \mathbb{R}^{n+k}_{\varphi(p)}$ of $\mathbf{X}$ with respect to $\mathbf{v} \in \mathbb{R}^n_p$, $p \in U$, is defined by

$$\begin{aligned}\nabla_{\mathbf{v}} \mathbf{X} &= \left(\varphi(p), \left.\frac{d}{dt}\right|_{t_0} (X \circ \alpha)\right) \\ &= (\varphi(p), \nabla_{\mathbf{v}} X_1, \ldots, \nabla_{\mathbf{v}} X_{n+k})\end{aligned}$$

where X is the vector part of $\mathbf{X}$ ($\mathbf{X}(q) = (\varphi(q), X(q))$ for $q \in U$), α is any parametrized curve in U with $\dot{\alpha}(t_0) = \mathbf{v}$, and the $X_i : U \to \mathbb{R}$ are the compo-

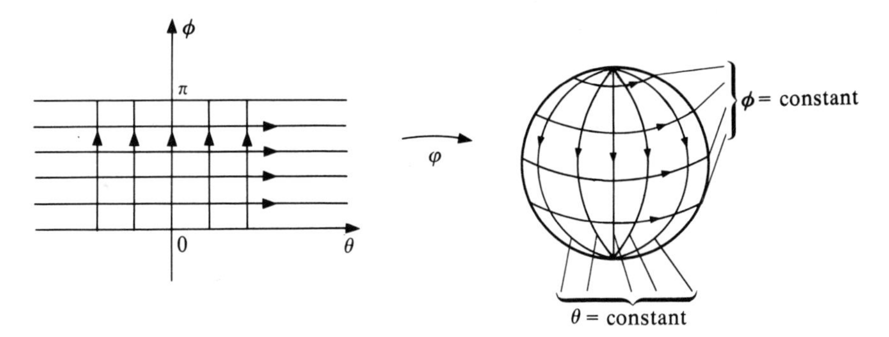

Figure 14.8 Coordinate curves on the parametrized 2-sphere (north and south poles deleted.) $\varphi(\theta, \phi) = (\cos\theta \sin\phi, \sin\theta \sin\phi, \cos\phi)$.

nents of $\mathbf{X}$ ($\mathbf{X}(q) = (\varphi(q), X_1(q), \ldots, X_{n+k}(q))$ for $q \in U$). Note that, when $\mathbf{v} = \mathbf{e}_i = (p, 0, \ldots, 1, \ldots, 0)$, we have

$$\nabla_{\mathbf{e}_i}\mathbf{X} = \left(\varphi(p), \frac{\partial X}{\partial u_i}(p)\right) = \left(\varphi(p), \frac{\partial X_1}{\partial u_i}(p), \ldots, \frac{\partial X_{n+k}}{\partial u_i}(p)\right).$$

Suppose now that $\varphi: U \to \mathbb{R}^{n+1}$ is a parametrized n-surface in $\mathbb{R}^{n+1}$. Then, for each $p \in U$, let $\mathbf{N}(p)$ denote the unique unit vector at $\varphi(p)$ such that $\mathbf{N}(p) \perp \text{Image } d\varphi_p$ and

$$\det\begin{pmatrix} \mathbf{E}_1(p) \\ \cdots \\ \mathbf{E}_n(p) \\ \mathbf{N}(p) \end{pmatrix} > 0$$

where the function det is defined as in Theorem 5 of Chapter 12. Then $\mathbf{N}$ is a smooth unit normal vector field along φ (Exercises 14.8 and 14.9). $\mathbf{N}$ is called the *orientation vector field* along φ. The linear map

$$L_p: (\text{Image } d\varphi_p) \to (\text{Image } d\varphi_p)$$

defined by

$$L_p(d\varphi_p(\mathbf{v})) = -\nabla_{\mathbf{v}}\mathbf{N}$$

is the *Weingarten map* at $p \in U$ of the parametrized n-surface $\varphi: U \to \mathbb{R}^{n+1}$. (Note that L_p is well defined because $d\varphi_p$ is one to one.) L_p is self-adjoint (Exercise 14.11). Its eigenvalues and unit eigenvectors are called the *principal curvatures* and *principal curvature directions* of φ at p. Its determinant is the *Gauss-Kronecker curvature* of φ at p (*Gaussian curvature* when $n = 2$) and $1/n$ times its trace is the *mean curvature* of φ at p.

EXAMPLE. Let φ be the parametrized torus in $\mathbb{R}^3$:

$$\varphi(\theta, \phi) = ((a + b\cos\phi)\cos\theta, (a + b\cos\phi)\sin\theta, b\sin\phi)$$

(Figure 14.5). The coordinate vector fields along φ have vector parts

$$E_1(\theta, \phi) = \frac{\partial\varphi}{\partial\theta} = (a + b\cos\phi)(-\sin\theta, \cos\theta, 0)$$

and

$$E_2(\theta, \phi) = \frac{\partial\varphi}{\partial\phi} = b(-\sin\phi\cos\theta, -\sin\phi\sin\theta, \cos\phi).$$

The orientation vector field $\mathbf{N}$ along φ has vector part

$$\begin{aligned} N(\theta, \phi) &= \frac{E_1(\theta, \phi) \times E_2(\theta, \phi)}{\|E_1(\theta, \phi) \times E_2(\theta, \phi)\|} \\ &= (\cos\theta\cos\phi, \sin\theta\cos\phi, \sin\phi). \end{aligned}$$

Hence, for $p = (\theta, \phi) \in \mathbb{R}^2$,

$$\begin{aligned} L_p(\mathbf{E}_1(p)) = L_p(d\varphi_p(p, 1, 0)) = -\nabla_{(p,1,0)}\mathbf{N} &= -\left(\varphi(p), \frac{\partial N}{\partial\theta}\right) \\ &= -(\varphi(p), -\sin\theta\cos\phi, \cos\theta\cos\phi, 0) \\ &= -\frac{\cos\phi}{a + b\cos\phi}\mathbf{E}_1(p) \end{aligned}$$

and

$$\begin{aligned} L_p(\mathbf{E}_2(p)) &= -\left(\varphi(p), \frac{\partial N}{\partial\phi}\right) \\ &= -(\varphi(p), -\cos\theta\sin\phi, -\sin\theta\sin\phi, \cos\phi) \\ &= -\frac{1}{b}\mathbf{E}_2(p). \end{aligned}$$

Therefore $\mathbf{E}_1(p)$ and $\mathbf{E}_2(p)$ are eigenvectors of L_p. The principal curvatures are $-(\cos\phi)/(a + b\cos\phi)$ and $-1/b$. The Gaussian curvature is $K(\theta, \phi) = (\cos\phi)/b(a + b\cos\phi)$. Note that $K > 0$ on the "outside" of the torus ($-\pi/2 < \phi < \pi/2$), $K < 0$ on the "inside" ($\pi/2 < \phi < 3\pi/2$), and $K = 0$ on the "top" ($\phi = \pi/2$) and on the "bottom" ($\phi = -\pi/2$) (see Figure 14.9).

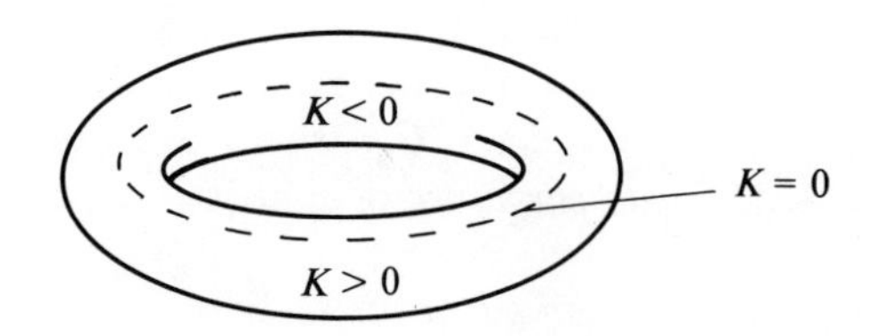

Figure 14.9 Gaussian curvature of a torus.

EXERCISES

14.1. Let S_1 be an n-surface in $\mathbb{R}^{n+1}$ and let S_2 be an m-surface in $\mathbb{R}^{m+1}$. Suppose $\varphi: S_1 \to \mathbb{R}^{m+1}$ is a smooth map such that $\varphi(S_1) \subset S_2$. Show that $d\varphi: T(S_1) \to T(S_2)$.

14.2. Let $\varphi: U_1 \to U_2$ and $\psi: U_2 \to \mathbb{R}^k$ be smooth, where $U_1 \subset \mathbb{R}^n$ and $U_2 \subset \mathbb{R}^m$. Verify the *chain rule* $d(\psi \circ \varphi) = d\psi \circ d\varphi$.

14.3. Verify that each of Examples 3 through 10 above satisfy the required condition that $d\varphi_p$ be non-singular for each $p \in$ domain φ.

14.4. Find the general formula describing the parametrized surface obtained by rotating about the x_3-axis a parametrized curve in the (x_1, x_3) plane. Verify that the parametrized surfaces of Examples 4 and 8 above are of this type.

14.5. Define $\varphi: U \to \mathbb{R}^4$, where $U = \{(\phi, \theta, \psi): \phi \in \mathbb{R}, 0 < \theta < \pi, 0 < \psi < \pi\}$, by $\varphi(\phi, \theta, \psi) = (\sin\phi \sin\theta \sin\psi, \cos\phi \sin\theta \sin\psi, \cos\theta \sin\psi, \cos\psi)$.

(a) Verify that φ is a parametrized 3-surface in $\mathbb{R}^4$.
(b) Show that the image of φ is contained in the unit 3-sphere in $\mathbb{R}^4$.

(ϕ, θ, and ψ are *spherical coordinates* on S^3.)

14.6. Let $\varphi: U \to \mathbb{R}^{n+1}$ be a parametrized n-surface in $\mathbb{R}^{n+1}$ and let $p = (a_1, \ldots, a_{n+2}) \in \mathbb{R}^{n+2}$, where $a_{n+2} \neq 0$. Define $\psi: U \times I \to \mathbb{R}^{n+2}$ where $I = \{t \in \mathbb{R}: 0 < t < 1\}$, by

$$\psi(t_1, \ldots, t_{n+1}) = (1 - t_{n+1})p + t_{n+1}(\varphi(t_1, \ldots, t_n), 0)$$

(see Figure 14.10). Show that ψ is a parametrized $(n+1)$-surface in $\mathbb{R}^{n+2}$. (ψ is the *cone over* φ *with vertex* p.)

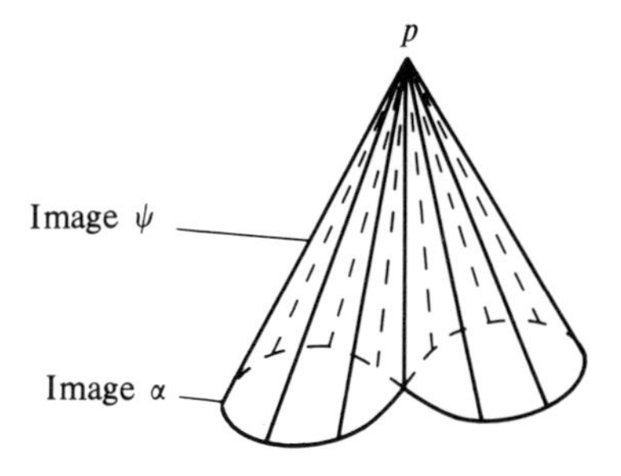

Figure 14.10 A cone over a parametrized curve α.

14.7. Let $\mathbf{X}$ be a smooth vector field on $\mathbb{R}^{n+k}$ and let $\varphi: U \to \mathbb{R}^{n+k}$ be a smooth map, U open in $\mathbb{R}^n$. Show that

$$\nabla_{\mathbf{v}}(\mathbf{X} \circ \varphi) = \nabla_{d\varphi(\mathbf{v})}\mathbf{X}$$

for all $\mathbf{v} \in T(U)$.

14.8. Let φ be a parametrized 2-surface in $\mathbb{R}^3$.

(a) Show that the orientation vector field $\mathbf{N}$ along φ is given by

$$\mathbf{N} = \frac{\mathbf{E}_1 \times \mathbf{E}_2}{\|\mathbf{E}_1 \times \mathbf{E}_2\|},$$

where $\mathbf{E}_1$ and $\mathbf{E}_2$ are the coordinate vector fields along φ.

(b) Conclude that $\mathbf{N}$ is smooth.

14.9. Let $\varphi: U \to \mathbb{R}^{n+1}$ be a parametrized n-surface in $\mathbb{R}^{n+1}$. Let $\mathbf{X}$ be the vector field along φ whose ith component is $(-1)^{n+1+i}$ times the determinant of the matrix obtained by deleting the ith column from the matrix

$$\begin{pmatrix} E_1 \\ \vdots \\ E_n \end{pmatrix},$$

where the $\mathbf{E}_i$ are the coordinate vector fields of φ. (Note that this matrix is just the transpose of the Jacobian matrix of φ.)

(a) Show that $\mathbf{X}(p) \neq 0$ for all $p \in U$.
(b) Show that $\mathbf{X}$ is a normal vector field along φ.
(c) Show that $\mathbf{N} = \mathbf{X}/\|\mathbf{X}\|$ is the orientation vector field along φ.
(d) Conclude that $\mathbf{N}$ is smooth.

14.10. Let $g: U \to \mathbb{R}$ be a smooth function on the open set U in $\mathbb{R}^n$ and let $\varphi: U \to \mathbb{R}^{n+1}$ be defined by $\varphi(u_1, \ldots, u_n) = (u_1, \ldots, u_n, g(u_1, \ldots, u_n))$. Show that the orientation vector field along φ is given by

$$\mathbf{N}(p) = \left(\varphi(p), -\frac{\partial g}{\partial u_1}(p), \ldots, -\frac{\partial g}{\partial u_n}(p), 1\right) \bigg/ \left[1 + \sum_{i=1}^{n} \left(\frac{\partial g}{\partial u_i}(p)\right)^2\right]^{1/2}.$$

14.11. Show that the Weingarten map at each point of a parametrized n-surface in $\mathbb{R}^{n+1}$ is self-adjoint.

14.12. Let $\varphi: U \to \mathbb{R}^{n+k}$ be a parametrized n-surface in $\mathbb{R}^{n+k}$. Show that $d\varphi: U \times \mathbb{R}^n \to \mathbb{R}^{n+k} \times \mathbb{R}^{n+k}$ is a parametrized $2n$-surface in $\mathbb{R}^{2n+2k}$.

14.13. Let $\varphi: U \to \mathbb{R}^{n+k}$ be a parametrized n-surface in $\mathbb{R}^{n+k}$. Let $\mathbf{E}_i$ denote the coordinate vector fields along φ and let $\mathbf{e}_i = (p, 0, \ldots, 1, \ldots, 0)$ for $p \in U$. Show that $\nabla_{\mathbf{e}_i} \mathbf{E}_j = \nabla_{\mathbf{e}_j} \mathbf{E}_i$ for all i and j.

14.14. Let φ be a parametrized n-surface in $\mathbb{R}^{n+1}$. Show that the Gauss-Kronecker curvature of φ is given by the formulas

$$K(p) = \frac{\det[L_p(\mathbf{E}_i(p)) \cdot \mathbf{E}_j(p)]}{\det[\mathbf{E}_i(p) \cdot \mathbf{E}_j(p)]} = \frac{\det[(\nabla_{\mathbf{e}_i} \mathbf{E}_j) \cdot \mathbf{N}(p)]}{\det[\mathbf{E}_i(p) \cdot \mathbf{E}_j(p)]}$$

where the $\mathbf{E}_i$ are the coordinate vector fields along φ, and $\mathbf{e}_i = (p, 0, \ldots, 1, \ldots, 0)$.

In Exercises 14.15–14.18, find the Gaussian curvature of the given parametrized 2-surface φ.

14.15. $\varphi(\theta, \phi) = (a \cos\theta \sin\phi, a \sin\theta \sin\phi, a \cos\phi)$ (sphere)

14.16. $\varphi(t, \theta) = (\cos\theta, \sin\theta, t)$ (right circular cylinder)

14.17. $\varphi(t, \theta) = (t\cos\theta, t\sin\theta, \theta)$ (helicoid)

14.18. $\varphi(t, \theta) = (\sinh t, \cosh t\cos\theta, \cosh t\sin\theta)$ (hyperboloid).

14.19. Find the Gauss-Kronecker curvature of the parametrized 3-surface φ, where

$$\varphi(x, y, z) = (x, y, z, x^2 + y^2 + z^2) \text{ (3-paraboloid in } \mathbb{R}^4).$$

14.20. Let $\varphi\colon I \times \mathbb{R} \to \mathbb{R}^3$ be the parametrized surface of revolution obtained by rotating the parametrized curve $\alpha(t) = (x(t), y(t))$ ($y(t) > 0$ for all $t \in I$) about the x_1-axis. Thus

$$\varphi(t, \theta) = (x(t), y(t)\cos\theta, y(t)\sin\theta)$$

for $t \in I$ and $\theta \in \mathbb{R}$.

(a) Show that the Gaussian curvature of φ is given by the formula

$$K = \frac{x'(x''y' - x'y'')}{y(x'^2 + y'^2)^2}$$

(b) Show that if α has unit speed this formula reduces to $K = -y''/y$.

14.21. Let $\alpha(t) = (x(t), y(t))$, where

$$x(t) = \int_0^t \sqrt{1 - e^{-2\tau}}\, d\tau, \qquad (t > 0)$$

$$y(t) = e^{-t}, \qquad (t > 0)$$

and let φ be the parametrized surface of revolution obtained by rotating α about the x_1-axis.

(a) Show that α has unit speed.

(b) Show that α has the property that for each $t > 0$ the segment between $\alpha(t)$

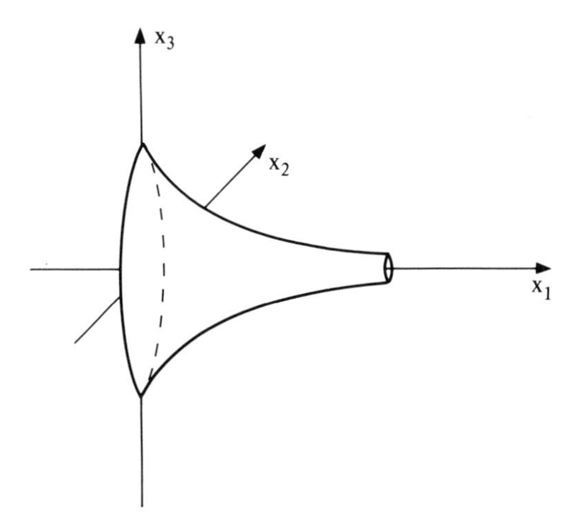

Figure 14.11 A pseudosphere.

and the x_1-axis of the tangent line to α at $\alpha(t)$ has constant length 1. (Hence Image α will be traced out by the end of a taut string one unit in length, initially vertical with the other end at the origin, being pulled along the x_1-axis.)

(c) Show that φ has constant Gaussian curvature $K = -1$. (φ is called a *parametrized pseudosphere* in $\mathbb{R}^3$; see Figure 14.11).

15 Local Equivalence of Surfaces and Parametrized Surfaces

In this chapter we shall establish two theorems which show that, locally, n-surfaces and parametrized n-surfaces are the same. In order to do this, we will need to use the following theorem from the calculus of several variables.

Inverse Function Theorem. *Let U be an open set in $\mathbb{R}^{n+1}$, let $\psi: U \to \mathbb{R}^{n+1}$ be smooth, and suppose $p \in U$ is such that $d\psi_p$ is non-singular. Then there exists an open set $V \subset U$ about p such that the restriction $\psi|_V$ of ψ to V maps V one to one onto an open set W in $\mathbb{R}^{n+1}$, and moreover the inverse map $(\psi|_V)^{-1}: W \to V$ is smooth.*

A proof of this theorem may be found in Fleming's *Functions of Several Variables* (Second Edition, Springer-Verlag, 1977). Note that, since the matrix for $d\psi_p$ with respect to the standard standard bases for $\mathbb{R}_p^{n+1}$ and $\mathbb{R}_{\psi(p)}^{n+1}$ is the Jacobian matrix $J_\psi(p)$ of ψ at p, the condition that $d\psi_p$ be non-singular says simply that $\det J_\psi(p) \neq 0$.

Theorem 1. *Let S be an n-surface in $\mathbb{R}^{n+1}$ and let $p \in S$. Then there exists an open set V about p in $\mathbb{R}^{n+1}$ and a parametrized n-surface $\varphi: U \to \mathbb{R}^{n+1}$ such that φ is a one to one map from U onto $V \cap S$* (see Figure 15.1).

Proof. Let $f: U_1 \to \mathbb{R}$ (U_1 open in $\mathbb{R}^{n+1}$) be a smooth function such that $S = f^{-1}(c)$ for some $c \in \mathbb{R}$ and $\nabla f(q) \neq 0$ for all $q \in S$. Choose $i \in \{1, \ldots, n+1\}$ such that $(\partial f/\partial x_i)(p) \neq 0$. Such an i exists since $\nabla f(p) \neq 0$. Define $\psi: U_1 \to \mathbb{R}^{n+1}$ by $\psi(x_1, \ldots, x_{n+1}) = (x_1, \ldots, x_{i-1}, f(x_1, \ldots, x_{n+1}), x_{i+1}, \ldots, x_{n+1})$. Thus ψ maps level sets of f into hyperplanes $x_i =$ constant, and in particular ψ maps S into the hyperplane $x_i = c$ (see Figure 15.2). The Jacobian matrix $J_\psi(p)$ is just the identity matrix with the ith column replaced by the components of $\nabla f(p)$. Hence $\det J_\psi(p) = (\partial f/\partial x_i)(p) \neq 0$. So, by the

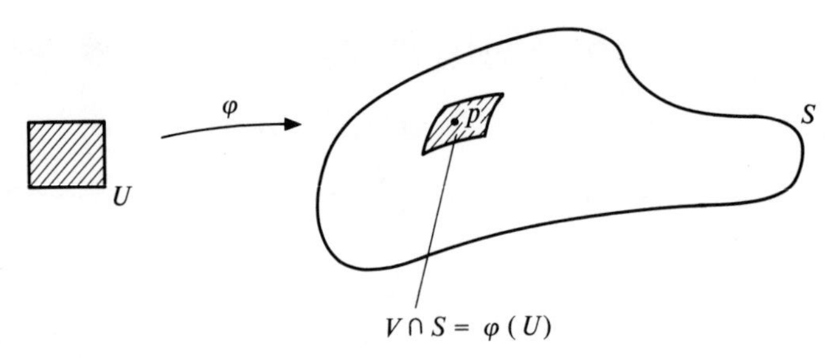

Figure 15.1 A parametrization of a portion of an n-surface.

inverse function theorem, there exists an open set $V_1 \subset U_1$ about p such that ψ maps V_1 one to one onto an open set W_1 about $\psi(p)$, and $(\psi|_{V_1})^{-1}: W_1 \to V_1$ is smooth. For each $j \in \{1, \ldots, n+1\}$, choose $a_j, b_j \in \mathbb{R}$ with $a_j < b_j$ such that

$$W = \{(x_1, \ldots, x_{n+1}): a_j < x_j < b_j \text{ for all } j\}$$

is a subset of W_1 and $\psi(p) \in W$. Finally, let $V = (\psi|_{V_1})^{-1}(W)$, let

$$U = \{(u_1, \ldots, u_n) \in \mathbb{R}^n: a_j < u_j < b_j \text{ for } j < i \\ \text{and } a_{j+1} < u_j < b_{j+1} \text{ for } j \geq i\},$$

and define $\varphi: U \to \mathbb{R}^{n+1}$ by (see Figure 15.2)

$$\varphi(u_1, \ldots, u_n) = (\psi|_V)^{-1}(u_1, \ldots, u_{i-1}, c, u_i, \ldots, u_n).$$

φ is the required parametrized n-surface. □

Remark. Note that the parametrized surface φ of Theorem 1 can be chosen so that the orientation vector fields $\mathbf{N}^\varphi$ of φ and $\mathbf{N}^S$ of S agree; that is, so that $\mathbf{N}^\varphi(q) = \mathbf{N}^S(\varphi(q))$ for all $q \in U$. Indeed, Image $d\varphi_q \subset S_{\varphi(q)}$ for all

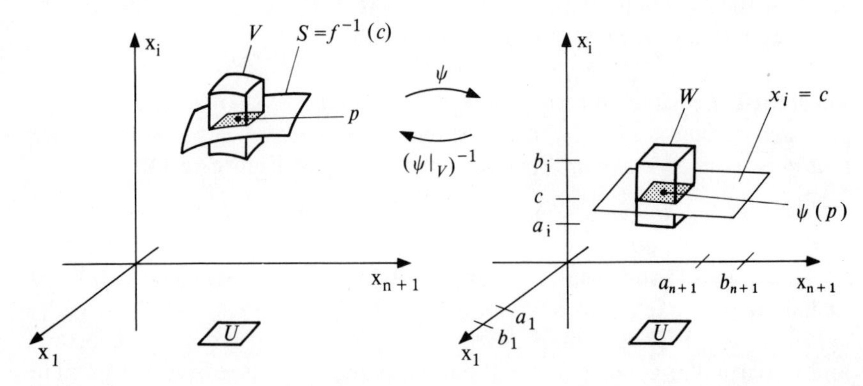

Figure 15.2 The map $\psi|_V$ maps the level set $V \cap f^{-1}(c)$ onto the intersection with W of the hyperplane $x_i = c$.

$q \in U$ since Image $\varphi \subset S$, so $\mathbf{N}^S(\varphi(q)) \perp$ Image $d\varphi_q$ for all $q \in U$. Moreover, the function $g\colon U \to \mathbb{R}$ defined by

$$g(q) = \det \begin{pmatrix} \mathbf{E}_1(q) \\ \vdots \\ \mathbf{E}_n(q) \\ \mathbf{N}^S(\varphi(q)) \end{pmatrix}$$

is continuous and nowhere zero. Since U is connected, g is either positive everywhere or negative everywhere. If $g(q) > 0$ for all $q \in U$ then $\mathbf{N}^\varphi = \mathbf{N}^S \circ \varphi$. If $g(q) < 0$ for all $q \in U$ then $\mathbf{N}^\varphi = -\mathbf{N}^S \circ \varphi$, so if we replace φ by the parametrized n-surface $\tilde{\varphi}\colon \tilde{U} \to \mathbb{R}^{n+1}$ defined by

$$\tilde{\varphi}(u_1, u_2, u_3, \ldots, u_n) = \varphi(u_2, u_1, u_3, \ldots, u_n)$$

we will have $\mathbf{N}^{\tilde{\varphi}} = \mathbf{N}^S \circ \tilde{\varphi}$. Here,

$$\tilde{U} = \{(u_1, u_2, u_3, \ldots, u_n) \in \mathbb{R}^n\colon (u_2, u_1, u_3, \ldots, u_n) \in U\}.$$

A parametrized n-surface $\varphi\colon U \to \mathbb{R}^{n+1}$ whose image is an open subset of the oriented n-surface S and whose orientation vector field agrees with that of S (i.e., $\mathbf{N}^\varphi = \mathbf{N}^S \circ \varphi$) is called a *local parametrization* of S. Theorem 1 guarantees the existence of a one to one local parametrization of S whose image is an open set in S about any given point of S. The inverse φ^{-1} of such a parametrization $\varphi\colon U \to S$ is often called a *chart* because through φ^{-1} the region Image $\varphi \subset S$ is "charted" on $U \subset \mathbb{R}^n$, just as a region of the earth is charted on a topographic or political map. φ^{-1} is also sometimes called a *coordinate system* because through φ^{-1} each point $p \in$ Image φ corresponds to an n-tuple of real numbers, the *coordinates* of p.

Example 1. Let φ be the map from the open square $0 < \theta < 2\pi$, $0 < \phi < 2\pi$ into $\mathbb{R}^3$ defined, for $a > b > 0$, by

$$\varphi(\theta, \phi) = ((a + b\cos\phi)\cos\theta, (a + b\cos\phi)\sin\theta, b\sin\phi).$$

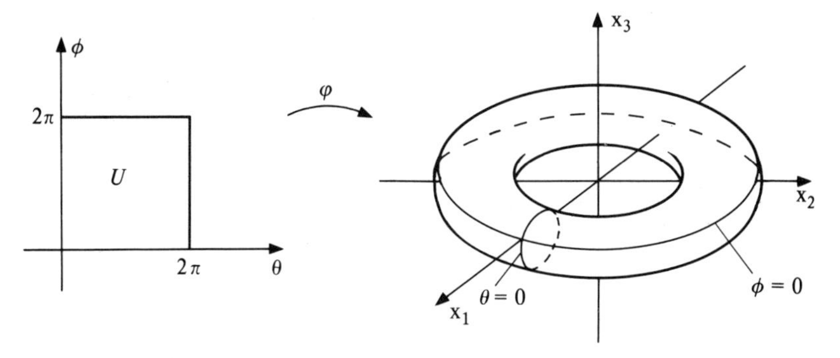

Figure 15.3 φ^{-1} is a chart on the portion of the torus obtained by deleting two circles ($\theta = 0$ and $\phi = 0$.)

Then φ^{-1} is a chart on the torus $(\sqrt{x_1^2 + x_2^2} - a)^2 + x_3^2 = b^2$, with two circles deleted (see Figure 15.3).

EXAMPLE 2. Let φ map the open rectangle $0 < \theta < 2\pi, 0 < \phi < \pi$ into $\mathbb{R}^3$ by $\varphi(\theta, \phi) = (\cos\theta \sin\phi, \sin\theta \sin\phi, \cos\phi)$. Then φ^{-1} is a chart on the unit sphere $\mathbf{S}^2$ with a semi-circle deleted (see Figure 15.4).

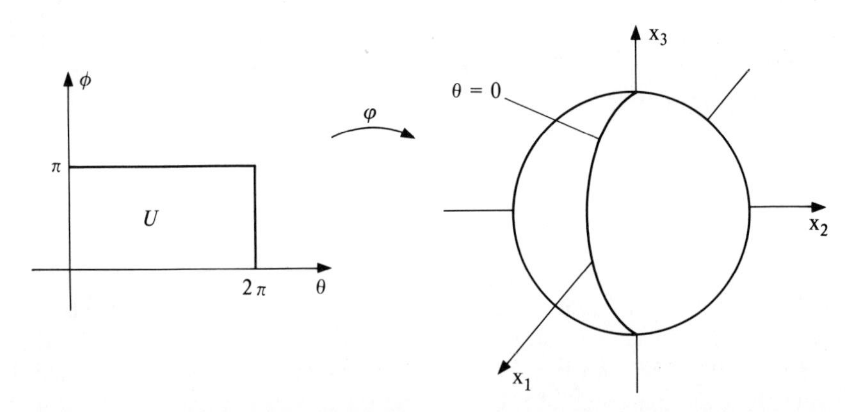

Figure 15.4 Spherical coordinates define a chart on the portion of the sphere $\mathbf{S}^2$ obtained by deleting the semi-circle $\theta = 0, 0 \le \phi \le \pi$.

EXAMPLE 3. A chart whose domain is the unit sphere with only one point deleted, and which is easily described for spheres of arbitrary dimension, is given by stereographic projection. Let $\mathbf{S}^n$ denote the unit n-sphere in $\mathbb{R}^{n+1}$ and let $q = (0, \ldots, 0, 1)$ denote the "north pole" of $\mathbf{S}^n$. Let $\varphi: \mathbb{R}^n \to \mathbf{S}^n$ be the map which sends each $p \in \mathbb{R}^n$ into the point different from q where the line through $(p, 0) \in \mathbb{R}^{n+1}$ and q cuts $\mathbf{S}^n$ (see Figure 15.5).

Since $\alpha(t) = t(p, 0) + (1 - t)q = (tp, 1 - t)$ is a parametrization of the line through $(p, 0)$ and q, and since $\|\alpha(t)\| = 1$ if and only if $t = 0$ or $t = 2/(\|p\|^2 + 1)$, the map φ is given by the formula

$$\varphi(x_1, \ldots, x_n) = (2x_1, \ldots, 2x_n, x_1^2 + \cdots + x_n^2 - 1)/(x_1^2 + \cdots + x_n^2 + 1).$$

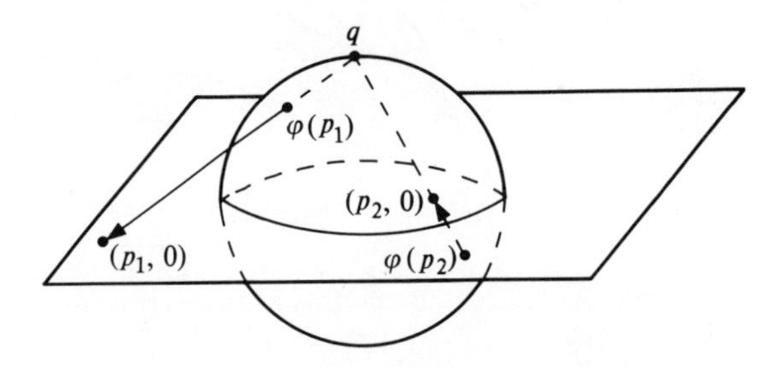

Figure 15.5 Stereographic projection.

The map φ is a parametrized surface which maps $\mathbb{R}^n$ one to one onto $\mathbf{S}^n - \{q\}$. The chart φ^{-1} is called *stereographic projection* from $\mathbf{S}^n - \{q\}$ onto the equatorial hyperplane. Note that φ leaves the equatorial $(n-1)$-sphere fixed, maps the unit ball $\{(x_1, \ldots, x_n) \in \mathbb{R}^n : x_1^2 + \cdots + x_n^2 < 1\}$ onto the "southern hemisphere" $\{(x_1, \ldots, x_{n+1}) \in \mathbf{S}^n : x_{n+1} < 0\}$, and maps the exterior $\{(x_1, \ldots, x_n) \in \mathbb{R}^n : x_1^2 + \cdots + x_n^2 > 1\}$ of the unit ball onto the "northern hemisphere" $\{(x_1, \ldots, x_{n+1}) \in \mathbf{S}^n : x_{n+1} > 0\}$ with the north pole deleted.

Theorem 2. *Let $\varphi: U \to \mathbb{R}^{n+1}$ be a parametrized n-surface in $\mathbb{R}^{n+1}$ and let $p \in U$. Then there exists an open set $U_1 \subset U$ about p such that $\varphi(U_1)$ is an n-surface in $\mathbb{R}^{n+1}$.*

Proof. Define $\psi: U \times \mathbb{R} \to \mathbb{R}^{n+1}$ by $\psi(q, s) = \varphi(q) + sN(q)$, where $N(q)$ is the vector part at q of the orientation vector field along φ. Then

$$J_\psi(p, 0) = \begin{pmatrix} \vdots & \vdots \\ J_\varphi(p) & N(p) \\ \vdots & \vdots \end{pmatrix} = \begin{pmatrix} \vdots & & \vdots & \vdots \\ E_1(p) & \cdots & E_n(p) & N(p) \\ \vdots & & \vdots & \vdots \end{pmatrix}$$

is the matrix whose columns are the vector parts at p of the coordinate vector fields $\mathbf{E}_i$ and of the unit normal vector field $\mathbf{N}$. Hence the columns of $J_\psi(p, 0)$ are linearly independent and $\det J_\psi(p, 0) \neq 0$. By the inverse function theorem, there exists an open set $V \subset U \times \mathbb{R}$ about $(p, 0)$ such that the restriction $\psi|_V$ of ψ to V maps V one to one onto the open set $\psi(V)$, and $(\psi|_V)^{-1}$ is smooth. By shrinking V if necessary, we may assume $V = U_1 \times I$ for some open set $U_1 \subset U$ containing p and some open interval $I \subset \mathbb{R}$ containing 0 (see Figure 15.6). Now define f: Image $\psi|_V \to \mathbb{R}$ by $f(\psi(q, s)) = s$; i.e., $f(\varphi(q) + sN(q)) = s$. Thus $f(\psi(q, s))$ is the perpendicular distance from $\psi(q, s)$ to Image φ. f is well defined and is smooth because f is the composition of the smooth map $(\psi|_V)^{-1}$ with the projection map $U_1 \times I \to I$. The level set $f^{-1}(0)$ is just $\varphi(U_1)$ because

$$f^{-1}(0) = \{\psi(q, s): q \in U_1, s = 0\} = \{\varphi(q): q \in U_1\}.$$

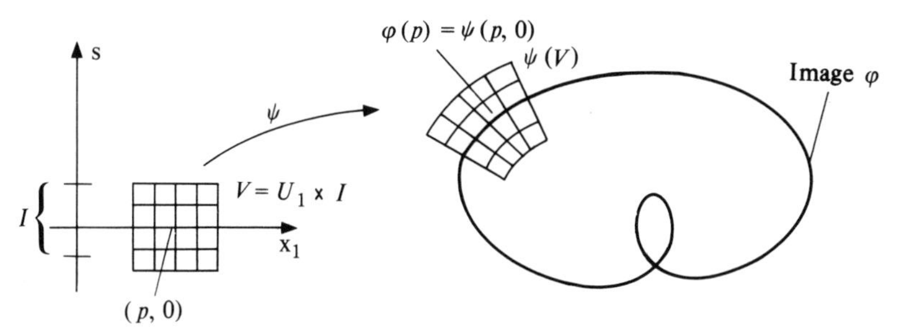

Figure 15.6 The inverse function theorem applied to a parametrized 1-surface φ. The straight lines in $\psi(V)$ are the lines $\beta_q(s) = \varphi(q) + sN(q)$ (q fixed). The transverse 1-surfaces are the images of the maps $\varphi_s: U_1 \to \mathbb{R}^2$ given by $\varphi_s(q) = \varphi(q) + sN(q)$ (s fixed).

Finally, $\nabla f(z) \neq 0$ for $z = \psi(q, 0) \in f^{-1}(0)$ because, letting $\alpha(s) = \psi(q, s) = \varphi(q) + sN(q)$, we have

$$\nabla f(z) \cdot \mathbf{N}(q) = \nabla f(\alpha(0)) \cdot \dot{\alpha}(0) = (f \circ \alpha)'(0) = 1 \neq 0.$$

Thus $\varphi(U_1) = f^{-1}(0)$ is an n-surface in $\mathbb{R}^{n+1}$. □

Theorem 1 says that about each point p of an n-surface S in $\mathbb{R}^{n+1}$ there is an open set V such that $S \cap V$ is the image of a one to one parametrized n-surface. Theorem 2 says that about each point p in the domain of a parametrized n-surface φ on $\mathbb{R}^{n+1}$ there exists an open set V such that Image $(\varphi|_V)$ is an n-surface. Whenever a subset S of $\mathbb{R}^{n+1}$ is described both as an n-surface $S = f^{-1}(c)$ and as the image of a parametrized n-surface $\varphi: U \to \mathbb{R}^{n+1}$ with $\mathbf{N}^\varphi(p) = \mathbf{N}^S(\varphi(p))$ for all $p \in U$, then φ and S have the same geometry at each point:

(i) The Weingarten map L_p^φ of φ at $p \in U$ is the same as the Weingarten map $L_{\varphi(p)}^S$ of S at $\varphi(p)$ because, for $\mathbf{v} \in \mathbb{R}_p^n$,

$$\begin{aligned} L_p^\varphi(d\varphi(\mathbf{v})) &= -\nabla_\mathbf{v} \mathbf{N}^\varphi = -\nabla_\mathbf{v}(\mathbf{N}^S \circ \varphi) = -(\mathbf{N}^S \circ \dot{\varphi} \circ \alpha)(t_0) \\ &= -\nabla_{\varphi \,\dot{\circ}\, \alpha(t_0)} \mathbf{N}^S = -\nabla_{d\varphi(\mathbf{v})} \mathbf{N}^S = L_{\varphi(p)}^S(d\varphi(\mathbf{v})) \end{aligned}$$

where $\alpha: I \to U$ is such that $\dot{\alpha}(t_0) = \mathbf{v}$.

(ii) The principal curvatures, Gauss-Kronecker curvature, and mean curvature of φ at $p \in U$ are equal to the corresponding quantities for S at $\varphi(p)$ since all are computed directly from the Weingarten map.

Remark. Theorem 2 establishes that if φ is a parametrized n-surface in $\mathbb{R}^{n+1}$ then, locally, Image φ is an n-surface; that is, a level set of a real valued function f with non-vanishing gradient. A natural question is whether a similar statement can be made about Image φ where φ is a parametrized n-surface in $\mathbb{R}^{n+k}$. The answer is affirmative. The statement is the same, with n-surface in $\mathbb{R}^{n+k}$ defined as follows.

A *surface of dimension n*, or *n-surface*, in $\mathbb{R}^{n+k} (k \geq 1)$ is a non-empty subset S of $\mathbb{R}^{n+k}$ of the form $S = f^{-1}(c)$ $(c \in \mathbb{R}^k)$ where $f: U \to \mathbb{R}^k$ (U open in $\mathbb{R}^{n+k}$) is a smooth function with the property that df_p has rank k for each $p \in S$. Since the matrix for df_p with respect to the standard bases for $\mathbb{R}_p^{n+k}$ and $\mathbb{R}_{\varphi(p)}^k$ is just the Jacobian matrix of f, whose columns are the vector parts of the gradient vectors $\nabla f_i(p)$, where $f(q) = (f_1(q), \ldots, f_k(q))$, $q \in U$, this definition can be rephased as follows: an n-surface in $\mathbb{R}^{n+k}$ is a non-empty subset of $\mathbb{R}^{n+k}$ of the form

$$S = f_1^{-1}(c_1) \cap \cdots \cap f_k^{-1}(c_k) = \bigcap_{i=1}^{k} f_i^{-1}(c_i)$$

where the $f_i: U \to \mathbb{R}$ (U open in $\mathbb{R}^{n+k}$) are smooth functions such that $\{\nabla f_1(p), \ldots, \nabla f_k(p)\}$ is linearly independent for each $p \in S$ (see Figure 15.7). Thus an n-surface in $\mathbb{R}^{n+k}$ is the intersection of k $(n+k-1)$-surfaces which

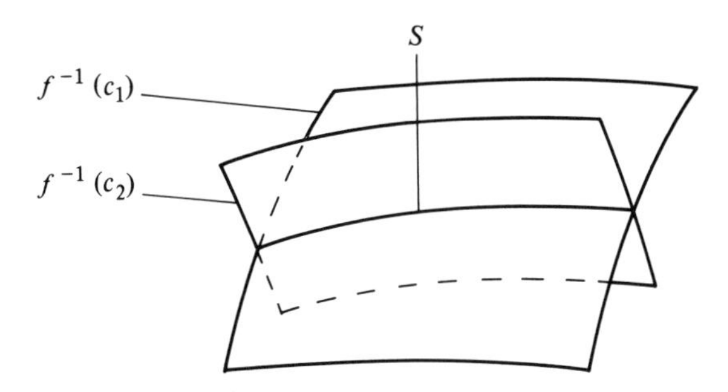

Figure 15.7 A 1-surface S in $\mathbb{R}^3$ is the intersection $f^{-1}(c_1) \cap f^{-1}(c_2)$ of two 2-surfaces.

meet "cleanly" in the sense that the normal directions are linearly independent at each point of the intersection.

The *tangent space* S_p at $p \in S$ to an n-surface $S = \bigcap_{i=1}^k f_i^{-1}(c_i)$ in $\mathbb{R}^{n+k}$ is the set of all vectors in $\mathbb{R}_p^{n+k}$ of the form $\dot{\alpha}(t_0)$ where α is any parametrized curve in S with $\alpha(t_0) = p$. Thus

$$\begin{aligned} S_p &= [f_1^{-1}(c_1)]_p \cap \cdots \cap (f_k^{-1}(c_k)]_p \\ &= \{\mathbf{v} \in \mathbb{R}_p^{n+k} : \nabla f_i(p) \cdot \mathbf{v} = 0 \text{ for } i \in \{1, \ldots, n\}\}. \end{aligned}$$

The k-dimensional subspace $S_p^\perp$ of $\mathbb{R}_p^{n+k}$ spanned by the vectors $\{\nabla f_1(p), \ldots, \nabla f_k(p)\}$ is the *normal space* to S at p (see Figure 15.8).

EXAMPLE 1. A 1-surface in $\mathbb{R}^3$ is usually called a *space curve* (see Figure 15.8).

EXAMPLE 2. Let $f_i: \mathbb{R}^4 \to \mathbb{R} (i \in \{1, 2\})$ be defined by

$$\begin{aligned} f_1(x_1, x_2, x_3, x_4) &= x_1^2 + x_2^2 \\ f_2(x_1, x_2, x_3, x_4) &= x_3^2 + x_4^2. \end{aligned}$$

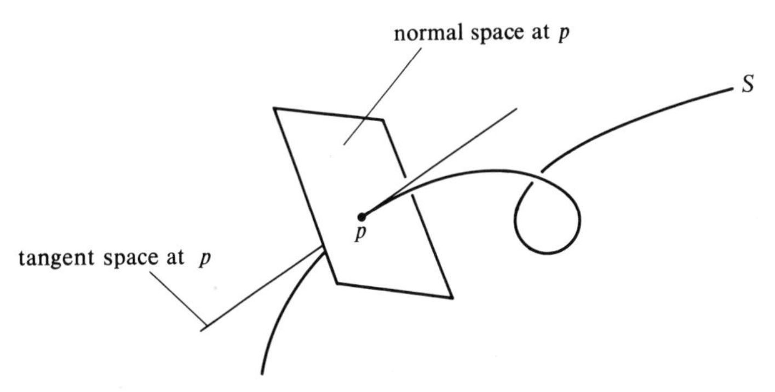

Figure 15.8 The tangent space and the normal space at a point p of a 1-surface (space curve) in $\mathbb{R}^3$.

Then $S = f_1^{-1}(1) \cap f_2^{-1}(1)$ is a 2-surface in $\mathbb{R}^4$ (a torus). Note that S is the Cartesian product of the unit circle in the (x_1, x_2)-plane and the unit circle in the (x_3, x_4)-plane. $S = \text{Image } \varphi$, where φ is the parametrized torus described in Example 9 of the previous chapter.

Remark. Just as the concept of n-surface in $\mathbb{R}^{n+1}$ as defined in Chapter 4 was not general enough to include surfaces like the Möbius band, the definition of n-surface in $\mathbb{R}^{n+k}$ given in this section is not general enough to include all subsets of $\mathbb{R}^{n+k}$ which might be called n-surfaces. The surfaces of Chapter 4 were "orientable" surfaces, so called because each such surface could be oriented by a choice of smooth unit normal vector field. The n-surfaces in $\mathbb{R}^{n+k}$ of this chapter are "normally frameable" n-surfaces, so called because each such surface can be "normally framed" by a choice of k smooth normal vector fields which form a basis for the normal space at each point of S. We shall not in this book consider more general types of surfaces.

Remark. We have defined a map from an n-surface $S \subset \mathbb{R}^{n+1}$ into $\mathbb{R}^k$ to be *smooth* if it is the restriction to S of a smooth function defined on some open set in $\mathbb{R}^{n+1}$ containing S. Using local parametrizations we can now give an alternate characterization of smoothness.

Theorem 3. *Let S be an n-surface in $\mathbb{R}^{n+1}$ and let $f\colon S \to \mathbb{R}^k$. Then f is smooth if and only if $f \circ \varphi\colon U \to \mathbb{R}^k$ is smooth for each local parametrization $\varphi\colon U \to S$.*

PROOF. If f is smooth then $f \circ \varphi$ is smooth for each φ since it is a composition of smooth functions.

Conversely, suppose $f \circ \varphi$ is smooth for each local parametrization φ of S. We must construct a smooth extension $\tilde{f}$ of f to an open set V in $\mathbb{R}^{n+1}$ containing S. For each $p \in S$, let $\varphi_p\colon U_p \to S$ be a local parametrization of S whose image contains p and let $\psi_p\colon U_p \times \mathbb{R} \to \mathbb{R}^{n+1}$ be defined by $\psi_p(q, s) = \varphi_p(q) + sN(\varphi_p(q))$, where $\mathbf{N}$ is an orientation on S. Then, as in the proof of Theorem 2, we can find an open set V_p about $(\varphi_p^{-1}(p), 0)$ in $U_p \times \mathbb{R}$ such that $\psi_p|_{V_p}$ maps V_p one to one onto an open set W_p in $\mathbb{R}^{n+1}$, and $(\psi_p|_{V_p})^{-1}\colon W_p \to V_p$ is smooth. Furthermore, by shrinking V_p if necessary, we may assume that $\psi_p(q, s) \in S$ for $(q, s) \in V_p$ if and only if $s = 0$. Now if we define $\tilde{f}_p\colon W_p \to \mathbb{R}^k$ by $\tilde{f}_p(\varphi_p(q) + sN(\varphi_p(q))) = f(\varphi_p(q))$ we will have constructed a smooth extension $\tilde{f}_p = (f \circ \varphi_p) \circ \pi \circ (\psi_p|_{V_p})^{-1}$ of $f|_{W_p \cap S}$ to the open set W_p in $\mathbb{R}^{n+1}$. Here, $\pi(q, s) = q$. We would like to piece these extensions together to get a smooth extension of f defined on the open set $\bigcup_{p \in S} W_p$. We can do this provided $\tilde{f}_{p_1} = \tilde{f}_{p_2}$ on $W_{p_1} \cap W_{p_2}$, for all $p_1, p_2 \in S$. But this may not be the case (see Figure 15.9). If, however, for each $p \in S$ we choose $\varepsilon_p > 0$ so that the ball of radius $2\varepsilon_p$ about p is contained in W_p and set $B_p =$ the ball of radius ε_p about p then

$$\varphi_{p_1}(q_1) + s_1 N(\varphi_{p_1}(q_1)) = \varphi_{p_2}(q_2) + s_2 N(\varphi_{p_2}(q_2))$$

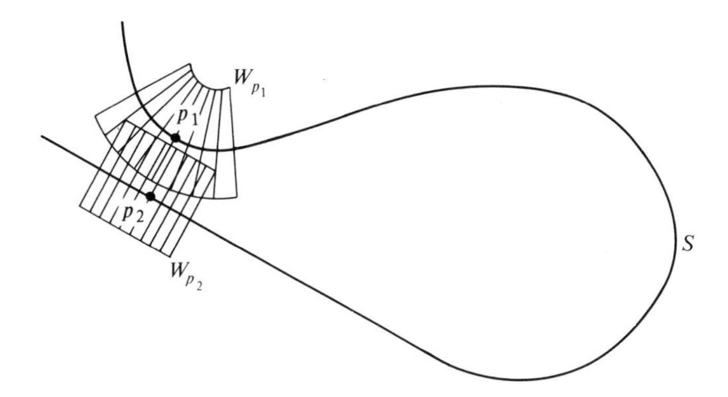

Figure 15.9 Construction of a smooth extension.

can only happen, for $(q_1, s_1) \in (\psi_{p_1}|_{V_{p_1}})^{-1}(B_{p_1})$ and $(q_2, s_2) \in (\psi_{p_2}|_{V_{p_2}})^{-1}(B_{p_2})$, when $\varphi_{p_1}(q_1) = \varphi_{p_2}(q_2)$ and $s_1 = s_2$. It follows that a function $\tilde{f}: \bigcup_{p \in S} B_p \to \mathbb{R}^k$ can be defined by $\tilde{f}(\varphi_p(q) + sN(\varphi_p(q))) = f(\varphi_p(q))$ and that this function is a smooth extension of f. $\square$

For $f: S \to \mathbb{R}^l$ a function defined on an n-surface S in $\mathbb{R}^{n+k}$ we can define *smoothness* of f either by the requirement that f be the restriction to S of a smooth function defined on some open set in $\mathbb{R}^{n+k}$ containing S or by the requirement that $f \circ \varphi$ be smooth for each local parametrization φ of S. These two requirements on f are equivalent by an argument analogous to that in the proof of Theorem 3.

A smooth map f with a smooth inverse is called a *diffeomorphism*. Thus, for example, the local parametrization $\varphi: U \to S$ constructed in the proof of Theorem 1 is a diffeomorphism from the open set $U \subset \mathbb{R}^{n+1}$ onto the open set $V \cap S$ about p in S.

Theorem 4. (*Inverse Function Theorem for n-surfaces*). *Let S and $\tilde{S}$ be n-surfaces, let $\psi: S \to \tilde{S}$ be a smooth map, and suppose $p \in S$ is such that $d\psi_p: S_p \to \tilde{S}_{\psi(p)}$ is nonsingular. Then there exists an open set V about p in S and an open set W about $\psi(p)$ in $\tilde{S}$ such that $\psi|_V$ is a diffeomorphism from V onto W.*

Proof. Let $\varphi_1: U_1 \to S$ and $\varphi_2: U_2 \to \tilde{S}$ be one to one local parametrizations of S and $\tilde{S}$, with $p \in \varphi_1(U_1)$ and $\psi(p) \in \varphi_2(U_2)$, constructed as in the proof of Theorem 1. Then $\varphi_2^{-1} \circ \psi \circ \varphi_1: U_1 \to U_2$ is smooth and $d(\varphi_2^{-1} \circ \psi \circ \varphi_1)_{\varphi_1^{-1}(p)} = (d\varphi_2^{-1})_{\psi(p)} \circ d\psi_p \circ (d\varphi_1)_{\varphi_1^{-1}(p)}$ is nonsingular so, by the inverse function theorem for $\mathbb{R}^n$, there exists an open set $V_1 \subset U_1$ containing $\varphi_1^{-1}(p)$ such that $(\varphi_2^{-1} \circ \psi \circ \varphi_1)|_{V_1}$ is a diffeomorphism from V_1 onto an open set $W_1 \subset U_2$ containing $\varphi_2^{-1} \circ \psi(p)$. Set $V = \varphi_1(V_1)$ and $W = \varphi_2(W_1)$. Then $\psi|_V = \varphi_2 \circ (\varphi_2^{-1} \circ \psi \circ \varphi_1) \circ \varphi_1^{-1}|_V$ is a diffeomorphism from V onto W. $\square$

Corollary. *Let S be a compact connected oriented n-surface in $\mathbb{R}^{n+1}$ whose Gauss-Kronecker curvature is nowhere zero. Then the Gauss map $N: S \to \mathbf{S}^n$ is a diffeomorphism.*

PROOF. By Theorem 5 of Chapter 13, N is one to one and onto, so we need only check that N^{-1} is smooth. But for each $p \in S$ and $\mathbf{v} \in S_p$, $dN(\mathbf{v})$ has the same vector part as $\nabla_{\mathbf{v}} \mathbf{N} = -L_p(\mathbf{v})$ and this can be zero only when $\mathbf{v} = \mathbf{0}$ since, by Theorem 6 of Chapter 12, the second fundamental form $\mathscr{S}_p$ of S at p is definite. Applying Theorem 4 we conclude that N^{-1} is smooth on an open set about each point of $\mathbf{S}^n$ and, by Theorem 3, this is sufficient. □

EXERCISES

15.1. Let $q = (0, \ldots, 0, -1)$ denote the "south pole" of the unit n-sphere $\mathbf{S}^n$. Find a formula for the parametrized n-surface $\varphi: \mathbb{R}^n \to \mathbf{S}^n - \{q\}$ which is the inverse of stereographic projection from $\mathbf{S}^n - \{q\}$ onto the equatorial hyperplane $x_{n+1} = 0$.

15.2. Find a formula for the parametrized n-surface $\varphi: \mathbb{R}^n \to \mathbf{S}^n - \{q\}$ where q is the north pole $(0, \ldots, 0, 1)$ of the unit n-sphere $\mathbf{S}^n$ and φ^{-1} is stereographic projection from $\mathbf{S}^n - \{q\}$ onto the tangent hyperplane $x_{n+1} = -1$ at the south pole $(0, \ldots, 0, -1)$. [Thus, for $p \in \mathbb{R}^n$, $\varphi(p)$ is the point of $\mathbf{S}^n$ different from q which lies on the line through q and $(p, -1) \in \mathbb{R}^{n+1}$.]

15.3. Let $\varphi: U \to \mathbb{R}^{n+1}$ be a parametrized n-surface in $\mathbb{R}^{n+1}$, let $\psi: U \times \mathbb{R} \to \mathbb{R}^{n+1}$ be defined by $\psi(q, s) = \varphi(q) + sN(q)$, let $V = U_1 \times I$ be such that $\psi|_V$ has a smooth inverse, and let $f(\psi(q, s)) = s$, as in the proof of Theorem 2.

(a) Show that the level sets $f^{-1}(c)$ $(c \in I)$ are everywhere orthogonal to the lines $\beta_q(s) = \varphi(q) + sN(q)$ $(q \in U_1$ fixed). [*Hint*: Note that each parametrized curve in $f^{-1}(c)$ is of the form $\varphi \circ \alpha + cN \circ \alpha$ where α is a parametrized curve in U_1.]

(b) Show that $\nabla f(z) = (z, N(q))$ for $z = \psi(q, s) \in \psi(U_1 \times I)$.

15.4. Show that in Theorem 2 it is not sufficient to simply restrict the domain of φ to an open subset U_1 of U on which φ is one to one in order to ensure that Image φ is an n-surface in $\mathbb{R}^{n+1}$: exhibit a one to one parametrized 1-surface in $\mathbb{R}^2$ whose image is not a 1-surface.

15.5. Let S be an oriented n-surface in $\mathbb{R}^{n+1}$ and let $T(S) = \{\mathbf{v} \in \mathbb{R}^{n+1}_p \subset \mathbb{R}^{2(n+1)}: p \in S \text{ and } \mathbf{v} \cdot \mathbf{N}(p) = 0\}$. Show that $T(S)$ is a $2n$-surface in $\mathbb{R}^{2n+2}$. ($T(S)$ is the *tangent bundle* of S.)

15.6. Let S be an oriented n-surface in $\mathbb{R}^{n+1}$ and let $T_1(S) = \{\mathbf{v} \in \mathbb{R}^{n+1}_p \subset \mathbb{R}^{2(n+1)}: p \in S,\ \mathbf{v} \cdot \mathbf{N}(p) = 0,\ \mathbf{v} \cdot \mathbf{v} = 1\}$. Show that $T_1(S)$ is a $(2n-1)$-surface in $\mathbb{R}^{2n+2}$. ($T_1(S)$ is the unit *sphere bundle* of S.)

15.7. (a) Viewing $\mathbb{R}^4$ as the set of all 2×2 matrices with real entries by identifying the 4-tuple $(x_1, \ldots, x_4)$ with the matrix

$$\begin{pmatrix} x_1 & x_2 \\ x_3 & x_4 \end{pmatrix},$$

show that the set $\mathrm{O}(2)$ of orthogonal 2×2 matrices is a 1-surface in $\mathbb{R}^4$. [Recall that a matrix A is orthogonal if A^{-1} is the transpose of A. This is equivalent to the condition that the rows of A form an orthonormal set.]

(b) Show that the tangent space $\mathrm{O}(2)_p$ to $\mathrm{O}(2)$ at $p = \left(\begin{smallmatrix} 1 & 0 \\ 0 & 1 \end{smallmatrix}\right)$ can be identified with the set of all skew-symmetric 2×2 matrices by showing that

$$\mathrm{O}(2)_p = \left\{\left(p, \begin{pmatrix} a & b \\ c & d \end{pmatrix}\right): a = d = 0, c = -b.\right\}$$

[*Hint*: Compute $(\alpha_i \cdot \alpha_j)'(t_0)$, where

$$\alpha(t) = \begin{pmatrix} \alpha_1(t) \\ \alpha_2(t) \end{pmatrix}$$

is an arbitrary parametrized curve in $\mathrm{O}(2)$ with $\alpha(t_0) = \left(\begin{smallmatrix} 1 & 0 \\ 0 & 1 \end{smallmatrix}\right)$.]

15.8. (a) Show that the set $\mathrm{O}(n)$ of orthogonal $n \times n$ matrices is an $n(n-1)/2$ surface in $\mathbb{R}^{n^2}$.

(b) What is the tangent space to $\mathrm{O}(n)$ at p = the identity matrix?

15.9. Show that if $S = f^{-1}(c)$ is an n-surface in $\mathbb{R}^{n+k}$ and $p \in S$ then the tangent space S_p to S at p is equal to the kernel of df_p.

15.10. Prove Theorem 1 with $n + 1$ replaced everywhere by $n + k$.

15.11. Prove Theorem 2 with $n + 1$ replaced by $n + k$. [Use $\psi: U \times \mathbb{R}^k \to \mathbb{R}^{n+k}$ defined by $\psi(q, t_1, \ldots, t_k) = \varphi(q) + \sum_{i=1}^{k} t_i N_i(q)$ where the $\mathbf{N}_i (i \in \{1, \ldots, k\})$ are vector fields along φ which span the normal space (Image $d\varphi_q)^\perp$ for each $q \in U$.]

15.12. Let $\varphi: \mathbb{R}^n \to S^n$ denote the inverse of stereographic projection from $S^n - \{(0, \ldots, 0, 1)\}$ onto the equatorial hyperplane $x_{n+1} = 0$.

(a) Show that for each $p \in \mathbb{R}^n$ there exists a real number $\lambda(p) > 0$ such that $\|d\varphi(\mathbf{v})\| = \lambda(p)\|\mathbf{v}\|$ for all $\mathbf{v} \in \mathbb{R}^n_p$. [*Hint*: Note that the vector part of $d\varphi(\mathbf{v})$ is just $(d/dt)|_0\, \varphi(p + tv)$, where $\mathbf{v} = (p, v)$.]

(b) Using the fact that $\mathbf{v} \cdot \mathbf{w} = (1/4)(\|\mathbf{v} + \mathbf{w}\|^2 - \|\mathbf{v} - \mathbf{w}\|^2)$, conclude that $d\varphi(\mathbf{v}) \cdot d\varphi(\mathbf{w}) = (\lambda(p))^2 \mathbf{v} \cdot \mathbf{w}$ for all $\mathbf{v}, \mathbf{w} \in \mathbb{R}^n_p$, and hence that $d\varphi$ preserves angles between vectors.

[This exercise thus shows that stereographic projection is a *conformal* (angle preserving) map.]

15.13. Let S be an n-surface in $\mathbb{R}^{n+1}$ and let $p \in S$. Show that the subset of S consisting of all points $q \in S$ which can be joined to p by a continuous curve in S is a connected n-surface.

15.14. Let $C = f_1^{-1}(c_1) \cap f_2^{-1}(c_2)$ be a 1-surface in $\mathbf{R}^3$ and let $\mathbf{X} = \nabla f_1 \times \nabla f_2$. Show that the restriction of $\mathbf{X}$ to C is a tangent vector field on C and that the maximal integral curve of $\mathbf{X}$ through $p \in C$ is a one to one or periodic map $\alpha: I \to C$. When does α map I onto C?

16 Focal Points

The construction in the proof of Theorem 2 of the previous chapter surrounds the parametrized n-surface $\varphi\colon U \to \mathbb{R}^{n+1}$ with a family of smooth maps $\varphi_s\colon U \to \mathbb{R}^{n+1} (s \in \mathbb{R})$ given by

$$\varphi_s(q) = \psi(q, s) = \varphi(q) + sN^{\varphi}(q)$$

(Figure 15.6). When $s = 0$, $\varphi_s = \varphi$ is a parametrized n-surface in $\mathbb{R}^{n+1}$. For $s \neq 0$, however, φ_s may fail to be a parametrized n-surface because there may be points $p \in U$ at which φ_s fails to be regular. At each such point there will be a direction $\mathbf{v} \in \mathbb{R}^n_p$ $(\|\mathbf{v}\| = 1)$ such that $d\varphi_s(\mathbf{v}) = \mathbf{0}$. If α is a parametrized curve in U with $\dot{\alpha}(t_0) = \mathbf{v}$, it follows that

$$\varphi_s \dot{\circ}\, \alpha(t_0) = d\varphi_s(\dot{\alpha}(t_0)) = \mathbf{0};$$

that is, the curve $\varphi_s \circ \alpha(t) = \varphi(\alpha(t)) + sN^{\varphi}(\alpha(t))$ (s fixed) pauses (has velocity zero) at $t = t_0$. Geometrically, this says that the normal lines which start along the curve $\varphi \circ \alpha$ near $\varphi(p) = \varphi(\alpha(t_0))$ tend to focus at $\mathscr{f} = \varphi_s(\alpha(t_0)) = \varphi_s(p)$ (see Figure 16.1). Such points $\mathscr{f}$ are called *focal points* of φ. Note that the normal lines along α need not actually meet at a focal point.

Given a parametrized n-surface $\varphi\colon U \to \mathbb{R}^{n+1}$ and a point $p \in U$, let $\beta\colon \mathbb{R} \to \mathbb{R}^{n+1}$ be defined by $\beta(s) = \varphi(p) + sN^{\varphi}(p)$. Thus β is a unit speed parametrization of the line normal to Image φ at $\varphi(p)$. A point $\mathscr{f} \in$ Image β is said to be a *focal point of* φ *along* β if $\mathscr{f} = \beta(s_0)$ where s_0 is such that the map $\varphi_{s_0}\colon U \to \mathbb{R}^{n+1}$ defined by $\varphi_{s_0}(q) = \varphi(q) + s_0 N^{\varphi}(q)$ is singular (not regular) at p.

Theorem 1. *Let $\varphi\colon U \to \mathbb{R}^{n+1}$ be a parametrized n-surface, let $p \in U$, and let $\beta\colon \mathbb{R} \to \mathbb{R}^{n+1}$ be the normal line given by $\beta(s) = \varphi(p) + sN^{\varphi}(p)$. Then the focal points of φ along β are the points $\beta(1/k_i(p))$, where the $k_i(p)$ are the non-zero*

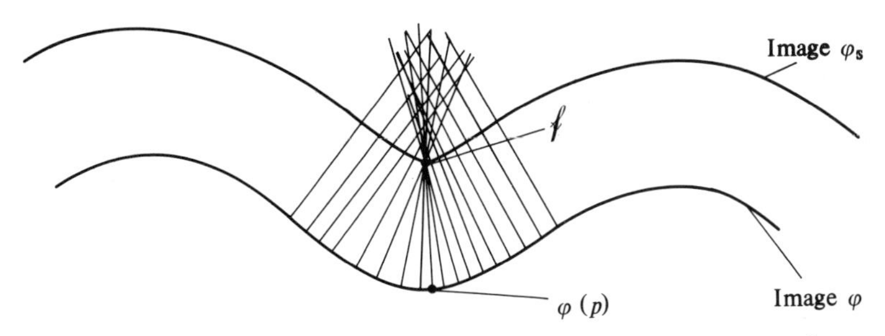

Figure 16.1 f is a focal point of the parametrized 1-surface φ in $\mathbb{R}^2$.

principal curvatures of φ at p. In particular, there exist at most n focal points of φ along β.

PROOF. $f = \beta(s) = \varphi_s(p)$ is a focal point of φ along β if and only if $d\varphi_s(\mathbf{v}) = \mathbf{0}$ for some $\mathbf{v} \in \mathbb{R}^n_p$, $\mathbf{v} \neq \mathbf{0}$. Letting $\alpha: I \to U$ be such that $\dot{\alpha}(t_0) = \mathbf{v}$, the vector part of $d\varphi_s(\mathbf{v})$ is

$$\left.\frac{d}{dt}\right|_{t_0} (\varphi_s \circ \alpha) = \left.\frac{d}{dt}\right|_{t_0} (\varphi \circ \alpha + sN^\varphi \circ \alpha) \qquad (s \text{ fixed})$$

$$= \left.\frac{d}{dt}\right|_{t_0} (\varphi \circ \alpha) + s\left.\frac{d}{dt}\right|_{t_0} (N^\varphi \circ \alpha).$$

Since this last expression is the vector part of $\varphi \mathbin{\dot{\circ}} \alpha(t_0) + s\nabla_\mathbf{v} \mathbf{N}^\varphi$, it follows that $d\varphi_s(\mathbf{v}) = \mathbf{0}$ if and only if

$$\begin{aligned}\mathbf{0} &= \varphi \mathbin{\dot{\circ}} \alpha(t_0) + s\nabla_\mathbf{v} N^\varphi \\ &= d\varphi(\mathbf{v}) - sL_p(d\varphi(\mathbf{v})).\end{aligned}$$

Hence, $d\varphi_s(\mathbf{v}) = \mathbf{0}$ if and only if

$$L_p(d\varphi(\mathbf{v})) = \frac{1}{s}(d\varphi(\mathbf{v})).$$

Note that s cannot be zero since $d\varphi(\mathbf{v}) \neq \mathbf{0}$. Thus f is a focal point of φ along β if and only if $1/s$ is an eigenvalue of L_p; that is, if and only if $1/s$ is a principal curvature of φ at p. □

Given an oriented n-surface S in $\mathbb{R}^{n+1}$, a point $f \in \mathbb{R}^{n+1}$ is said to be a *focal point* of S along the normal line $\beta(s) = p + sN_S(p)$ $(p \in S)$ if f is a focal point along β of φ, where φ is any parametrization, with $N^\varphi = N^S \circ \varphi$, of an open set about p in S. Thus f is a point where normal lines along some curve through p in S tend to focus. By the previous theorem, the focal points of S along β are the points $\beta(1/k_i(p))$ where the $k_i(p)$ are the non-zero principal curvatures of S at p. Note that the location of the focal points does not depend on the choice of orientation on S since reversing $\mathbf{N}^S$ also causes the principal curvatures $k_i(p)$ to change sign, so the focal points

$p + (1/k_i(p))N^S(p)$ remain the same. Note further that the normal lines which tend to focus at $p + (1/k_i(p))N^S(p)$ are those which begin along a curve in S which moves out from p in the ith principal curvature direction.

The next theorem describes the most important property of focal points; namely, that distance from an n-surface is locally minimized along normal lines only up to the first focal point. We shall state and prove this theorem for oriented surfaces but of course it is also valid for parametrized surfaces.

Theorem 2. *Let S be an oriented n-surface in $\mathbb{R}^{n+1}$. Let $p \in S$ and let p_0 lie on the line normal to S through p. Define $h: S \to \mathbb{R}$ by $h(q) = \|q - p_0\|^2$. Then h attains a local minimum at p if and only if there are no focal points of S between p and p_0 along the normal line through p* (see Figure 16.2).

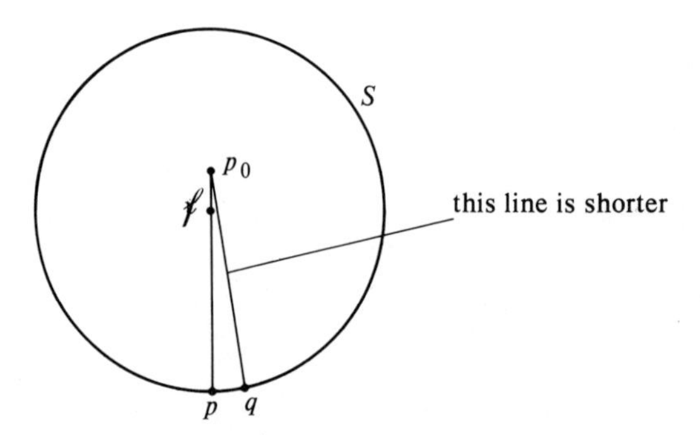

Figure 16.2 Distance to S is no longer minimized along the normal beyond the first focal point f.

PROOF. Since p_0 lies on the normal line to S at p, $p_0 = p + sN(p)$ for some $s \in \mathbb{R}$. We shall assume that $s > 0$; if not, we can change the sign of s by reversing the orientation on S. Define $\tilde{h}: \mathbb{R}^{n+1} \to \mathbb{R}$ by

$$\tilde{h}(q) = \|q - p_0\|^2 = (q - p_0) \cdot (q - p_0)$$

so that h is the restriction of $\tilde{h}$ to S. Then

$$\nabla\tilde{h}(q) = 2(q, q - p_0).$$

In particular,

$$\nabla\tilde{h}(p) = 2(p, p - p_0) = -2s\mathbf{N}(p)$$

so h is stationary at p (grad $h(p) = \mathbf{0}$). The Hessian of h at p on $\mathbf{v} \in S_p$ ($\mathbf{v} \neq \mathbf{0}$) is given by (see Chapter 13)

$$\begin{aligned}\mathscr{H}_p(\mathbf{v}) &= [\nabla_{\mathbf{v}}(\nabla\tilde{h} - ((\nabla\tilde{h}) \cdot \mathbf{N})\mathbf{N})] \cdot \mathbf{v} \\ &= [\nabla_{\mathbf{v}}(\nabla\tilde{h}) + ((\nabla\tilde{h}) \cdot \mathbf{N})(p)L_p(\mathbf{v})] \cdot \mathbf{v} \\ &= [2\mathbf{v} - 2sL_p(\mathbf{v})] \cdot \mathbf{v} = 2\|\mathbf{v}\|^2(1 - sk(\mathbf{v}/\|\mathbf{v}\|))\end{aligned}$$

where $k(\mathbf{v}/\|\mathbf{v}\|)$ is the normal curvature of S in the direction $\mathbf{v}/\|\mathbf{v}\|$. Hence

$$\mathscr{H}_p(\mathbf{v}) \geq 2\|\mathbf{v}\|^2(1 - sk_n(p))$$

where $k_n(p)$ is the maximum value of normal curvature at p; that is, $k_n(p)$ is the maximal principal curvature of S at p. It follows that if $k_n(p) < 0$, or if $k_n(p) > 0$ and $s < 1/k_n(p)$, then $\mathscr{H}_p$ is positive definite so h attains a local minimum at p. By continuity, h must also attain a local minimum at p when $k_n(p) > 0$ and $s = 1/k_n(p)$. Thus h attains a local minimum at p whenever either $k_n(p) < 0$ or $k_n(p) > 0$ and $s \leq 1/k_n(p)$; that is, whenever there are no focal points between p and $p_0 = p + sN(p)$ along the normal line to S through p (see Figure 16.3).

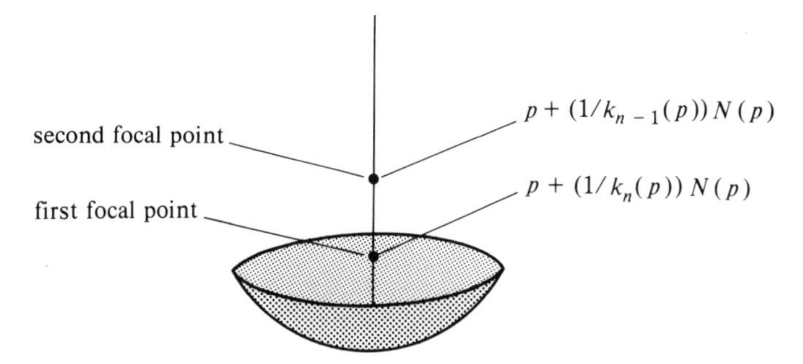

Figure 16.3 For $k_n(p) > 0$, the first focal point occurs at $p + (1/k_n(p))N(p)$.

Conversely, if $0 < 1/k_n(p) < s$ then

$$\mathscr{H}_p(\mathbf{v}_n) = 2(1 - sk(\mathbf{v}_n)) = 2(1 - sk_n(p)) < 0,$$

where $\mathbf{v}_n$ is a principal curvature direction at p corresponding to the principal curvature k_n, so $\mathscr{H}_p$ is not positive semi-definite and h does not attain a local minimum at p. In fact, if α is any parametrized curve in S with $\dot{\alpha}(t_0) = \mathbf{v}_n$ then

$$(h \circ \alpha)'(t_0) = (\text{grad } h)(p) \cdot \dot{\alpha}(t_0) = 0$$

and

$$(h \circ \alpha)''(t_0) = \nabla_{\mathbf{v}_n}(\text{grad } h) \cdot \dot{\alpha}(t_0) = \mathscr{H}_p(\mathbf{v}_n) < 0$$

so distance from p_0 decreases as one moves out from p in S in the direction $\mathbf{v}_n$.

□

The set of all focal points along all normal lines to an n-surface S in $\mathbb{R}^{n+1}$ is called the *focal locus* of S. This set may be visualized as follows. For each $s \in \mathbb{R}$, the set

$$S_s = \{q \in \mathbb{R}^{n+1} : q = p + sN(p) \text{ for some } p \in S\}$$

of all points obtained by moving out from S along the normals a distance s looks like an n-surface with singularities (points where the tangent space has dimension $< n$) at points which are focal points of S. These singularities usually appear as cusps or folds in S_s. The set S_s may be viewed as the position at time s/c (c = speed of light) of an advancing wavefront caused by a flash of light along S at time $s = 0$. By watching these wavefronts we can watch the singularities trace out the focal locus (see Figure 16.4).

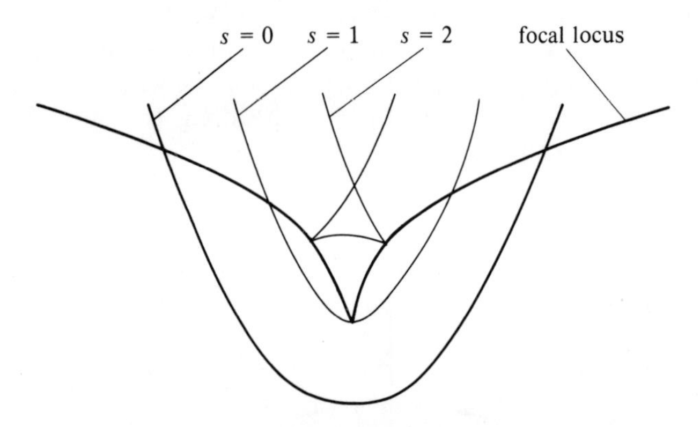

Figure 16.4 The focal locus of a parabola as the set of singularities of an advancing wavefront.

Exercises

16.1. Let $\varphi: \mathbb{R} \to \mathbb{R}^2$ be the ellipse $\varphi(t) = (a \cos t, b \sin t)$ $(a > 0, b > 0)$.

(a) Show that the focal locus of φ is the image of the parametrized curve

$$\alpha(t) = \left(\frac{a^2 - b^2}{a} \cos^3 t, \frac{b^2 - a^2}{b} \sin^3 t\right).$$

(b) Sketch the focal locus of φ.

16.2. Let C be an oriented plane curve and let $p \in C$ be such that the curvature $\kappa(p)$ of C at p is not zero.

(a) Show that for $q \in C$ sufficiently close to p, the normal lines to C at p and at q intersect at some point $h(q) \in \mathbb{R}^2$.
(b) Show that as q approaches p along C, the point $h(q)$ approaches the focal point of C along the normal line through p. [Thus the focal locus of C is the "envelope" of the family of normal lines of C (see Figure 16.5).]

16.3. Let $\varphi: I \to \mathbb{R}^2$ be a regular parametrized curve in $\mathbb{R}^2$ with curvature κ nowhere zero. For $t \in I$, let

$$\alpha(t) = \varphi(t) + (1/\kappa(t))N^{\varphi}(t),$$

so that α is a parametrization of the focal locus of φ.

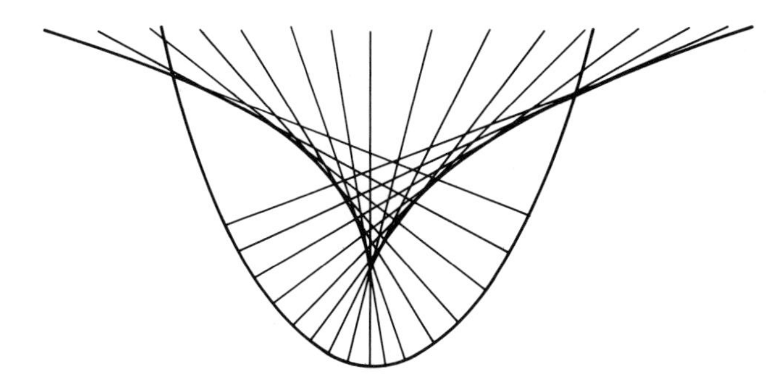

Figure 16.5 The focal locus of a parabola as the envelope of its normal lines.

(a) Show that α is regular ($\dot{\alpha}(t) \neq \mathbf{0}$) at $t \in I$ if and only if $\kappa'(t) \neq 0$.
(b) Show that, for each $t \in I$ with $\kappa'(t) \neq 0$, the normal line to Image φ at $\varphi(t)$ is tangent at $\alpha(t)$ to the focal locus of φ (see Figure 16.5).
(c) Show that on any subinterval $[t_1, t_2] \subset I$ such that $\kappa'(t) \neq 0$ for $t_1 < t < t_2$, the length of the line segment from $\varphi(t)$ to $\alpha(t)$ plus the length of α from $\alpha(t)$ to $\alpha(t_2)$ is constant as a function of t (see Figure 16.6).

[*Remark.* Exercise 16.3 shows that any portion of a plane curve which has regular focal locus can be traced out by unwinding a string from the focal locus (Figure 16.6). The focal locus of a plane curve C is the locus of the centers of curvature and is often called the *evolute* of C. A curve which is obtained from another by unwinding a string is called an *involute*.]

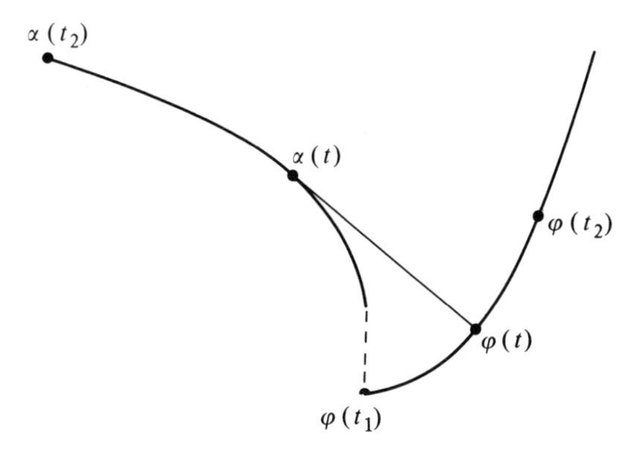

Figure 16.6 Half a parabola as an involute of its focal locus.

16.4. Let $\varphi\colon I \to \mathbb{R}^2$ be a regular parametrized curve in $\mathbb{R}^2$ and let $t_0 \in I$. For each $s \in \mathbb{R}$, define $\varphi_s\colon I \to \mathbb{R}^2$ by $\varphi_s(t) = \varphi(t) + sN^{\varphi}(t)$, let I_s denote the largest interval about t_0 on which φ_s is regular, and let $\kappa_s\colon I_s \to \mathbb{R}$ denote the curvature of the restriction of φ_s to I_s.

(a) Show that I_s is an open interval about t_0 for each $s < 1/\kappa(t_0)$, where κ is the curvature of φ.

(b) Show that, for $s < 1/\kappa(t_0)$,

$$\kappa_s(t_0) = \frac{1}{\dfrac{1}{\kappa(t_0)} - s}$$

and conclude that $\lim_{s \to 1/\kappa(t_0)} |\kappa_s(t_0)| = \infty$. [*Hint*: You may assume for ease of computation that φ is a unit speed curve.]

16.5. Let S be an oriented n-surface in $\mathbb{R}^{n+1}$. For $p \in S$ and $\mathbf{v} \in S_p$ ($\mathbf{v} \neq 0$), let $\alpha: I \to S$ be a parametrized curve in S such that $\dot{\alpha}(t_0) = \mathbf{v}$ and define $\psi: \mathbb{R} \times I \to \mathbb{R}^{n+1}$ by $\psi(s, t) = \alpha(t) + sN^S(\alpha(t))$. Let $\mathbf{X}$ denote the vector field, along the line β normal to S at p ($\beta(s) = p + sN^S(p)$), defined by $\mathbf{X}(s) = d\psi(s, 0, 0, 1)$ (see Figure 16.7).

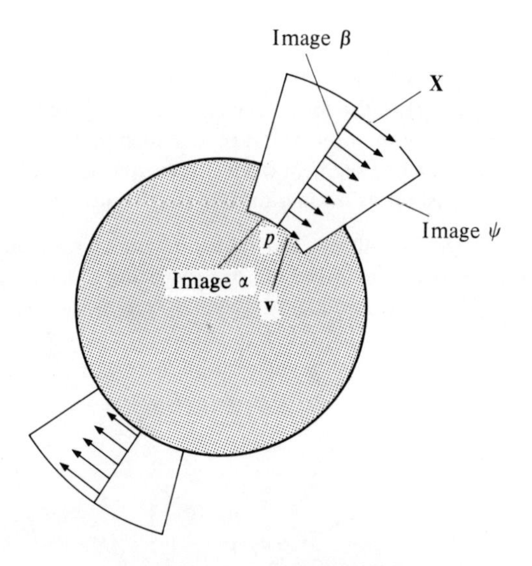

Figure 16.7 A Jacobi field.

(a) Show that $\mathbf{X}(0) = \mathbf{v}$ and that $\dot{\mathbf{X}}(0) = -L_p(\mathbf{v})$. [*Hint*: The vector part of $\dot{\mathbf{X}}$ is $(\partial/\partial s)(\partial\psi/\partial t)$.]

(b) Show that $\ddot{\mathbf{X}} = \mathbf{0}$ and conclude that $\mathbf{X}(s) = (\beta(s), v + sw)$ where v and w are the vector parts of $\mathbf{v}$ and of $-L_p(\mathbf{v})$ respectively.

(c) Show that $\mathbf{X}(s) = 0$ if and only if $\beta(s)$ is a focal point of S along β and $\mathbf{v}$ points in a principal curvature direction corresponding to the principal curvature $1/s$.

Remark. It follows from (b) that the vector field $\mathbf{X}$ does not depend on the choice of parametrized curve α with initial velocity $\mathbf{v}$. X is called the *Jacobi field* along β generated by $\mathbf{v}$.

Surface Area and Volume 17

We consider now the problem of how to find the volume (area when $n = 2$) of an n-surface in $\mathbb{R}^{n+1}$. As with the length of plane curves, this is done in two steps. First we define the volume of a parametrized n-surface and then we define the volume of an n-surface in terms of local parametrizations.

Recall that the length of a parametrized curve $\alpha\colon I \to \mathbb{R}^2$ is defined by the formula

$$l(\alpha) = \int_I \|\dot{\alpha}\| = \int_a^b \|\dot{\alpha}(t)\|\, dt$$

where a and b are the endpoints of I. In the language of parametrized surfaces, if α is regular then the velocity field $\dot{\alpha}$ is just the coordinate vector field $\mathbf{E}_1$ along the parametrized 1-surface α in $\mathbb{R}^2$, and

$$\|\dot{\alpha}(t)\| = \|\mathbf{E}_1(t)\| = \|\mathbf{E}_1(t)\| \det\begin{pmatrix} \mathbf{E}_1(t)/\|\mathbf{E}_1(t)\| \\ \mathbf{N}(t) \end{pmatrix} = \det\begin{pmatrix} \mathbf{E}_1(t) \\ \mathbf{N}(t) \end{pmatrix}$$

where $\mathbf{N}$ is the orientation vector field along α. The second equality here is a consequence of the fact that the vectors $\mathbf{E}_1(t)/\|\mathbf{E}_1(t)\|$ and $\mathbf{N}(t)$ form an orthonormal basis for $\mathbb{R}^2_{\alpha(t)}$ so

$$\det\begin{pmatrix} \mathbf{E}_1(t)/\|\mathbf{E}_1(t)\| \\ \mathbf{N}(t) \end{pmatrix}$$

is the determinant of an orthogonal matrix and hence is equal to ± 1; the sign is $+$ because the basis $\mathbf{E}_1(t)/\|\mathbf{E}_1(t)\|$ is consistant with the orientation $\mathbf{N}$. The formula for the length of α can now be rewritten as

$$l(\alpha) = \int_I \det\begin{pmatrix} \mathbf{E}_1 \\ \mathbf{N} \end{pmatrix}.$$

This integral is clearly a special case of an integral defined for parametrized n-surfaces.

The *volume* of a parametrized n-surface $\varphi\colon U \to \mathbb{R}^{n+1}$ is defined to be the integral

$$V(\varphi) = \int_U \det\begin{pmatrix} \mathbf{E}_1 \\ \vdots \\ \mathbf{E}_n \\ \mathbf{N} \end{pmatrix} = \int_U \det\begin{pmatrix} \mathbf{E}_1(u_1, \ldots, u_n) \\ \vdots \\ \mathbf{E}_n(u_1, \ldots, u_n) \\ \mathbf{N}(u_1, \ldots, u_n) \end{pmatrix} du_1 \cdots du_n$$

where $\mathbf{E}_1, \ldots, \mathbf{E}_n$ are the coordinate vector fields along φ and $\mathbf{N}$ is the orientation vector field along φ. When $n = 1$, the volume of φ is usually called the *length* of φ and is denoted $l(\varphi)$. When $n = 2$, the volume of φ is usually called the *area* of φ and is denoted $A(\varphi)$. Note that, since the volume integrand

$$\det\begin{pmatrix} \mathbf{E}_1 \\ \vdots \\ \mathbf{E}_n \\ \mathbf{N} \end{pmatrix}$$

is everywhere positive, $V(\varphi) > 0$. $V(\varphi)$ may be $+\infty$.

An intuitive explanation of why this particular integral should measure volume is that the integrand measures "volume magnification" along φ (see Figure 17.1).

It is convenient to have a formula for volume which does not require calculation of the orientation vector field $\mathbf{N}$.

Theorem 1. *Let $\varphi\colon U \to \mathbb{R}^{n+1}$ be a parametrized n-surface. Then*

$$V(\varphi) = \int_U (\det(\mathbf{E}_i \cdot \mathbf{E}_j))^{1/2}.$$

PROOF.

$$\left[\det\begin{pmatrix} \mathbf{E}_1 \\ \vdots \\ \mathbf{E}_n \\ \mathbf{N} \end{pmatrix}\right]^2 = \det\begin{pmatrix} E_1 \\ \vdots \\ E_n \\ N \end{pmatrix} \det\begin{pmatrix} E_1 \\ \vdots \\ E_n \\ N \end{pmatrix}^t \qquad (t = \text{transpose})$$

$$= \det\left[\begin{pmatrix} E_1 \\ \cdots \\ E_n \\ N \end{pmatrix} (E_1^t \cdots E_n^t\ N^t)\right]$$

$$= \det\begin{pmatrix} E_1 \cdot E_1 & \cdots & E_1 \cdot E_n & E_1 \cdot N \\ \vdots & & \vdots & \vdots \\ E_n \cdot E_1 & \cdots & E_n \cdot E_n & E_n \cdot N \\ N \cdot E_1 & \cdots & N \cdot E_n & N \cdot N \end{pmatrix}$$

$$= \det \begin{pmatrix} \mathbf{E}_1 \cdot \mathbf{E}_1 & \cdots & \mathbf{E}_1 \cdot \mathbf{E}_n & 0 \\ \vdots & & \vdots & \vdots \\ \mathbf{E}_n \cdot \mathbf{E}_1 & \cdots & \mathbf{E}_n \cdot \mathbf{E}_n & 0 \\ 0 & \cdots & 0 & 1 \end{pmatrix}$$

$$= \det(\mathbf{E}_i \cdot \mathbf{E}_j).$$

Taking the square root of both sides and integrating over U yields the required formula. □

Remark. The functions $g_{ij} = \mathbf{E}_i \cdot \mathbf{E}_j\colon\ U \to \mathbb{R}$ are called the *metric coefficients* along φ. The determinant $\det(g_{ij})$ is usually denoted, in the differential geometry literature, by the letter g. The volume integral then takes the form $V(\varphi) = \int_U \sqrt{g}$.

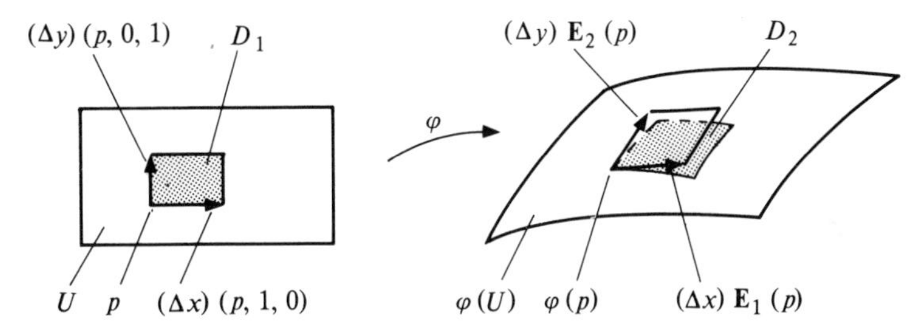

Figure 17.1 Area magnification along a parametrized 2-surface φ in $\mathbb{R}^3$. The small shaded rectangle D_1 in U with area $(\Delta x)(\Delta y)$ is mapped by φ to the small shaded region D_2 in $\varphi(U)$. The area of D_2 is closely approximated by the area of the parallelogram in $\mathbb{R}^3_{\varphi(p)}$ spanned by $d\varphi((\Delta x)(p, 1, 0)) = (\Delta x)\mathbf{E}_1(p)$ and $d\varphi((\Delta y)(p, 0, 1)) = (\Delta y)\mathbf{E}_2(p)$. The area of this latter parallelogram is

$$\begin{aligned} \|(\Delta x)\mathbf{E}_1(p) \times (\Delta y)\mathbf{E}_2(p)\| &= (\Delta x)(\Delta y)\mathbf{E}_1(p) \times \mathbf{E}_2(p) \cdot \mathbf{N}(p) \\ &= (\Delta x)(\Delta y)\det\begin{pmatrix} \mathbf{E}_1(p) \\ \mathbf{E}_2(p) \\ \mathbf{N}(p) \end{pmatrix}. \end{aligned}$$

The ratio of areas of the two shaded regions is

$$\frac{A(D_2)}{A(D_1)} = \det\begin{pmatrix} \mathbf{E}_1(p) \\ \mathbf{E}_2(p) \\ \mathbf{N}(p) \end{pmatrix} + \varepsilon(p, \Delta x, \Delta y)$$

where $\lim_{(\Delta x, \Delta y) \to 0} \varepsilon(p, \Delta x, \Delta y) = 0$. The limiting ratio

$$\lim_{(\Delta x, \Delta y) \to 0} \frac{A(D_2)}{A(D_1)} = \det\begin{pmatrix} \mathbf{E}_1(p) \\ \mathbf{E}_2(p) \\ \mathbf{N}(p) \end{pmatrix}$$

measures area magnification at p under φ; its integral over U gives the area of φ.

EXAMPLE. Let $\varphi\colon U \to \mathbb{R}^3$ be defined by

$$\varphi(\theta, \phi) = (r\cos\theta\sin\phi,\ r\sin\theta\sin\phi,\ r\cos\phi)$$

where $U = \{(\theta, \phi) \in \mathbb{R}^2\colon -\pi < \theta < \pi,\ 0 < \phi < \pi\}$. Thus φ is the spherical coordinate parametrization of the sphere of radius r in $\mathbb{R}^3$ with half of a great circle removed. Its area can be found by either of the above formulae:

$$A(\varphi) = \int_U \det\begin{pmatrix} \mathbf{E}_1 \\ \mathbf{E}_2 \\ \mathbf{N} \end{pmatrix}$$

$$= \int_0^\pi \int_{-\pi}^\pi \begin{vmatrix} -r\sin\theta\sin\phi & r\cos\theta\sin\phi & 0 \\ r\cos\theta\cos\phi & r\sin\theta\cos\phi & -r\sin\phi \\ -\cos\theta\sin\phi & -\sin\theta\sin\phi & -\cos\phi \end{vmatrix} d\theta\, d\phi$$

$$= \int_0^\pi \int_{-\pi}^\pi r^2 \sin\phi\, d\theta\, d\phi = 4\pi r^2$$

or

$$A(\varphi) = \int_U (\det(\mathbf{E}_i \cdot \mathbf{E}_j))^{1/2} = \int_0^\pi \int_{-\pi}^\pi \begin{vmatrix} r^2\sin^2\phi & 0 \\ 0 & r^2 \end{vmatrix}^{1/2} d\theta\, d\phi$$

$$= \int_0^\pi \int_{-\pi}^\pi r^2 \sin\phi\, d\theta\, d\phi = 4\pi r^2.$$

The formula of Theorem 1 allows us also to define the volume of parametrized n-surfaces in $\mathbb{R}^{n+k}$ for all $k \geq 0$. Even more generally, it enables us to attach an n-dimensional volume to any smooth map $\varphi\colon U \to \mathbb{R}^{n+k}$, U open in $\mathbb{R}^n$. A smooth map $\varphi\colon U \to \mathbb{R}^{n+k}$, U open in $\mathbb{R}^n$, is called a *singular n-surface* in $\mathbb{R}^{n+k}$. The adjective "singular" is used to emphasize that φ is not required to be regular; i.e., $d\varphi_p$ may be singular for some (or, even, for all) $p \in U$. Note that each parametrized n-surface is a singular n-surface but that singular n-surfaces are not, in general, parametrized n-surfaces. The *volume* $V(\varphi)$ of a singular n-surface $\varphi\colon U \to \mathbb{R}^{n+k}$ is defined by

$$V(\varphi) = \int_U (\det(\mathbf{E}_i \cdot \mathbf{E}_j))^{1/2}$$

where, as usual, the $\mathbf{E}_i$ are the coordinate vector fields along φ.

EXAMPLE. Let $\varphi\colon U \to \mathbb{R}^3$ be the parametrized 3-surface defined by

$$\varphi(r, \theta, \phi) = (r\cos\theta\sin\phi,\ r\sin\theta\sin\phi,\ r\cos\phi)$$

where $U = \{(r, \theta, \phi)\colon 0 < r < a,\ 0 < \theta < 2\pi,\ 0 < \phi < \pi\}$. φ maps U one to one onto the open ball of radius a about the origin in $\mathbb{R}^3$ with a half disc

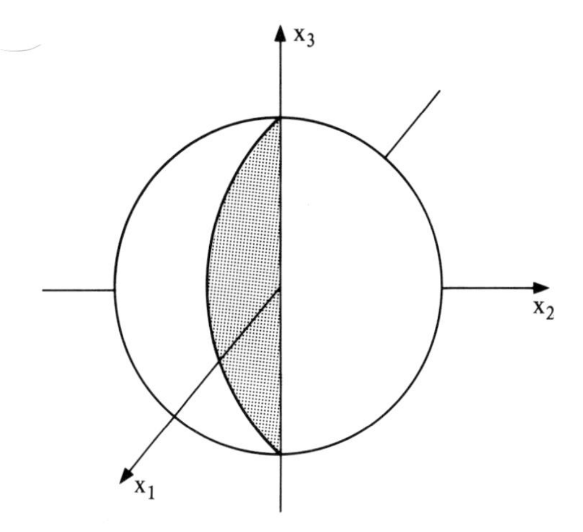

Figure 17.2 The open ball $B = \{(x_1, x_2, x_3) \in \mathbb{R}^3 : x_1^2 + x_2^2 + x_3^2 < a^2\}$ with the half disc $\{(x_1, x_2, x_3) \in B : x_1 \geq 0, x_2 = 0\}$ deleted.

removed (see Figure 17.2). For $p = (r, \theta, \phi) \in U$, we find

$$\mathbf{E}_1(p) = \left(p, \frac{\partial \varphi}{\partial r}(p)\right) = (p, \cos\theta \sin\phi, \sin\theta, \sin\phi, \cos\phi)$$

$$\mathbf{E}_2(p) = \left(p, \frac{\partial \varphi}{\partial \theta}(p)\right) = (p, -r\sin\theta\sin\phi, r\cos\theta\sin\phi, 0)$$

$$\mathbf{E}_3(p) = \left(p, \frac{\partial \varphi}{\partial \phi}(p)\right) = (p, r\cos\theta\cos\phi, r\sin\theta\cos\phi, -r\sin\phi)$$

so

$$\begin{aligned} V(\varphi) &= \int_U (\det(\mathbf{E}_i \cdot \mathbf{E}_j))^{1/2} \\ &= \int_0^\pi \int_0^{2\pi} \int_0^a \begin{vmatrix} 1 & 0 & 0 \\ 0 & r^2 \sin^2 \phi & 0 \\ 0 & 0 & r^2 \end{vmatrix}^{1/2} dr\, d\theta\, d\phi \\ &= \int_0^\pi \int_0^{2\pi} \int_0^a r^2 \sin\phi \, dr\, d\theta\, d\phi = \frac{4}{3}\pi a^3. \end{aligned}$$

Theorem 2. *Let $\varphi: U \to \mathbb{R}^{n+1}$ be a parametrized n-surface and let $N: U \to \mathbf{S}^n$ denote its Gauss map ($\mathbf{N}(p) = (\varphi(p), N(p))$ for all $p \in U$, where $\mathbf{N}$ is the orientation vector field along φ). Then*

$$V(N) = \int_U |K| \det(\mathbf{E}_i^\varphi \cdot \mathbf{E}_j^\varphi)^{1/2}$$

where $K: U \to \mathbb{R}$ is the Gauss-Kronecker curvature of φ and the $\mathbf{E}_i^\varphi$ are the coordinate vector fields along φ.

PROOF. The coordinate vector field $\mathbf{E}_i^N$ of the singular n-surface N has vector part at $p \in U$ equal to $\partial N/\partial x_i(p)$, which is the same as the vector part of

$$\nabla_{\mathbf{e}_i}\mathbf{N} = -L_p(d\varphi_p(\mathbf{e}_i)) = -L_p(\mathbf{E}_i^\varphi(p)) = -\sum_{k=1}^{n} a_{ki}(p)\mathbf{E}_k^\varphi(p)$$

where $\mathbf{e}_i = (p, 0, \ldots, 1, \ldots, 0)$ (the 1 in the $(i+1)$th spot), L_p is the Weingarten map of φ at p, and $(a_{ij}(p))$ is the matrix for L_p with respect to the basis $\{\mathbf{E}_i^\varphi(p)\}$ for the tangent space Image $d\varphi_p$. Hence

$$\begin{aligned}
\det(\mathbf{E}_i^N \cdot \mathbf{E}_j^N) &= \det\left(\sum_k a_{ki}\mathbf{E}_k^\varphi \cdot \sum_l a_{lj}\mathbf{E}_l^\varphi\right) \\
&= \det\left(\sum_{k,l} a_{ki}a_{lj}\mathbf{E}_k^\varphi \cdot \mathbf{E}_l^\varphi\right) \\
&= (\det(a_{ij}))^2 \det(\mathbf{E}_i^\varphi \cdot \mathbf{E}_j^\varphi) \\
&= K^2 \det(\mathbf{E}_i^\varphi \cdot \mathbf{E}_j^\varphi).
\end{aligned}$$

Taking the square root of both sides and integrating, we obtain

$$V(N) = \int_U (\det(\mathbf{E}_i^N \cdot \mathbf{E}_j^N))^{1/2} = \int_U |K| \det(\mathbf{E}_i^\varphi \cdot \mathbf{E}_j^\varphi)^{1/2}. \qquad \square$$

Corollary. *The Gauss-Kronecker curvature at $p \in U$ of a parametrized n-surface $\varphi: U \to \mathbb{R}^{n+1}$ has absolute value*

$$|K(p)| = \lim_{\varepsilon \to 0} V(N|_{B_\varepsilon})/V(\varphi|_{B_\varepsilon})$$

where $B_\varepsilon = \{q \in U: \|q - p\| < \varepsilon\}$.

PROOF. By Theorem 2 and the mean value theorem for integrals,

$$\frac{V(N|_{B_\varepsilon})}{V(\varphi|_{B_\varepsilon})} = \frac{|K(p_1)| \det(\mathbf{E}_i^\varphi(p_1) \cdot \mathbf{E}_j^\varphi(p_1))^{1/2} \int_{B_\varepsilon} 1}{(\det(\mathbf{E}_i^\varphi(p_2) \cdot \mathbf{E}_j^\varphi(p_2)))^{1/2} \int_{B_\varepsilon} 1}$$

for some $p_1, p_2 \in B_\varepsilon$. Taking the limit as $\varepsilon \to 0$ completes the proof. $\square$

Remark. This Corollary provides us with a geometric interpretation of the *magnitude* of the Gauss-Kronecker curvature K of an oriented n-surface $S \subset \mathbb{R}^{n+1}$ in terms of volume magnification under the Gauss map N. The significance of the *sign* of K is as follows. Taking $\mathbf{Z} = \mathbf{N}$ in Theorem 5 of Chapter 12 and using the fact that $dN(\mathbf{v})$ and $\nabla_{\mathbf{v}}\mathbf{N}$ have the same vector part for all $\mathbf{v} \in S_p$, $p \in S$, we find that

$$K(p) = (-1)^n \det\begin{pmatrix} \nabla_{\mathbf{v}_1}\mathbf{N} \\ \vdots \\ \nabla_{\mathbf{v}_n}\mathbf{N} \\ \mathbf{N}(p) \end{pmatrix} \Big/ \det\begin{pmatrix} \mathbf{v}_1 \\ \vdots \\ \mathbf{v}_n \\ \mathbf{N}(p) \end{pmatrix} = \det\begin{pmatrix} dN(\mathbf{v}_1) \\ \vdots \\ dN(\mathbf{v}_n) \\ \mathbf{N}^{S^n}(N(p)) \end{pmatrix} \Big/ \det\begin{pmatrix} \mathbf{v}_1 \\ \vdots \\ \mathbf{v}_n \\ \mathbf{N}(p) \end{pmatrix}$$

where $\mathbf{v}_1, \ldots, \mathbf{v}_n$ is any basis for S_p and $\mathbf{N}^{S^n}$ is the *standard orientation* on S^n, defined by $\mathbf{N}^{S^n}(q) = (q, (-1)^n q)$. Thus $K(p) > 0$ if and only if dN maps each basis for S_p consistent with the orientation on S to a basis for $S^n_{N(p)}$ consistent with the standard orientation on S^n. Given any smooth map $f\colon S \to \tilde{S}$ from one oriented n-surface in $\mathbb{R}^{n+1}$ to another and any $p \in S$ with $df_p\colon S_p \to \tilde{S}_{f(p)}$ non-singular, f is said to be *orientation preserving* at p if df maps bases for S_p consistent with the orientation on S to bases for $\tilde{S}_{f(p)}$ consistent with the orientation on $\tilde{S}$; otherwise f is said to be *orientation reversing* at p. Thus, the *Gauss-Kronecker curvature $K(p)$ at a point p of an oriented n-surface $S \subset \mathbb{R}^{n+1}$ is positive if and only if the Gauss map $N\colon S \to S^n$ is orientation preserving at p, and $K(p)$ is negative if and only if N is orientation reversing at p.*

Just as the length of a parametrized curve is unchanged under reparametrization, the volume of a singular n-surface is unchanged under reparametrization. A *reparametrization* of a singular n-surface $\varphi\colon U_1 \to \mathbb{R}^{n+k}$ is a singular n-surface $\psi\colon U_2 \to \mathbb{R}^{n+k}$ of the form $\psi = \varphi \circ h$ where $h\colon U_2 \to U_1$ is a smooth map with smooth inverse and Jacobian determinant everywhere positive (see Figure 17.3). Note that any pair of such singular n-surfaces will always have the same images. Furthermore, if φ is a parametrized n-surface and ψ is a reparametrization of φ then ψ is a parametrized n-surface and, if $k = 1$, the orientation vector fields $\mathbf{N}^\varphi$ along φ and $\mathbf{N}^\psi$ along ψ are related by $\mathbf{N}^\psi(p) = \mathbf{N}^\varphi(h(p))$, for all $p \in U_2$ (Exercise 17.11).

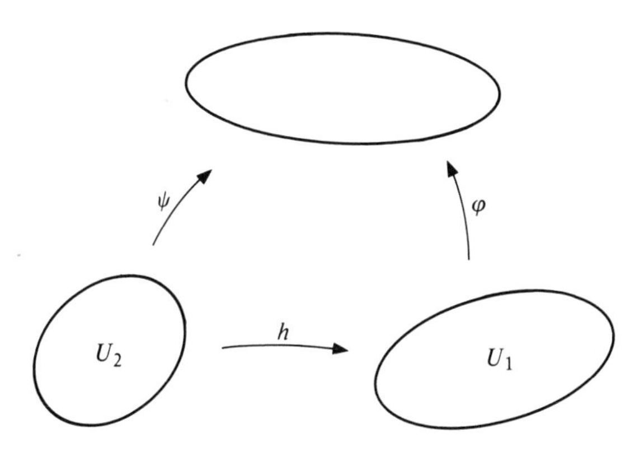

Figure 17.3 ψ is a reparametrization of φ.

Theorem 3. *Let* $\varphi\colon U_1 \to \mathbb{R}^{n+k}$ *be a singular n-surface and let* $\psi = \varphi \circ h\colon U_2 \to \mathbb{R}^{n+k}$ *be a reparametrization of* φ. *Then* $V(\psi) = V(\varphi)$.

PROOF. Let $\mathbf{E}_i^\varphi$ and $\mathbf{E}_i^\psi$ denote, respectively, the coordinate vector fields along φ and along ψ. Then, letting $\mathbf{X}_i$ denote the vector field on $\mathbb{R}^n$ defined by $\mathbf{X}_i(p) = (p, 0, \ldots, 1, \ldots, 0)$ where the 1 is in the $(i+1)$th spot we have, for $p \in U_2$,

$$\begin{aligned}\mathbf{E}_i^\psi(p) &= d\psi(\mathbf{X}_i(p)) = d(\varphi \circ h)(\mathbf{X}_i(p)) = d\varphi(dh(\mathbf{X}_i(p)))\\ &= d\varphi\left(\sum_{k=1}^n h_{ki}(p)\mathbf{X}_k(h(p))\right) = \sum_{k=1}^n h_{ki}(p)\, d\varphi(\mathbf{X}_k(h(p)))\\ &= \sum_{k=1}^n h_{ki}(p)\mathbf{E}_k^\varphi(h(p)),\end{aligned}$$

where the $h_{ij} = \partial h_i/\partial x_j$ are the entries in the Jacobian matrix for h. Hence

$$\mathbf{E}_i^\psi \cdot \mathbf{E}_j^\psi = \sum_{k,\,l=1}^n h_{ki} h_{lj} \mathbf{E}_k^\varphi \circ h \cdot \mathbf{E}_l^\varphi \circ h,$$

and

$$\det(\mathbf{E}_i^\psi \cdot \mathbf{E}_j^\psi) = (\det(h_{ij}))^2 \det(\mathbf{E}_i^\varphi \circ h \cdot \mathbf{E}_j^\varphi \circ h) = J_h^2 \det(\mathbf{E}_i^\varphi \circ h \cdot \mathbf{E}_j^\varphi \circ h),$$

where J_h is the Jacobian determinant of h. By the change of variables theorem for multiple integrals,

$$\begin{aligned}V(\psi) &= \int_{U_2} (\det(\mathbf{E}_i^\psi \cdot \mathbf{E}_j^\psi))^{1/2}\\ &= \int_{h^{-1}(U_1)} (\det(\mathbf{E}_i^\varphi \circ h \cdot \mathbf{E}_j^\varphi \circ h))^{1/2} J_h\\ &= \int_{U_1} \det(\mathbf{E}_i^\varphi \cdot \mathbf{E}_j^\varphi)^{1/2} = V(\varphi).\end{aligned}$$

□

Passing now to oriented n-surfaces S in $\mathbb{R}^{n+1}$, we would like to obtain the volume of S by adding up the volumes of subsets which are images of one to one singular n-surfaces. However, it is generally not possible to express S as a disjoint union of such sets. It is possible to express S as the union of images of closed rectangles which overlap only along the boundaries (see Figure 17.4); the set of boundary points will then contribute nothing to the volume integral so the overlap will not matter and the volume of S can then be defined as the sum of volumes of these "singular rectangles." This procedure, although intuitively quite attractive, is difficult to carry out rigorously. We shall adopt an alternate approach.

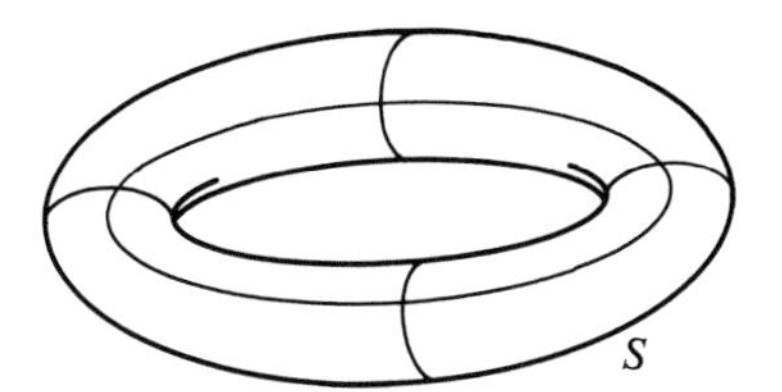

Figure 17.4 The surface S can be expressed as a union of rectangular regions which overlap only along the boundaries of the rectangles.

Consider the volume integrand

$$\det\begin{pmatrix} \mathbf{E}_1 \\ \vdots \\ \mathbf{E}_n \\ \mathbf{N} \end{pmatrix}$$

of a one to one local parametrization $\varphi\colon U \to S$. If we replace φ by a reparametrization $\psi = \varphi \circ h\colon U_2 \to S$, the volume integrand changes in that the coordinate vector fields of φ are replaced by the coordinate vector fields of ψ, but the volume integral does not change. This suggests that essential part of the volume integrand at $p \in S$ is the function ζ which assigns to each ordered set $\{\mathbf{v}_1, \ldots, \mathbf{v}_n\}$ of n vectors in S_p the number

$$\zeta(\mathbf{v}_1, \ldots, \mathbf{v}_n) = \det\begin{pmatrix} \mathbf{v}_1 \\ \vdots \\ \mathbf{v}_n \\ \mathbf{N}(p) \end{pmatrix}.$$

ζ is called the *volume form* on S. We shall see that the volume form can be integrated over a compact oriented n-surface S in $\mathbb{R}^{n+1}$; its integral will be the volume of S. ζ is an example of a differential n-form.

A *differential k-form*, usually called simply a *k-form*, on an n-surface S is a function ω which assigns to each ordered set $\{\mathbf{v}_1, \ldots, \mathbf{v}_k\}$ of k vectors in S_p, $p \in S$, a real number $\omega(\mathbf{v}_1, \ldots, \mathbf{v}_k)$ such that

(i) for each $i \in \{1, \ldots, k\}$ and $\mathbf{v}_1, \ldots, \mathbf{v}_{i-1}, \mathbf{v}_{i+1}, \ldots, \mathbf{v}_k \in S_p$, $p \in S$, the function which sends $\mathbf{v} \in S_p$ to $\omega(\mathbf{v}_1, \ldots, \mathbf{v}_{i-1}, \mathbf{v}, \mathbf{v}_{i+1}, \ldots, \mathbf{v}_k) \in \mathbb{R}$ is linear, and

(ii) for each $\mathbf{v}_1, \ldots, \mathbf{v}_k \in S_p$ and for each permutation σ of the integers $\{1, \ldots, k\}$,

$$\omega(\mathbf{v}_{\sigma(1)}, \ldots, \mathbf{v}_{\sigma(k)}) = (\operatorname{sign} \sigma)\omega(\mathbf{v}_1, \ldots, \mathbf{v}_k).$$

Note that the conditions (i) and (ii) simply express familiar properties of the determinant when $k = n$ and ω is taken to be the volume form ζ. Property (i) is called *multilinearity* and property (ii) is called *skewsymmetry*.

Given a k-form ω and tangent vector fields $\mathbf{X}_1, \ldots, \mathbf{X}_k$ on S, we can define a real valued function $\omega(\mathbf{X}_1, \ldots, \mathbf{X}_k)$ on S by

$$[\omega(\mathbf{X}_1, \ldots, \mathbf{X}_k)](p) = \omega(\mathbf{X}_1(p), \ldots, \mathbf{X}_k(p)).$$

The k-form ω is said to be *smooth* if $\omega(\mathbf{X}_1, \ldots, \mathbf{X}_k)\colon S \to \mathbb{R}$ is smooth whenever the vector fields $\mathbf{X}_1, \ldots, \mathbf{X}_k$ are smooth.

Example 1. A 1-form on S is simply a function $\omega\colon T(S) = \bigcup_{p \in S} S_p \to \mathbb{R}$ such that the restriction of ω to S_p is linear for each $p \in S$. Thus, for example, if $\mathbf{X}$ is a tangent vector field on S then the function $\omega_{\mathbf{X}}\colon T(S) \to \mathbb{R}$ defined by

$$\omega_{\mathbf{X}}(\mathbf{v}) = \mathbf{X}(p) \cdot \mathbf{v} \qquad (\mathbf{v} \in S_p,\, p \in S)$$

is a 1-form on S. $\omega_{\mathbf{X}}$ is called the 1-form *dual* to $\mathbf{X}$.

Example 2. Let ω_1 and ω_2 be 1-forms on S. Then the function $\omega_1 \wedge \omega_2$ defined by

$$(\omega_1 \wedge \omega_2)(\mathbf{v}_1, \mathbf{v}_2) = \omega_1(\mathbf{v}_1)\omega_2(\mathbf{v}_2) - \omega_1(\mathbf{v}_2)\omega_2(\mathbf{v}_1)$$

is a 2-form on S. $\omega_1 \wedge \omega_2$ is the *exterior product* of ω_1 and ω_2.

Example 3. The volume form ζ on an oriented n-surface S in $\mathbb{R}^{n+1}$ is a smooth n-form on S.

Example 4. Let ω be a k-form $(k > 1)$ on S and let $\mathbf{X}$ be a tangent vector field on S. Then the function $\mathbf{X} \lrcorner\, \omega$ defined by

$$(\mathbf{X} \lrcorner\, \omega)(\mathbf{v}_1, \ldots, \mathbf{v}_{k-1}) = \omega(\mathbf{X}(p), \mathbf{v}_1, \ldots, \mathbf{v}_{k-1})$$

is a $(k-1)$-form on S. $\mathbf{X} \lrcorner\, \omega$ is the *interior product* of $\mathbf{X}$ and ω.

Example 5. Let ω be a k-form on an m-surface $\tilde{S}$ and let $f\colon S \to \tilde{S}$ be a smooth map from the n-surface S into $\tilde{S}$. Then the function $f^*\omega$ defined by

$$(f^*\omega)(\mathbf{v}_1, \ldots, \mathbf{v}_k) = \omega(df(\mathbf{v}_1), \ldots, df(\mathbf{v}_k))$$

is a k-form on S. $f^*\omega$ is called the *pull-back* of ω under f.

The *sum* of two k-forms ω_1 and ω_2 on an n-surface S is the k-form $\omega_1 + \omega_2$ defined by

$$(\omega_1 + \omega_2)(\mathbf{v}_1, \ldots, \mathbf{v}_k) = \omega_1(\mathbf{v}_1, \ldots, \mathbf{v}_k) + \omega_2(\mathbf{v}_1, \ldots, \mathbf{v}_k).$$

The *product* of a function $f\colon S \to \mathbb{R}$ and a k-form ω on S is the k-form $f\omega$ on S defined by

$$(f\omega)(\mathbf{v}_1, \ldots, \mathbf{v}_k) = f(p)\omega(\mathbf{v}_1, \ldots, \mathbf{v}_k),$$

for $\mathbf{v}_1, \ldots, \mathbf{v}_k \in S_p$, $p \in S$. Note that the sum of two smooth k-forms is smooth and that the product of a smooth function and a smooth k-form is smooth.

For ω a smooth k-form on an n-surface S and φ a singular k-surface in S (i.e., with Image $\varphi \subset S$), the *integral* of ω over φ is defined by

$$\int_{\varphi} \omega = \int_{U} \omega(\mathbf{E}_1^{\varphi}, \ldots, \mathbf{E}_k^{\varphi}),$$

provided the latter integral exists. The vector fields $\mathbf{E}_i^{\varphi}$ are of course the coordinate vector fields along φ. The integral will exist, in particular, whenever the function $\omega(\mathbf{E}_1^{\varphi}, \ldots, \mathbf{E}_k^{\varphi})$ is identically zero outside some compact subset of U.

EXAMPLE. For φ a local parametrization of S and ζ the volume form on S, $\int_{\varphi} \zeta = V(\varphi)$.

Note that if ψ is a reparametrization of the singular k-surface φ in S and ω is a smooth k-form on S such that $\int_{\varphi} \omega$ exists then $\int_{\psi} \omega$ exists and $\int_{\psi} \omega = \int_{\varphi} \omega$. Indeed, if $\varphi\colon U_1 \to S$ and $\psi = \varphi \circ h\colon U_2 \to S$ then, as in the proof of Theorem 3, $\mathbf{E}_i^{\psi} = \sum_{j=1}^{k} h_{ji} \mathbf{E}_j^{\varphi} \circ h$ so

$$\begin{aligned} \int_{\psi} \omega &= \int_{U_2} \omega(\mathbf{E}_1^{\psi}, \ldots, \mathbf{E}_k^{\psi}) \\ &= \int_{U_2} \omega\left(\sum_{j_1=1}^{k} h_{j_1 1} \mathbf{E}_{j_1}^{\varphi} \circ h, \ldots, \sum_{j_k=1}^{k} h_{j_k k} \mathbf{E}_{j_k}^{\varphi} \circ h \right) \\ &= \int_{U_2} \sum_{j_1 \cdots j_k} h_{j_1 1} \cdots h_{j_k k} \omega(\mathbf{E}_{j_1}^{\varphi} \circ h, \ldots, \mathbf{E}_{j_k}^{\varphi} \circ h), \end{aligned}$$

where the last equality is a consequence of the multilinearity of ω. By the skewsymmetry of ω, $\omega(\mathbf{E}_{j_1}^{\varphi} \circ h, \ldots, \mathbf{E}_{j_k}^{\varphi} \circ h)$ is zero whenever two or more of the j_i's are equal, and $\omega(\mathbf{E}_{j_1}^{\varphi}, \ldots, \mathbf{E}_{j_k}^{\varphi}) = (\operatorname{sign} \sigma)\omega(\mathbf{E}_1^{\varphi}, \ldots, \mathbf{E}_k^{\varphi})$ whenever $j_1 = \sigma(1), \ldots, j_k = \sigma(k)$ for some permutation σ of $\{1, \ldots, k\}$. Hence

$$\begin{aligned} \int_{\psi} \omega &= \int_{U_2} \sum_{\sigma} (\operatorname{sign} \sigma) h_{\sigma(1)1} \cdots h_{\sigma(k)k} \omega(\mathbf{E}_1^{\varphi} \circ h, \ldots, \mathbf{E}_k^{\varphi} \circ h) \\ &= \int_{h^{-1}(U_1)} (\omega(\mathbf{E}_1^{\varphi}, \ldots, \mathbf{E}_k^{\varphi}) \circ h) J_h \\ &= \int_{U_1} \omega(\mathbf{E}_1^{\varphi}, \ldots, \mathbf{E}_k^{\varphi}) = \int_{\varphi} \omega, \end{aligned}$$

as claimed.

We now proceed to define the integral of an n-form ω over a compact oriented n-surface S. This is done by expressing the n-form as a sum of n-forms ω_i each of which is identically zero outside the image of some one to one local parametrization φ_i and then defining $\int_S \omega = \sum_i \int_{\varphi_i} \omega_i$. The n-forms ω_i will be obtained as products of ω with functions f_i with the property that $\sum_i f_i = 1$.

A *partition of unity* on an n-surface S is a finite collection of smooth functions $f_i: S \to \mathbb{R}$ $(i \in \{1, \ldots, m\})$ such that

(i) $f_i(q) \geq 0$ for all $i \in \{1, \ldots, m\}$ and all $q \in S$,
(ii) for each $i \in \{1, \ldots, m\}$ there exists a one to one local parametrization $\varphi_i: U_i \to S$ such that f_i is identically zero outside the image under φ_i of a compact subset of U_i, and
(iii) $\sum_{i=1}^{m} f_i(q) = 1$ for all $q \in S$.

If $\{\varphi_i\}$ is any collection of one to one local parametrizations satisfying (ii), the partition of unity $\{f_i\}$ is said to be *subordinate* to $\{\varphi_i\}$.

Remark. The requirement that the collection of functions $\{f_i\}$ be finite can be replaced by a less restrictive requirement called local finiteness, but finite partitions of unity will be adequate for our needs.

Theorem 4. *Let S be a compact n-surface in $\mathbb{R}^{n+1}$. Then there exists a partition of unity on S.*

PROOF. First we construct, for each $p \in S$, a "bump function" $g_p: S \to \mathbb{R}$ as follows. Given $p \in S$, let $\varphi_p: U_p \to S$ be a one to one local parametrization of S whose image is an open set about p in S, as constructed in the proof of Theorem 1 of Chapter 15. Choose $r_p > 0$ such that

$$B_p = \{x \in \mathbb{R}^n: \|x - \varphi_p^{-1}(p)\| \leq r_p\} \subset U_p$$

and define the bump function $g_p: S \to \mathbb{R}$ by

$$g_p(q) = \begin{cases} e^{-1/(r_p^2 - \|\varphi_p^{-1}(q) - \varphi_p^{-1}(p)\|^2)} & \text{for } q \in V_p \\ 0 & \text{for } q \notin V_p \end{cases}$$

where

$$V_p = \{q \in \varphi_p(U_p): \|\varphi_p^{-1}(q) - \varphi_p^{-1}(p)\| < r_p\}$$

(see Figure 17.5). Then g_p is smooth (Exercise 17.17), g_p is identically zero outside the image under φ_p of the compact set $B_p \subset U_p$, and $g_p(q) > 0$ for all q in the open set V_p about p in S.

Assuming for the moment that we can find a finite set $\{p_1, \ldots, p_m\}$ of points in S such that $\bigcup_{i=1}^{m} V_{p_i} = S$, we can define $f_i: S \to \mathbb{R}$ for each $i \in \{1, \ldots, m\}$ by

$$f_i(q) = g_{p_i}(q) \Big/ \sum_{j=1}^{m} g_{p_j}(q).$$

Note that the denominator $\sum_{j=1}^{m} g_{p_j}(q)$ is nowhere zero, since each $g_{p_j}(q) \geq 0$ and $g_{p_j}(q) > 0$ when j is such that $q \in V_{p_j}$. $\{f_i\}$ is then a partition of unity on S, subordinate to $\{\varphi_{p_i}\}$.

Finally, the fact that there exists a finite set $\{p_1, \ldots, p_m\}$ of points in S such that $\bigcup_{i=1}^{m} V_{p_i} = S$ is a consequence of the Heine-Borel theorem: the sets

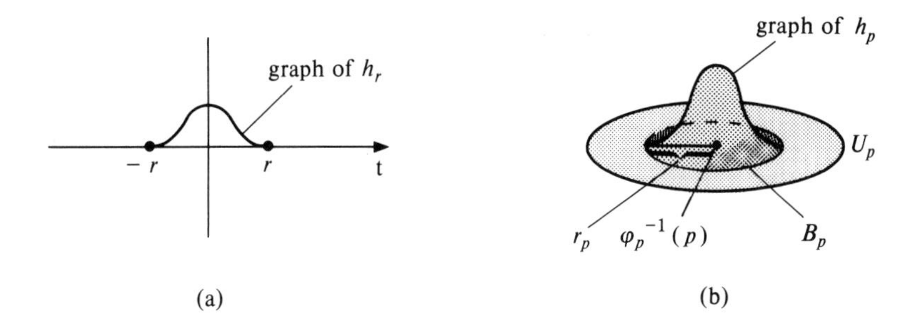

Figure 17.5 Construction of a smooth bump function.

(a) The function $h_r: \mathbb{R} \to \mathbb{R}$ given by

$$h_r(t) = \begin{cases} e^{-1/(r^2-t^2)} & \text{if } |t| < r \\ 0 & \text{if } |t| \geq r \end{cases}$$

is smooth.

(b) The bump function $g_p: S \to \mathbb{R}$ is such that g_p is identically zero outside the image of the local parametrization $\varphi_p: U_p \to S$, and $g_p \circ \varphi_p = h_p$ where $h_p(x) = h_{r_p}(\|x - \varphi_p^{-1}(p)\|)$ for $x \in U_p$.

$\{V_p: p \in S\}$ are open sets in S which cover S ($\bigcup_{p \in S} V_p = S$); the Heine-Borel Theorem (see, e.g., Fleming, *Functions of Several Variables*, Springer-Verlag 1977 (Second Edition)) states that every covering of a compact set S by open sets has a finite subcovering; i.e., there is a finite collection $\{V_{p_1}, \ldots, V_{p_m}\}$ of these sets with $\bigcup_{i=1}^m V_{p_i} = S$. □

The *integral* of a smooth n-form ω over a compact oriented n-surface $S \subset \mathbb{R}^{n+1}$ is defined to be the real number

$$\int_S \omega = \sum_i \int_{\varphi_i} (f_i \omega),$$

where $\{f_i\}$ is any partition of unity on S subordinate to a collection $\{\varphi_i\}$ of one to one local parametrizations of S. Note that $\int_S \omega$ does not depend on the choice of partition of unity because if $\{\tilde{f}_j\}$ is another, subordinate to local parametrizations $\{\tilde{\varphi}_j\}$, then

$$\begin{aligned}
\sum_j \int_{\tilde{\varphi}_j} (\tilde{f}_j \omega) &= \sum_j \int_{\tilde{\varphi}_j} \Big(\sum_i f_i\Big) \tilde{f}_j \omega = \sum_j \sum_i \int_{\tilde{\varphi}_j} (f_i \tilde{f}_j \omega) \\
&\overset{(1)}{=} \sum_{i,j} \int_{\tilde{\varphi}_{ji}} (f_i \tilde{f}_j \omega) \overset{(2)}{=} \sum_{i,j} \int_{\varphi_{ij}} (\tilde{f}_j f_i \omega) \\
&\overset{(3)}{=} \sum_i \sum_j \int_{\varphi_i} (\tilde{f}_j f_i \omega) = \sum_i \int_{\varphi_i} \Big(\sum_j \tilde{f}_j\Big) f_i \omega = \sum_i \int_{\varphi_i} (f_i \omega),
\end{aligned}$$

where $\tilde{\varphi}_{ji}$ is the restriction of $\tilde{\varphi}_j$ to the open set $\tilde{\varphi}_j^{-1}(V_{ij})$, $V_{ij} = (\text{Image } \tilde{\varphi}_j) \cap (\text{Image } \varphi_i)$, and φ_{ij} is the restriction of φ_i to $\varphi_i^{-1}(V_{ij})$. Equalities (1) and (3) hold because $f_i \tilde{f}_j$ is identically zero outside V_{ij}, and equality (2) holds because φ_{ij} is a reparametrization of $\tilde{\varphi}_{ji}$ (see Exercise 17.18).

Using integration of forms, we can now define the *volume* of a compact oriented n-surface $S \subset \mathbb{R}^{n+1}$ to be the integral over S of its volume form ζ:

$$V(S) = \int_S \zeta.$$

We can also now define the *integral* over S of any smooth function $f\colon S \to \mathbb{R}$ by

$$\int_S f = \int_S f\zeta.$$

EXERCISES

17.1. Find the area of the parametrized "cylinder in $\mathbb{R}^3$ with a line removed" given by $\varphi(\theta, t) = (r \cos \theta, r \sin \theta, t)$, $0 < \theta < 2\pi$, $0 < t < h$.

17.2. Find the area of the parametrized "cone in $\mathbb{R}^3$ with a line removed" given by $\varphi(\theta, t) = (tr \cos \theta, tr \sin \theta, (1 - t)h)$, $0 < \theta < 2\pi$, $0 < t < 1$.

17.3. Find the area of the parametrized "torus in $\mathbb{R}^3$ with two circles removed" given by

$$\varphi(\theta, \phi) = ((a + b \cos \phi)\cos \theta, (a + b \cos \phi)\sin \theta, b \sin \phi),$$
$$0 < \theta < 2\pi,\ 0 < \phi < 2\pi.$$

17.4. Find the area of the parametrized "torus in $\mathbb{R}^4$ with two circles removed" given by

$$\varphi(\theta, \phi) = (a \cos \theta, a \sin \theta, b \cos \phi, b \sin \phi), \qquad 0 < \theta < 2\pi,\ 0 < \phi < 2\pi.$$

17.5. Find the volume of the parametrized "unit 3-sphere in $\mathbb{R}^4$ with part of a 2-sphere removed" given by

$$\varphi(\phi, \theta, \psi) = (\sin \phi \sin \theta \sin \psi, \cos \phi \sin \theta \sin \psi, \cos \theta \sin \psi, \cos \psi),$$
$$0 < \phi < 2\pi,\ 0 < \theta < \pi,\ 0 < \psi < \pi.$$

17.6. Show that the area of the parametrized surface of revolution $\varphi(t, \theta) = (x(t), y(t)\cos \theta, y(t)\sin \theta)$, $a < t < b$, $0 < \theta < 2\pi$, where $y(t) > 0$ for $a < t < b$, is given by the formula

$$A(\varphi) = \int_a^b 2\pi y(t)((x'(t))^2 + (y'(t))^2)^{1/2}\, dt.$$

17.7. Let $g\colon U \to \mathbb{R}$ be a smooth function on the open set $U \subset \mathbb{R}^n$. Define $\varphi\colon U \to \mathbb{R}^{n+1}$ by $\varphi(u_1, \ldots, u_n) = (u_1, \ldots, u_n, g(u_1, \ldots, u_n))$. Show that $V(\varphi) = \int_U (1 + \sum (\partial g/\partial u_i)^2)^{1/2}$.

17.8. Let $\varphi_n: U \to \mathbb{R}^{n+1}$ be defined by

$$\varphi_n(\theta_1, \ldots, \theta_n) = (\sin\theta_1 \sin\theta_2 \cdots \sin\theta_n, \cos\theta_1 \sin\theta_2 \cdots \sin\theta_n, \cos\theta_2 \sin\theta_3 \cdots \sin\theta_n, \ldots, \cos\theta_{n-1}\sin\theta_n, \cos\theta_n),$$

where $U = \{(\theta_1, \ldots, \theta_n) \in \mathbb{R}^n : 0 < \theta_1 < 2\pi, 0 < \theta_i < \pi \text{ for } 2 \le i \le n\}$.

(a) Show that φ_n is a parametrized n-surface.
(b) Show that φ_n maps U one to one onto a subset of the unit n-sphere S^n.
(c) Show that S^n − Image φ_n is contained in the $(n-1)$-sphere $\{(x_1, \ldots, x_{n+1}) \in S^n : x_1 = 0\}$. (It will then follow that $V(\varphi_n) = V(S^n)$ since S^n − Image φ_n has n-dimensional volume zero.)
(d) Find a formula expressing the volume of φ_n as a multiple of the volume of φ_{n-1}. [*Hint*: Introduce a zero into the last corner entry of the matrix

$$\begin{pmatrix} E_1 \\ \vdots \\ E_n \\ N \end{pmatrix}$$

by adding to the last row a suitable multiple of the next to the last row. Then expand the determinant by minors of the last column. Finally pull out all factors of $\sin\theta_n$ and integrate with respect to θ_n.]
(e) Find $V(\varphi_n)$.

17.9. Let $\varphi: U \to \mathbb{R}^3$ be a parametrized 2-surface in $\mathbb{R}^3$. Show that the area of φ is given by the formula $A(\varphi) = \int_U \|\partial\varphi/\partial u_1 \times \partial\varphi/\partial u_2\|$.

17.10. Let $\varphi: U \to \mathbb{R}^{n+1}$ be a parametrized n-surface in $\mathbb{R}^{n+1}$. Let $\mathbf{W}$ be the vector field along φ whose ith component at $p \in U$ is $(-1)^{n+i}$ times the determinant of the matrix obtained by deleting the ith row from the Jacobian matrix of φ at p (or the ith column from the matrix

$$\begin{pmatrix} E_1(p) \\ \vdots \\ E_n(p) \end{pmatrix}).$$

(a) Show that $\mathbf{W}$ is a normal vector field along φ, and that $\mathbf{W}/\|\mathbf{W}\|$ is the orientation vector field along φ.
(b) Show that $V(\varphi) = \int_U \|\mathbf{W}\|$.

17.11. Let $\varphi: U_1 \to \mathbb{R}^{n+1}$ be a parametrized n-surface and let $\psi = \varphi \circ h: U_2 \to \mathbb{R}^{n+1}$ be a reparametrization of φ. Show that $\mathbf{N}^\psi = \mathbf{N}^\varphi \circ h$, where $\mathbf{N}^\psi$ and $\mathbf{N}^\varphi$ are, respectively, the orientation vector fields along φ and along ψ. [*Hint*: Show that

$$\det\begin{pmatrix} \mathbf{E}_1^\psi \\ \vdots \\ \mathbf{E}_n^\psi \\ \mathbf{N}^\psi \end{pmatrix} = J_h \det\begin{pmatrix} \mathbf{E}_1^\varphi \circ h \\ \vdots \\ \mathbf{E}_n^\varphi \circ h \\ \mathbf{N}^\psi \circ h \end{pmatrix}$$

where J_h is the Jacobian determinant of h.]

17.12. Let ω be a k-form on the n-surface S.

(a) Show that if $\{\mathbf{v}_1, \ldots, \mathbf{v}_k\}$ is a linearly dependent set of vectors in S_p, $p \in S$, then $\omega(\mathbf{v}_1, \ldots, \mathbf{v}_k) = 0$.
(b) Show that if $k > n$ then ω is identically zero.

17.13. Let S be an oriented n-surface in $\mathbb{R}^{n+1}$ and let ζ be the volume form on S.

(a) Show that if $\{\mathbf{v}_1, \ldots, \mathbf{v}_n\}$ is an orthonormal basis for S_p, $p \in S$, then $\zeta(\mathbf{v}_1, \ldots, \mathbf{v}_n) = \pm 1$ and $\zeta(\mathbf{v}_1, \ldots, \mathbf{v}_n) = +1$ if and only if the basis $\{\mathbf{v}_1, \ldots, \mathbf{v}_n\}$ is consistent with the orientation $\mathbf{N}$ of S.
(b) Show that if ω is any n-form on S then there exists a function $f: S \to \mathbb{R}$ such that $\omega = f\zeta$. [*Hint*: Set $f(p) = \omega(\mathbf{v}_1, \ldots, \mathbf{v}_n)$ where $\{\mathbf{v}_1, \ldots, \mathbf{v}_n\}$ is any orthonormal basis for S_p consistent with the orientation $\mathbf{N}$ on S. Then compute the values of ω and of $f\zeta$ on $\mathbf{w}_1, \ldots, \mathbf{w}_n$ where $\mathbf{w}_j = \sum_{i=1}^n a_{ij}\mathbf{v}_i$.]

17.14. Let ω_1 be a k-form and let ω_2 be an l-form on the oriented n-surface $S \subset \mathbb{R}^{n+1}$. Define the *exterior product* $\omega_1 \wedge \omega_2$ by

$$(\omega_1 \wedge \omega_2)(\mathbf{v}_1, \ldots, \mathbf{v}_{k+l}) = \frac{1}{k!\,l!} \sum (\text{sign } \sigma)\omega_1(\mathbf{v}_{\sigma(1)}, \ldots, \mathbf{v}_{\sigma(k)})\omega_2(\mathbf{v}_{\sigma(k+1)}, \ldots, \mathbf{v}_{\sigma(k+l)})$$

for $\mathbf{v}_1, \ldots, \mathbf{v}_{k+l} \in S_p$, where the sum is over all permutations σ of $\{1, \ldots, k+l\}$.

(a) Show that $\omega_1 \wedge \omega_2$ is a $(k+l)$-form on S.
(b) Show that $\omega_2 \wedge \omega_1 = (-1)^{kl}\omega_1 \wedge \omega_2$.
(c) Show that if ω_3 is another l-form on S then $\omega_1 \wedge (\omega_2 + \omega_3) = \omega_1 \wedge \omega_2 + \omega_1 \wedge \omega_3$.
(d) Show that if ω_3 is an m-form on S then $(\omega_1 \wedge \omega_2) \wedge \omega_3 = \omega_1 \wedge (\omega_2 \wedge \omega_3)$.
(e) Show that if $\mathbf{X}_1, \ldots, \mathbf{X}_{n+1}$ are tangent vector fields on S such that for each $p \in S$ $\{\mathbf{X}_1(p), \ldots, \mathbf{X}_{n+1}(p)\}$ is an orthonormal basis for S_p consistent with the orientation on S, and if, for each i, ω_i is the 1-form on S dual to $\mathbf{X}_i$ then

$$\omega_1 \wedge \cdots \wedge \omega_n = \zeta$$

where ζ is the volume form on S. [*Hint*: Use mathematical induction to prove that for $1 \leq k \leq n$, $(\omega_1 \wedge \cdots \wedge \omega_k)(\mathbf{X}_1, \ldots, \mathbf{X}_k) = 1$ and that $\mathbf{X}_i \lrcorner (\omega_1 \wedge \cdots \wedge \omega_k) = 0$ for all $i > k$. Then use Exercise 17.13.]

17.15. Let S be an oriented n-surface in $\mathbb{R}^{n+1}$, let $\tilde{S}$ be an oriented m-surface in $\mathbb{R}^{m+1}$, let $f: S \to \tilde{S}$ be a smooth map, and let ω be a smooth k-form on $\tilde{S}$.

(a) Show that $f^*\omega$ is a smooth k-form on S.
(b) Show that if φ is a singular k-surface in S then

$$\int_\varphi f^*\omega = \int_{f \circ \varphi} \omega.$$

(c) Show that if $m = n$ and if f is an orientation preserving diffeomorphism then

$$\int_S f^*\omega = \int_{\tilde{S}} \omega.$$

17.16. Show that the antipodal map $f: S^n \to S^n$ on the n-sphere S^n, defined by $f(p) = -p$, is orientation preserving if and only if n is odd.

17.17. (a) Show that the function $h: \mathbb{R} \to \mathbb{R}$ defined by

$$h(t) = \begin{cases} e^{-1/t} & \text{if } t > 0 \\ 0 & \text{if } t \leq 0 \end{cases}$$

is smooth.

(b) Conclude that for each $r > 0$ the function $h_r: \mathbb{R} \to \mathbb{R}$ given by

$$h_r(t) = \begin{cases} e^{-1/(r^2 - t^2)} & \text{if } |t| < r \\ 0 & \text{if } |t| \geq r \end{cases}$$

is smooth and that the bump function g_p defined in the proof of Theorem 4 is smooth.

17.18. Let S be an oriented n-surface in $\mathbb{R}^{n+1}$ and let $\varphi: U \to S$ and $\psi: V \to S$ be one to one local parametrizations of S. Show that if $W = \varphi(U) \cap \psi(V) \neq \varnothing$ then $h = \varphi^{-1} \circ \psi|_{\psi^{-1}(W)}$ is an orientation preserving diffeomorphism from $\psi^{-1}(W)$ to $\varphi^{-1}(W)$. Conclude that $\psi|_{\psi^{-1}(W)}$ is a reparametrization of $\varphi|_{\varphi^{-1}(W)}$.

17.19. Let $\mathbf{X}$ and $\mathbf{Y}$ be vector fields on $\mathbb{R}^3$ and let $\omega_\mathbf{X}$ and $\omega_\mathbf{Y}$ be their dual 1-forms. Show that, for $\mathbf{v}$ and $\mathbf{w}$ in $\mathbb{R}^3_p$, $p \in \mathbb{R}^3$,

$$(\omega_\mathbf{X} \wedge \omega_\mathbf{Y})(\mathbf{v}, \mathbf{w}) = (\mathbf{X} \times \mathbf{Y})(p) \cdot (\mathbf{v} \times \mathbf{w}).$$

[Hint: By multilinearity, it suffices to check this equation when $\mathbf{v}$ and $\mathbf{w}$ are standard basis vectors.]

18 Minimal Surfaces

Let $\varphi\colon U \to \mathbb{R}^{n+1}$ be a parametrized n-surface in $\mathbb{R}^{n+1}$. A *variation* of φ is a smooth map $\psi\colon U \times (-\varepsilon, \varepsilon) \to \mathbb{R}^{n+1}$ with the property that $\psi(p, 0) = \varphi(p)$ for all $p \in U$. Thus a variation surrounds the n-surface φ with a family of singular n-surfaces $\varphi_s\colon U \to \mathbb{R}^{n+1}(-\varepsilon < s < \varepsilon)$ defined by $\varphi_s(p) = \psi(p, s)$.

A variation ψ of the form

$$\psi(p, s) = \varphi(p) + sf(p)N(p),$$

where f is a smooth function along φ and N is the Gauss map of φ, is called a *normal variation* of φ. If f is the constant function 1, this variation ψ is of the type considered in Chapter 16 which, for each s, pushes φ out a distance s along the normal. If f is a bump function similar to the one whose graph appears in Figure 17.5(b), the variation ψ introduces a bump in φ by pushing φ out along the normal only in a ball B_p about $p \in U$ (see Figure 18.1). If φ already has a bump at p, then the normal variation ψ may tend to remove the bump.

A variation ψ with the property that $\psi(p, s) = \psi(p, 0)$ for $-\varepsilon < s < \varepsilon$ whenever p lies outside some compact subset C of U is said to be *compactly*

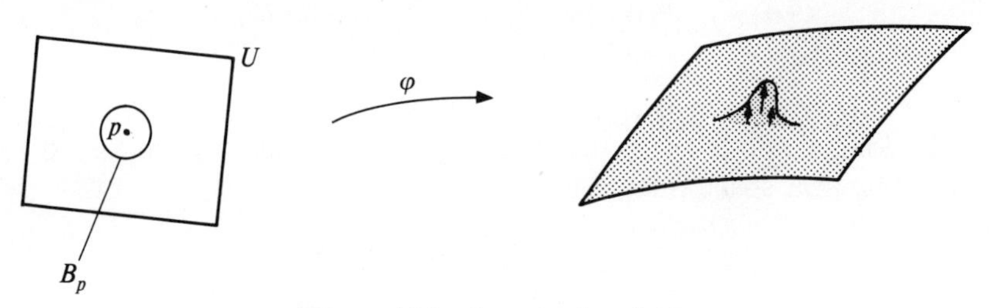

Figure 18.1 A normal variation.

supported. Note that if $\psi: U \times (-\varepsilon, \varepsilon) \to \mathbb{R}^{n+1}$ is a compactly supported variation of φ there is an $\varepsilon_1 > 0$ such that each φ_s, for $|s| < \varepsilon_1$, is a parametrized n-surface. One way to see this is to observe that the function $\delta: U \times (-\varepsilon, \varepsilon) \to \mathbb{R}$ defined by

$$\delta(p, s) = \det \begin{pmatrix} E_1^s(p) \\ \vdots \\ E_n^s(p) \\ N(p) \end{pmatrix}$$

where the E_i^s are the vector parts of the coordinate vector fields $\mathbf{E}_i^s$ along φ_s and N is the Gauss map along φ, is continuous. Hence, the set

$$C_1 = \{(p, s) \in \mathbb{R}^{n+1}: p \in C, \ |s| \leq \varepsilon/2, \quad \text{and} \quad \delta(p, s) = 0\}$$

is compact. If C_1 is empty, let $\varepsilon_1 = \varepsilon/2$; otherwise let ε_1 be the minimum value of g on C_1, where $g: \mathbb{R}^{n+1} \to \mathbb{R}$ is given by $g(p, s) = |s|$. Then

(i) $\varepsilon_1 \neq 0$ since $\delta(p, 0) \neq 0$ for all $p \in U$,
(ii) $\delta(p, s) \neq 0$ whenever $p \in C$ and $|s| < \varepsilon_1$, and
(iii) $\delta(p, s) \neq 0$ for all $s(|s| < \varepsilon)$ whenever $p \notin C$ (since then the $\mathbf{E}_i^s$ are equal to the coordinate vector fields $\mathbf{E}_i$ along φ).

Thus $\delta(p, s) \neq 0$ whenever $p \in U$ and $|s| < \varepsilon_1$ so the coordinate vector fields $\mathbf{E}_i^s$ of φ_s are linearly independent; i.e., φ_s is regular as required.

We shall analyse the effect that compactly supported normal variations have on volume. Let $\varphi: U \to \mathbb{R}^{n+1}$ be a parametrized n-surface with finite volume and let $\psi: U \times (-\varepsilon, \varepsilon) \to \mathbb{R}^{n+1}$ be a compactly supported normal variation of φ,

$$\psi(p, s) = \varphi(p) + sf(p)N(p).$$

Then the coordinate vector field $\mathbf{E}_i^s$ of φ_s has vector part

$$E_i^s = \frac{\partial \varphi_s}{\partial u_i} = \frac{\partial \varphi}{\partial u_i} + s\frac{\partial f}{\partial u_i} N + sf\frac{\partial N}{\partial u_i}.$$

But, for each $p \in U$, $\partial N/\partial u_i(p)$ is the vector part of

$$\nabla_{\mathbf{E}_i(p)}\mathbf{N} = -L_p(\mathbf{E}_i(p)) = -\sum_{j=1}^{n} c_{ji}(p)\mathbf{E}_j(p),$$

where L_p is the Weingarten map of φ at p, the $\mathbf{E}_i$ are the coordinate vector fields along φ, and $(c_{ij}(p))$ is the matrix for L_p with respect to the basis $\{\mathbf{E}_i(p)\}$ for the tangent space Image $d\varphi_p$. Hence

$$E_i^s = E_i + s\left(\frac{\partial f}{\partial u_i} N - f\sum_{j=1}^{n} c_{ji} E_j\right).$$

The volume of φ_s is

$$V(\varphi_s) = \int_U \det \begin{pmatrix} \mathbf{E}_1^s \\ \vdots \\ \mathbf{E}_n^s \\ \mathbf{N}^s \end{pmatrix}$$

where $\mathbf{N}^s$ is the orientation vector field along φ_s. Its rate of change at $s = 0$ is

$$\frac{d}{ds}\bigg|_0 V(\varphi_s) = \int_U \frac{\partial}{\partial s}\bigg|_0 \det \begin{pmatrix} \mathbf{E}_1^s \\ \vdots \\ \mathbf{E}_n^s \\ \mathbf{N}^s \end{pmatrix}.$$

But, since $\mathbf{E}_i^0 = \mathbf{E}_i$ and $\mathbf{N}^0 = \mathbf{N}$, we have

$$\frac{\partial}{\partial s}\bigg|_0 \det \begin{pmatrix} E_1^s \\ \vdots \\ E_n^s \\ N^s \end{pmatrix} = \sum_{i=1}^n \det \begin{pmatrix} E_1 \\ \vdots \\ \dfrac{\partial E_i^s}{\partial s}\bigg|_{s=0} \\ \vdots \\ E_n \\ N \end{pmatrix} + \det \begin{pmatrix} E_1 \\ \vdots \\ E_n \\ \dfrac{\partial N^s}{\partial s}\bigg|_{s=0} \end{pmatrix}$$

$$\overset{(1)}{=} \sum_{i=1}^n \det \begin{pmatrix} E_1 \\ \vdots \\ \dfrac{\partial f}{\partial u_i} N - f \sum_{j=1}^n c_{ji} E_j \\ \vdots \\ E_n \\ N \end{pmatrix} = -f \sum_{i,\, j=1}^n c_{ji} \det \begin{pmatrix} E_1 \\ \vdots \\ E_j \\ \vdots \\ E_n \\ N \end{pmatrix} \quad (E_j \text{ in the } i\text{th row})$$

$$\overset{(2)}{=} -f \sum_{i=1}^n c_{ii} \det \begin{pmatrix} E_1 \\ \vdots \\ E_n \\ N \end{pmatrix}.$$

To obtain equality (1) we have used the fact that $(\partial N^s/\partial s)|_{s=0}$, being perpendicular to $N = N^0$, is a linear combination of $\{E_1, \ldots, E_n\}$ so the determinant of the matrix with these $n + 1$ vectors as rows must be zero. To obtain equality (2) we have used the fact that when $j \neq i$ the coefficient of c_{ji} is the determinant of a matrix with two equal rows and is therefore zero. We conclude, then, that

$$\left.\frac{d}{ds}\right|_0 V(\varphi_s) = -n \int_U fH \det\begin{pmatrix} \mathbf{E}_1 \\ \vdots \\ \mathbf{E}_n \\ \mathbf{N} \end{pmatrix}$$

where $H(p) = (1/n)$ trace L_p is the mean curvature of φ at $p \in U$.

Remark. It can be shown (see Exercise 18.7) that this formula is valid for all compactly supported variations ψ of φ (not just for normal ones), where the function $f\colon U \to \mathbb{R}$ is defined by $f = \mathbf{X} \cdot \mathbf{N}$, $\mathbf{X}$ being the *variation vector field* along φ defined by $\mathbf{X}(p) = \mathbf{E}_{n+1}^{\psi}(p, 0)$, $\mathbf{E}_{n+1}^{\psi}$ the $(n + 1)$th coordinate vector field along ψ. The formula is also valid for all normal variations (not necessarily compactly supported) as long as all the necessary integrals are defined and the interchange of $(d/ds)|_0$ and $\int_U$ can be justified.

The volume integral is said to be *stationary* at the parametrized n-surface $\varphi\colon U \to \mathbb{R}^{n+1}$ if $V(\varphi) < \infty$ and $(d/ds)|_0 V(\varphi_s) = 0$ for all compactly supported normal variations ψ of φ. This is the case, for example, when the volume of φ is smaller than that of each parametrized n-surface φ_s which can be obtained from φ via a small compactly supported normal variation.

Theorem. *Let $\varphi\colon U \to \mathbb{R}^{n+1}$ be a parametrized n-surface with finite volume in $\mathbb{R}^{n+1}$. Then the volume integral is stationary at φ with respect to compactly supported normal variations if and only if the mean curvature of S is identically zero.*

PROOF. Certainly if $H = 0$ then, for every compactly supported normal variation ψ of φ,

$$\left.\frac{d}{ds}\right|_0 V(\varphi_s) = -n \int_U fH \det\begin{pmatrix} \mathbf{E}_1 \\ \vdots \\ \mathbf{E}_n \\ \mathbf{N} \end{pmatrix} = 0.$$

Conversely, if $H(p) \neq 0$ for some $p \in U$, choose an $\varepsilon > 0$ such that the closed ball about p of radius ε is contained in U, let $h\colon U \to \mathbb{R}$ be a smooth bump function with $h(p) = 1$, $h(q) \geq 0$ for all $q \in U$, and $h(q) = 0$ for all q with $\|q - p\| \geq \varepsilon$, and let ψ be the normal variation of φ with $f = hH$. Then ψ is compactly supported and $fH = hH^2$ is non-negative on U and positive at p, so

$$\left.\frac{d}{ds}\right|_0 V(\varphi_s) = -n \int_U hH^2 \det\begin{pmatrix} \mathbf{E}_1 \\ \vdots \\ \mathbf{E}_n \\ \mathbf{N} \end{pmatrix} < 0. \qquad \square$$

A parametrized or oriented n-surface in $\mathbb{R}^{n+1}$ with mean curvature identically zero is called a *minimal surface*. The adjective "minimal" is used here because a minimal surface usually arises as a surface whose volume is minimal among all surfaces obtainable from it via small normal variations. Minimal 2-surfaces in $\mathbb{R}^3$ are found in nature as soap films: if a soap film takes the shape of a surface spanning a wire frame then (assuming air pressure is the same on both sides of the film) in order for the surface to be stable its area must be minimal among all nearby surfaces spanning the given frame (see Figure 18.2).

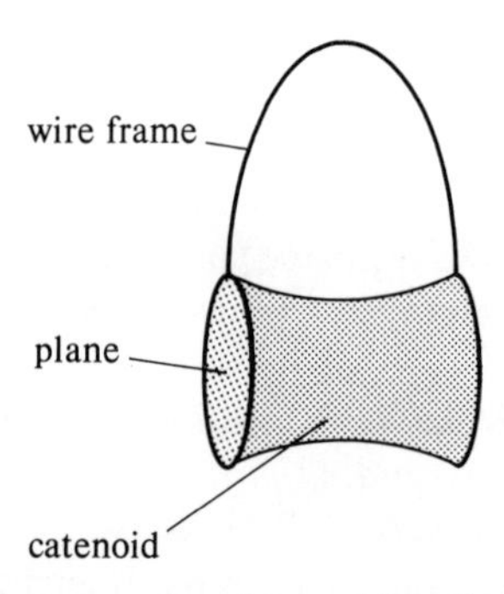

Figure 18.2 Minimal surfaces can be obtained by dipping a wire frame in a soap solution. (For best results, keep the distance between the parallel circles small.)

Clearly an n-plane $a_1 x_1 + \cdots + a_{n+1} x_{n+1} = b$ is a minimal surface in $\mathbb{R}^{n+1}$ since all principal curvatures, and hence also the mean curvature, are identically zero. On the other hand, there are no compact minimal surfaces in $\mathbb{R}^{n+1}$ since, by Theorem 4 of Chapter 12, each compact n-surface in $\mathbb{R}^{n+1}$ must contain a point where all principal curvatures are different from zero and have the same sign.

We shall find all 2-surfaces of revolution in $\mathbb{R}^3$ which are minimal. Suppose first that $\alpha\colon I \to \mathbb{R}^2$ is a parametrized curve of the form $\alpha(t) = (t, y(t))$ for some smooth function $y\colon I \to \mathbb{R}$ with $y(t) > 0$ for all $t \in I$. The 2-surface obtained by rotating α about the x_1-axis is given by

$$\varphi(t, \theta) = (t, y(t)\cos\theta, y(t)\sin\theta).$$

A straight-forward computation (Exercise 18.1) shows that the principal curvatures of φ are given by

$$\kappa_1(t, \theta) = -y''(t)/(1 + (y'(t))^2)^{3/2}$$
$$\kappa_2(t, \theta) = 1/y(t)(1 + (y'(t))^2)^{1/2}.$$

Thus the mean curvature of φ will be zero if and only if $y(t)$ satisfies the differential equation

$$\frac{y''}{(1 + y'^2)^{3/2}} = \frac{1}{y(1 + y'^2)^{1/2}}.$$

Multiplying both sides by $y'(1 + y'^2)^{1/2}$ we obtain

$$\frac{y'y''}{1 + y'^2} = \frac{y'}{y}$$

which integrates to

$$\tfrac{1}{2} \log(1 + y'^2) = \log y + \log c = \log(cy),$$

or

$$1 + y'^2 = (cy)^2$$

where $c > 0$ is a constant of integration. Solving for y' gives

$$y' = \pm((cy)^2 - 1)^{1/2} \qquad \text{or} \qquad y'/((cy)^2 - 1)^{1/2} = \pm 1.$$

Integrating again, we obtain

$$(1/c)(\cosh^{-1}(cy) - c_2) = \pm t$$

or

$$y = (1/|c_1|)\cosh(c_1 t + c_2)$$

where $c_1 = \pm c$ and c_2 is another constant of integration.

A curve in $\mathbb{R}^2$ of the form $x_2 = (1/|c_1|)\cosh(c_1 x_1 + c_2)$ is called a *catenary*; a surface of revolution obtained by rotating such a curve about the x_1-axis is called a *catenoid* (Figure 18.2). The above argument shows that each minimal surface in $\mathbb{R}^3$ which can be obtained by rotating the graph of a smooth function about the x_1-axis is a portion of a catenoid.

If we drop the requirement that the parametrized curve α have image the graph of a function, we obtain in addition to catenoids only portions of planes. Indeed, if $\alpha(t) = (x(t), y(t))$ then on any interval where $x' \neq 0$ there exists a reparametrization β of α of the form $\beta(t) = (t, y \circ x^{-1}(t))$ so on that interval Image α is the graph of a function. On any interval where x' is identically zero, α must be of the form $\alpha(t) = (c, y(t))$ for some $c \in \mathbb{R}$ so the surface of revolution obtain by rotating this portion of α about the $x_1$axis is contained in the plane $x_1 = c$. Since two catenoids, defined by choosing different values of c_1 and c_2, do not fit together smoothly, and a portion of a catenoid cannot be glued smoothly onto a portion of a plane, we conclude that *the only connected minimal surfaces of revolution in $\mathbb{R}^3$ are portions of catenoids and portions of planes.*

EXERCISES

18.1. Find the principal curvatures of the parametrized surface of revolution obtained by rotating about the x_1-axis the parametrized curve $\alpha(t) = (t, y(t))$, where $y(t) > 0$ for all $t \in I$.

18.2. Show that the parametrized helicoid $\varphi: \mathbb{R}^2 \to \mathbb{R}^3$ defined by $\varphi(t, \theta) = (t \cos \theta, t \sin \theta, \theta)$ is a minimal surface.

18.3. Let S be a connected minimal 1-surface in $\mathbb{R}^2$. Show that S is a segment of a straight line.

18.4. Show that the Gaussian curvature of a minimal 2-surface in $\mathbb{R}^3$ is everywhere ≤ 0.

18.5. Show that an oriented 2-surface S in $\mathbb{R}^3$ is a minimal surface if and only if for each $p \in S$ there exist orthogonal directions $\mathbf{v}$ and $\mathbf{w}$ in S_p on which the normal curvature of S is zero. (Directions $\mathbf{v} \in S_p$ for which the normal curvature $k(\mathbf{v})$ is zero are called *asymptotic directions*.)

18.6. Show that if the Gauss map of a minimal 2-surface S in $\mathbb{R}^3$ is regular then it is conformal; i.e., show that if $dN_p: S_p \to S^2_{N(p)}$ is non-singular ($p \in S$) then there exists $\lambda(p) > 0$ such that $\|dN_p(\mathbf{v})\| = \lambda(p)\|\mathbf{v}\|$ for all $\mathbf{v} \in S_p$.

18.7. Let $\alpha: [a, b] \to \mathbb{R}^2$ be a unit speed curve and let $\psi: [a, b] \times (-\varepsilon, \varepsilon) \to \mathbb{R}^2$ be a variation of α. Show that

$$\frac{d}{ds}\bigg|_0 l(\alpha_s) = (\mathbf{X} \cdot \mathbf{N})\bigg|_a^b - \int_a^b (\mathbf{X} \cdot \mathbf{N})(t)\kappa(t)\, dt$$

where $\alpha_s(t) = \psi(t, s)$, $\mathbf{X}(t) = \mathbf{E}_2^\psi(t, 0)$, $\mathbf{N}$ is the orientation vector field along α, and κ is the curvature along α. Conclude that if $\psi(a, s) = \alpha(a)$ and $\psi(b, s) = \alpha(b)$ for all s then

$$\frac{d}{ds}\bigg|_0 l(\alpha_s) = -\int_a^b (\mathbf{X} \cdot \mathbf{N})(t)\kappa(t)\, dt.$$

[*Hint*: Verify that $(\partial/\partial s)|_0 \|\alpha_s\| = ((\partial^2\psi/\partial t\, \partial s) \cdot (\partial\psi/\partial t))|_{s=0}$ and integrate by parts.]

The Exponential Map 19

In Chapter 7 we defined geodesics as "straightest curves" in an n-surface. In this chapter we shall examine the role of geodesics as "shortest curves." We begin by using a technique of the calculus of variations analogous to the one we used in Chapter 18 to study minimal surfaces. Now, however, we shall vary parametrized curves rather than parametrized surfaces.

Let $\alpha: [a, b] \to S$ be a parametrized curve in an n-surface $S \subset \mathbb{R}^{n+1}$. A *variation* of α is a smooth map $\psi: [a, b] \times (-\varepsilon, \varepsilon) \to S$ $(\varepsilon > 0)$ such that $\psi(t, 0) = \alpha(t)$ for all $t \in I$ (see Figure 19.1). The two coordinate vector fields $\mathbf{E}_1$ and $\mathbf{E}_2$ along ψ, defined by

$$\mathbf{E}_1(t, s) = d\psi(t, s, 1, 0)$$

$$\mathbf{E}_2(t, s) = d\psi(t, s, 0, 1),$$

are then tangent to S along ψ. Note that $\mathbf{E}_1(t, 0) = \dot{\alpha}(t)$ for all $t \in I$. The vector field $\mathbf{X}$ along α defined by $\mathbf{X}(t) = \mathbf{E}_2(t, 0)$ is called the *variation vector field* along α associated with the variation ψ.

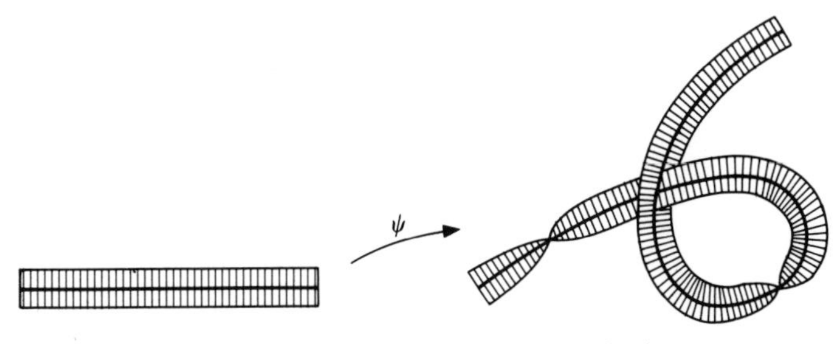

Figure 19.1 A variation of a parametrized curve.

A variation ψ of a parametrized curve α defines a family of parametrized curves α_s: $[a, b] \to S$ by $\alpha_s(t) = \psi(t, s)$. The length of α_s is given by the length integral

$$l(\alpha_s) = \int_a^b \|\dot{\alpha}_s(t)\| \, dt = \int_a^b \|\mathbf{E}_1(t, s)\| \, dt.$$

The derivative of this function of s is

$$\begin{aligned}\frac{d}{ds} l(\alpha_s) &= \int_a^b \frac{\partial}{\partial s} (\mathbf{E}_1 \cdot \mathbf{E}_1)^{1/2} \, dt \\ &= \int_a^b \left[\left(\frac{\partial}{\partial s}\left(\frac{\partial \psi}{\partial t}\right) \cdot \frac{\partial \psi}{\partial t} \right) \Big/ \|\mathbf{E}_1\| \right] dt \\ &= \int_a^b \left[\left(\frac{\partial}{\partial t}\left(\frac{\partial \psi}{\partial s}\right) \cdot \frac{\partial \psi}{\partial t} \right) \Big/ \|\mathbf{E}_1\| \right] dt.\end{aligned}$$

If we assume now that α is a unit speed curve then $\|\mathbf{E}_1\|_{s=0} = \|\dot{\alpha}\| = 1$ so

$$\begin{aligned}\left.\frac{d}{ds}\right|_0 l(\alpha_s) &= \int_a^b \dot{\mathbf{X}} \cdot \dot{\alpha} \, dt \\ &= \int_a^b [(\mathbf{X} \cdot \dot{\alpha})' - \mathbf{X} \cdot \ddot{\alpha}] \, dt\end{aligned}$$

or

$$\boxed{\left.\frac{d}{ds}\right|_0 l(\alpha_s) = (\mathbf{X} \cdot \dot{\alpha})(b) - (\mathbf{X} \cdot \dot{\alpha})(a) - \int_a^b (\mathbf{X} \cdot \ddot{\alpha}) \, dt.}$$

This boxed formula is called the *first variation formula* for the length integral. It is valid for any variation ψ of any unit speed curve α in S. Note that the right hand side depends only on the variation vector field $\mathbf{X}$; any two variations of α with the same variation vector field will yield the same value of $(d/ds)|_0 \, l(\alpha_s)$.

A variation ψ: $[a, b] \times (-\varepsilon, \varepsilon) \to S$ is said to be a *fixed endpoint variation* of $\alpha(t) = \psi(t, 0)$ if $\psi(a, s) = \alpha(a)$ and $\psi(b, s) = \alpha(b)$ for all $s \in (-\varepsilon, \varepsilon)$. The variation ψ is said to be a *normal variation* if the variation vector field $\mathbf{X}$ is everywhere orthogonal to α ($\mathbf{X}(t) \perp \dot{\alpha}(t)$ for all $t \in [a, b]$). Specializing the first variation formula to these situations yields the following.

Theorem 1. *Let* α: $[a, b] \to S$ *be a unit speed curve in an n-surface* $S \subset \mathbb{R}^{n+1}$. *Then the following three conditions are equivalent:*

(i) *The length integral is stationary at* α *with respect to fixed endpoint variations.*
(ii) *The length integral is stationary at* α *with respect to normal variations.*
(iii) α *is a geodesic in* S.

In particular, if α is a shortest curve in S joining two points of S then α is a geodesic.

PROOF. If $\psi\colon [a, b] \times (-\varepsilon, \varepsilon) \to S$ is a fixed endpoint variation of α then $\psi(a, s) = \alpha(a)$ for $|s| < \varepsilon$ so $\mathbf{X}(a) = (\alpha(a), (\partial\psi/\partial s)(a, 0)) = \mathbf{0}$ and similarly $\mathbf{X}(b) = \mathbf{0}$. If ψ is a normal variation of α then $\mathbf{X}(a) \cdot \dot{\alpha}(a) = 0$ and $\mathbf{X}(b) \cdot \dot{\alpha}(b) = 0$. In either case, the first variation formula reduces to

$$\frac{d}{ds}\bigg|_0 l(\alpha_s) = -\int_a^b (\mathbf{X} \cdot \ddot{\alpha})\, dt.$$

If α is a geodesic in S then $\ddot{\alpha}(t) \perp S_{\alpha(t)}$ for all $t \in [a, b]$ so $\mathbf{X} \cdot \ddot{\alpha} = 0$ along α and therefore $(d/ds)|_0\, l(\alpha_s) = 0$, for all fixed endpoint or normal variations ψ of α. Thus (iii) $\Rightarrow$ (i) and (iii) $\Rightarrow$ (ii).

On the other hand, if α is not a geodesic then there is a $t_0 \in [a, b]$ such that $\ddot{\alpha}(t_0) \notin S^{\perp}_{\alpha(t_0)}$; i.e., such that the tangential component $\dot{\alpha}'(t_0)$ of $\ddot{\alpha}(t_0)$ is not zero. (Recall from Chapter 8 that $\dot{\alpha}'$ is the covariant acceleration of α.) We will construct a fixed endpoint normal variation ψ of α whose variation vector field along α is $f\dot{\alpha}'$ where f is a nonnegative smooth function along α with $f(a) = f(b) = 0$ and $f(t_0) > 0$. This will be a normal variation of α since, $\dot{\alpha}$ being a unit vector field along α, $\dot{\alpha}' \perp \dot{\alpha}$. The first variation formula, for this variation, becomes

$$\frac{d}{ds}\bigg|_0 l(\alpha_s) = -\int_a^b f\dot{\alpha}' \cdot \ddot{\alpha} = -\int_a^b f\|\dot{\alpha}'\|^2 < 0,$$

proving that (i) $\Rightarrow$ (iii) and (ii) $\Rightarrow$ (iii).

To construct the variation ψ, let $\varphi\colon U \to S$ be a one to one parametrized n-surface whose image is an open set in S containing $\alpha(t_0)$. Choose a_1, b_1 with $a < a_1 < b_1 < b$ such that $\alpha([a_1, b_1]) \subset \text{Image } \varphi$. Define $\beta\colon [a_1, b_1] \to U$ by $\beta(t) = \varphi^{-1} \circ \alpha(t)$, let $f\colon [a, b] \to \mathbb{R}$ be a smooth bump function with $f(t_0) > 0$ and $f(t) = 0$ for all $t \notin [a_1, b_1]$, and let $\mathbf{Y}$ be the smooth vector field along β defined by $\mathbf{Y}(t) = f(t)(d\varphi_{\beta(t)})^{-1}(\dot{\alpha}'(t))$ (see Figure 19.2). Now define $\psi\colon [a, b] \times (-\varepsilon, \varepsilon) \to S$ by

$$\psi(t, s) = \begin{cases} \varphi(\beta(t) + sY(t)) & \text{for } t \in [a_1, b_1],\ s \in (-\varepsilon, \varepsilon) \\ \varphi \circ \beta(t) = \alpha(t) & \text{for } t \notin [a_1, b_1],\ s \in (-\varepsilon, \varepsilon) \end{cases}$$

where $\varepsilon > 0$ is chosen small enough so that $\beta(t) + sY(t) \in U$ for all $(t, s) \in [a_1, b_1] \times (-\varepsilon, \varepsilon)$. Then ψ is a fixed endpoint variation of α with variation vector field

$$\mathbf{X}(t) = d\psi(t, 0, 0, 1) = \begin{cases} d\varphi(\mathbf{Y}(t)) & \text{for } t \in [a_1, b_1] \\ \mathbf{0} & \text{for } t \notin [a_1, b_1] \end{cases}$$

$$= f(t)\dot{\alpha}'(t)$$

as required.

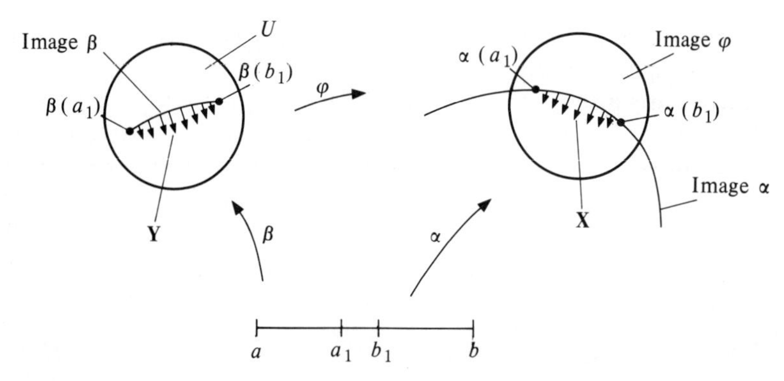

Figure 19.2 Construction of a length decreasing variation along a non-geodesic.

Finally, if α is shortest among all curves in S joining $\alpha(a)$ to $\alpha(b)$ then $l(\alpha_s)$ is minimum at $s = 0$, for all fixed endpoint variations of α, so the length integral is stationary at α and α must be a geodesic in S. □

Remark 1. The proof above not only shows that if α is not a geodesic then α does not minimize length but in fact describes how to obtain a shorter curve from $\alpha(a)$ to $\alpha(b)$: simply deform α, keeping the endpoints fixed, in the direction of the tangential component $\dot{\alpha}'$ of acceleration of α (see Figure 19.3).

Remark 2. A review of the proof of Theorem 1 will show that replacing the hypothesis that α be a unit speed curve in S with the hypothesis that α be a constant speed curve in S will not alter the validity of the theorem, although the first variation formula will change slightly.

Remark 3. The first variation formula can be rewritten in terms of the covariant acceleration $\dot{\alpha}'$ of α as follows:

$$\frac{d}{ds}\bigg|_0 l(\alpha_s) = (\mathbf{X} \cdot \dot{\alpha})(b) - (\mathbf{X} \cdot \dot{\alpha})(a) - \int_a^b (\mathbf{X} \cdot \dot{\alpha}')\, dt.$$

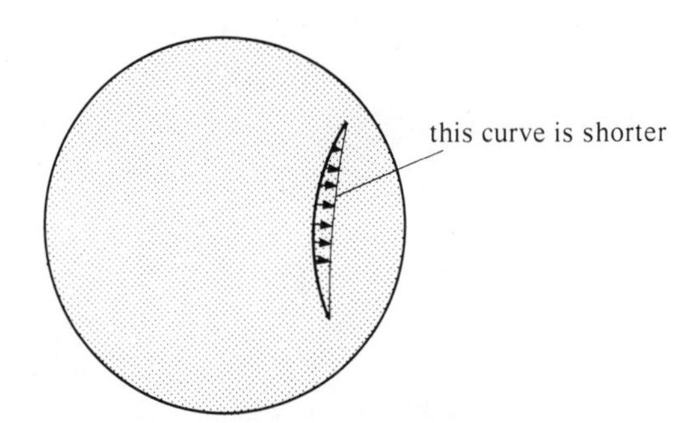

Figure 19.3 Decreasing lengths of curves on S^2.

Theorem 1 establishes that a shortest unit speed curve between two points p and q in an n-surface $S \subset \mathbb{R}^{n+1}$ must be a geodesic. It does not show that there exists a shortest curve between two points (in fact, there may be none: consider a 2-plane in $\mathbb{R}^3$ with a point removed) and it does not show that a geodesic $\alpha: [a, b] \to S$ is a shortest curve (even locally) between $\alpha(a)$ and $\alpha(b)$ (in fact, it may not be; see Figure 19.4). But we will show that, if p and $q \in S$ are close enough, there does exist a geodesic connecting p to q which is in fact shortest among all curves in S joining p to q. To prove these facts, we shall use the exponential map of an n-surface.

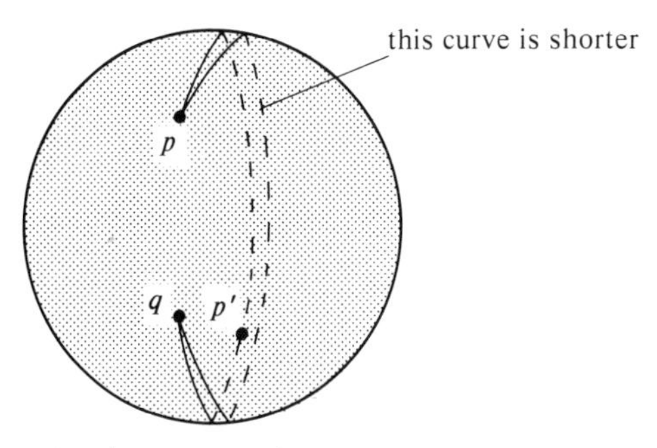

Figure 19.4 Geodesics (great circles) on the sphere do not minimize the length integral, even locally, beyond the conjugate (antipodal) point p'.

For $\mathbf{v} \in T(S) = \bigcup_{p \in S} S_p$, let $\alpha_{\mathbf{v}}$ denote the unique maximal geodesic in S with $\dot{\alpha}_{\mathbf{v}}(0) = \mathbf{v}$. Let

$$U = \{\mathbf{v} \in T(S)\colon 1 \in \text{domain } \alpha_{\mathbf{v}}\}$$

and let $\exp: U \to S$ be defined by $\exp(\mathbf{v}) = \alpha_{\mathbf{v}}(1)$. exp is called the *exponential map* of S.

Note that the zero vector in S_p is in U for each $p \in S$ and that its image under exp is p.

Example. The maximal geodesic in the unit circle $\mathbf{S}^1 \subset \mathbb{R}^2$ with initial velocity $\mathbf{v} = (1, 0, 0, \theta)$ is the constant speed global parametrization $\alpha_{\mathbf{v}}(t) = (\cos \theta t, \sin \theta t)$ of $\mathbf{S}^1$. Hence

$$\exp(1, 0, 0, \theta) = \alpha_{\mathbf{v}}(1) = (\cos \theta, \sin \theta).$$

Viewing $\mathbb{R}^2$ as the set $\mathbb{C}$ of complex numbers by identifying (a, b) with $a + bi$, this formula may be rewritten in the form $\exp(1, 0, 0, \theta) = \cos \theta + i \sin \theta = e^{i\theta}$.

Theorem 2. *The exponential map* $\exp: U \to S$ *of an n-surface S in $\mathbb{R}^{n+1}$ has the following properties:*

(i) *The domain* U *of* exp *is an open set in* $T(S)$.
(ii) *If* $\mathbf{v} \in U$ *then* $t\mathbf{v} \in U$ *for* $0 \leq t \leq 1$.
(iii) exp *is a smooth map.*
(iv) *For each* $p \in S$ *there exists a set* U_p, *open in* S_p *and containing* $\mathbf{0} \in S_p$, *such that* $U_p \subset U$ *and* $\exp|_{U_p}$ *is a diffeomorphism from* U_p *onto an open subset of* S *containing* p.
(v) *For each* $p \in S$ *and* $\mathbf{v} \in S_p$, *the maximal geodesic* $\alpha_{\mathbf{v}}$ *with* $\dot{\alpha}_{\mathbf{v}}(0) = \mathbf{v}$ *is given by the formula*

$$\alpha_{\mathbf{v}}(t) = \exp(t\mathbf{v}).$$

PROOF. (v) is immediate from the fact that for each $t \in \mathbb{R}$ the parametrized curve $\alpha(s) = \alpha_{\mathbf{v}}(ts)$, defined on the interval $\{s \in \mathbb{R}\colon ts \in I\}$ where I is the domain of $\alpha_{\mathbf{v}}$, is a geodesic with $\dot{\alpha}(0) = t\dot{\alpha}_{\mathbf{v}}(0) = t\mathbf{v}$. By uniqueness of geodesics, $\alpha_{t\mathbf{v}}(s) = \alpha(s) = \alpha_{\mathbf{v}}(ts)$ for all s such that $ts \in I$. Taking $s = 1$ yields, for each $t \in I$, $\alpha_{\mathbf{v}}(t) = \alpha_{t\mathbf{v}}(1) = \exp(t\mathbf{v})$.

(ii) follows from (v) because if $\mathbf{v} \in U$ then 1 is in the domain of $\alpha_{\mathbf{v}}$ so $\alpha_{\mathbf{v}}(t) = \exp(t\mathbf{v})$ is defined for $0 \leq t \leq 1$.

(i) Recall (Exercise 15.5) that $T(S)$ is a $2n$-surface in $\mathbb{R}^{2n+2}$. Consider the smooth vector field $\mathbf{X}$ on $T(S)$ defined by

$$\mathbf{X}(\mathbf{v}) = (p, v, v, -(\mathbf{v} \cdot \nabla_{\mathbf{v}} \mathbf{N})N(p))$$

for $\mathbf{v} = (p, v) \in T(S)$. $\mathbf{X}$ is called the *geodesic spray* on $T(S)$. We shall relate the integral curves of $\mathbf{X}$ to the geodesics of S.

For $\alpha\colon I \to S$ any parametrized curve in S, the *natural lift* of α to $T(S)$ is the parametrized curve $\tilde{\alpha}\colon I \to T(S)$ given by

$$\tilde{\alpha}(t) = \dot{\alpha}(t) = \left(\alpha(t), \frac{d\alpha}{dt}(t)\right).$$

The velocity of $\tilde{\alpha}$ is

$$\dot{\tilde{\alpha}}(t) = \left(\alpha(t), \frac{d\alpha}{dt}(t), \frac{d\alpha}{dt}(t), \frac{d^2\alpha}{dt^2}(t)\right)$$

so $\tilde{\alpha}$ is an integral curve of $\mathbf{X}$ if and only if

$$\frac{d^2\alpha}{dt^2} = -(\dot{\alpha} \cdot \nabla_{\dot{\alpha}} \mathbf{N})N \circ \alpha.$$

But this is just the differential equation (G) (see Chapter 7) of a geodesic in S. Thus $\alpha\colon I \to S$ *is a geodesic in* S *if and only if its natural lift* $\tilde{\alpha}$ *to* $T(S)$ *is an integral curve of the geodesic spray* $\mathbf{X}$. Furthermore, for each $\mathbf{v} \in T(S)$, the maximal geodesic $\alpha_{\mathbf{v}}$ with initial velocity $\mathbf{v}$ has natural lift $\tilde{\alpha}_{\mathbf{v}}$ with $\mathbf{v} = \tilde{\alpha}_{\mathbf{v}}(0)$ and $\mathbf{X}(\mathbf{v}) = \dot{\tilde{\alpha}}(0)$ so $\mathbf{X}$ is a tangent vector field on $T(S)$ whose maximal integral curve through $\mathbf{v} \in T(S)$ is $\tilde{\alpha}_{\mathbf{v}}$. It follows that, for each $\mathbf{v} \in T(S)$, *the maximal geodesic* $\alpha_{\mathbf{v}}$ *in* S *with* $\dot{\alpha}_{\mathbf{v}}(0) = \mathbf{v}$ *is given by the formula*

$\alpha_{\mathbf{v}} = \pi \circ \beta_{\mathbf{v}}$ *where* $\beta_{\mathbf{v}}$ *is the maximal integral curve of* $\mathbf{X}$ *with* $\beta_{\mathbf{v}}(0) = \mathbf{v}$, *and* $\pi: T(S) \to S$ *is defined by* $\pi(p, v) = p$.

Now, if $\mathbf{v} \in T(S)$ is in the domain U of the exponential map then the geodesic $\alpha_{\mathbf{v}}$ has domain containing the interval $[0, 1]$ and hence the maximal integral curve $\beta_{\mathbf{v}} = \tilde{\alpha}_{\mathbf{v}}$ of $\mathbf{X}$ through $\mathbf{v}$ has domain containing $[0, 1]$. As in the proof of the Corollary to Theorem 4 of Chapter 13, we can choose $\bar{\varepsilon} > 0$ such that for each t in the compact set $[0, 1]$ there is an open set V_t in $T(S)$ containing $\beta_{\mathbf{v}}(t)$ such that the integral curve of $\mathbf{X}$ through each point of V_t has domain containing the interval $(-\bar{\varepsilon}, \bar{\varepsilon})$. Setting $V = \bigcup_{t \in [0, 1]} V_t$ we obtain an open set V in $T(S)$, containing $\beta_{\mathbf{v}}([0, 1])$, such that through each point $\mathbf{w}$ of V there passes an integral curve $\beta_{\mathbf{w}}$ of $\mathbf{X}$ with $\dot{\beta}_{\mathbf{w}}(0) = \mathbf{w}$ and such that domain $(\beta_{\mathbf{w}})$ contains $(-\bar{\varepsilon}, \bar{\varepsilon})$. By Theorem 4 of Chapter 13, the map $\psi: (-\bar{\varepsilon}, \bar{\varepsilon}) \times V \to T(S)$ defined by $\psi(t, \mathbf{w}) = \beta_{\mathbf{w}}(t)$ is smooth. Moreover, by uniqueness of integral curves $\beta_{\beta_{\mathbf{w}}(t)}(s) = \beta_{\mathbf{w}}(t + s)$ for all t and s such that $t, s \in (-\bar{\varepsilon}, \bar{\varepsilon})$ and such that $\beta_{\mathbf{w}}(t) \in V$. Choosing k a positive integer such that $1/k < \bar{\varepsilon}$ and defining $\psi_{1/k}: V \to T(S)$ by $\psi_{1/k}(\mathbf{w}) = \psi(1/k, \mathbf{w}) = \beta_{\mathbf{w}}(1/k)$, it follows that

$$(\psi_{1/k} \circ \psi_{1/k})(\mathbf{w}) = \beta_{\beta_{\mathbf{w}}(1/k)}(1/k) = \beta_{\mathbf{w}}(2/k)$$

for all $\mathbf{w} \in V$ such that $\psi_{1/k}(\mathbf{w}) = \beta_{\mathbf{w}}(1/k) \in V$ and, iterating k times,

$$(\psi_{1/k} \circ \cdots \circ \psi_{1/k})(\mathbf{w}) = \beta_{\mathbf{w}}(k/k) = \beta_{\mathbf{w}}(1)$$

for all $\mathbf{w}$ in the open set

$$W = \{\mathbf{w} \in V: \psi_{1/k}(\mathbf{w}) \in V, \psi_{1/k} \circ \psi_{1/k}(\mathbf{w}) \in V, \ldots, \\ (\psi_{1/k} \circ \cdots \circ \psi_{1/k})(\mathbf{w}) \in V \text{ (composition } k - 1 \text{ times)}\}.$$

Thus $1 \in \text{domain}(\beta_{\mathbf{w}}) = \text{domain}(\pi \circ \beta_{\mathbf{w}}) = \text{domain } \alpha_{\mathbf{w}}$ for all $\mathbf{w} \in W$. In other words, $W \subset U$. Since $\mathbf{v} \in W$, we have succeeded in finding, for each $\mathbf{v} \in U$, an open set W in $T(S)$ such that $\mathbf{v} \in W \subset U$. Hence U is an open set in $T(S)$.

(iii) Since, in the notation of the previous paragraph,

$$\exp(\mathbf{w}) = \alpha_{\mathbf{w}}(1) = \pi \circ \beta_{\mathbf{w}}(1) = (\pi \circ \psi_{1/k} \circ \cdots \circ \psi_{1/k})(\mathbf{w})$$

for all $\mathbf{w} \in W$, exp is smooth.

(iv) We need only check that $(d \exp)_{\mathbf{0}}: (S_p)_{\mathbf{0}} \to S_p$ is non-singular, for then the inverse function theorem applies. But each element of $(S_p)_{\mathbf{0}}$ is of the form $\dot{\alpha}(0)$ where $\alpha(t) = t\mathbf{v}$ for some $\mathbf{v} \in S_p$ and, by (v),

$$(d \exp)(\dot{\alpha}(0)) = (\exp \dot{\circ}\, \alpha)(0) = \dot{\alpha}_{\mathbf{v}}(0) = \mathbf{v}$$

so $(d \exp)(\dot{\alpha}(0)) = \mathbf{0}$ only if $\dot{\alpha}(0) = \mathbf{0}$. This says that $(d \exp)_{\mathbf{0}}$ is non-singular.

□

According to Theorem 2, the geodesics in S through $p \in S$ can be described as the images under exp of the rays $\alpha(t) = t\mathbf{v}$ in S_p (see Figure 19.5).

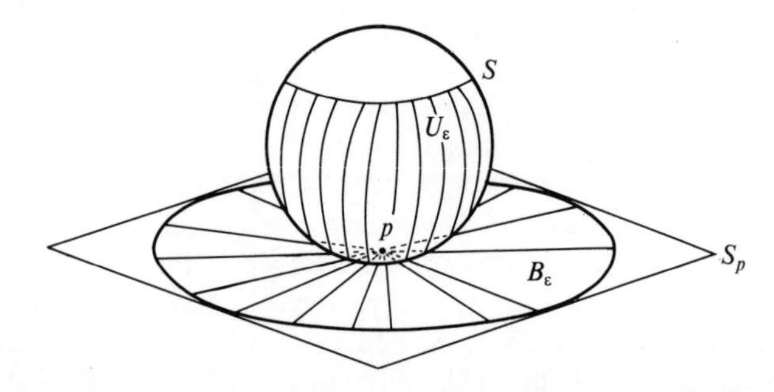

Figure 19.5 The geodesics in S through p are the images under the exponential map of the rays through $\mathbf{0}$ in S_p. Moreover, exp maps the ε-ball B_ε about $\mathbf{0}$ diffeomorphically onto the open set U_ε in S, for sufficiently small $\varepsilon > 0$.

Moreover, for $\varepsilon > 0$ sufficiently small, exp maps the ε-ball $B_\varepsilon = \{\mathbf{v} \in S_p : \|\mathbf{v}\| < \varepsilon\}$ diffeomorphically onto an open set U_ε in S. For $q \in U_\varepsilon$ it follows that there exists a geodesic in U_ε joining p to q; namely, the geodesic $\alpha_{\mathbf{v}}(t) = \exp(t\mathbf{v})$ $(0 \le t \le 1)$ where $\mathbf{v} \in B_\varepsilon$ is such that $\exp(\mathbf{v}) = q$. Furthermore, this geodesic is the unique (up to reparametrization) geodesic in U_ε joining p to q. We shall show that in fact this geodesic has length less than or equal to that of every parametrized curve in S joining p to q. The proof depends on two facts about the differential of the exponential map.

Lemma. *Let S be an n-surface in $\mathbb{R}^{n+1}$ and let $U \subset T(S)$ be the domain of the exponential map of S. For $p \in S$, and $\mathbf{v} \in S_p \cap U$, d exp has the following effect on vectors tangent to S_p at $\mathbf{v}$:*

(i) *If $\mathbf{w} \in (S_p)_{\mathbf{v}}$ is tangent at $\mathbf{v}$ to the ray $\alpha(t) = t\mathbf{v}$ through $\mathbf{v}$ (i.e., if $\mathbf{w}$ is a multiple of $\dot{\alpha}(1)$) then $\|(d \exp)(\mathbf{w})\| = \|\mathbf{w}\|$.*

(ii) *If $\mathbf{w} \in (S_p)_{\mathbf{v}}$ is orthogonal to the ray $\alpha(t) = t\mathbf{v}$ through $\mathbf{v}$ (i.e., if $\dot{\alpha}(1) \cdot \mathbf{w} = 0$) then $(d \exp)(\mathbf{w})$ is orthogonal to the geodesic $(\exp \circ \alpha)(t) = \exp(t\mathbf{v})$.*

Remark. Statement (ii) is usually called the *Gauss lemma* (see Figure 19.6).

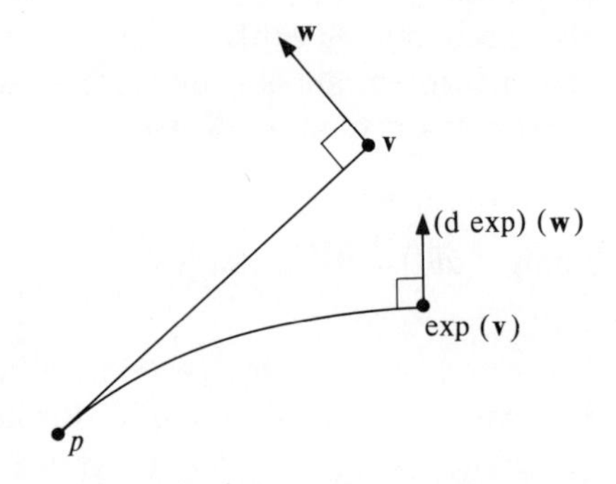

Figure 19.6 The Gauss lemma: d exp preserves orthogonality to radial geodesics.

PROOF. (i) $(\exp \circ \alpha)(t) = \exp(t\mathbf{v})$ is the maximal geodesic in S with initial velocity $\mathbf{v}$. Since geodesics have constant speed,

$$\|(d\exp)(\dot{\alpha}(1))\| = \|(\exp \dot{\circ} \alpha)(1)\| = \|(\exp \dot{\circ} \alpha)(0)\| = \|\mathbf{v}\| = \|\dot{\alpha}(1)\|.$$

Since $d\exp$ is linear on $T(S)_{\mathbf{v}} \supset (S_p)_{\mathbf{v}}$, it follows that if $\mathbf{w} = c\dot{\alpha}(1)$ for some $c \in \mathbb{R}$ then

$$\|(d\exp)(\mathbf{w})\| = |c|\,\|(d\exp)(\dot{\alpha}(1))\| = |c|\,\|\dot{\alpha}(1)\| = \|\mathbf{w}\|.$$

(ii) Each $\mathbf{w} \in (S_p)_{\mathbf{v}}$ is of the form $\mathbf{w} = \dot{\beta}(0)$ where $\beta(s) = \mathbf{v} + s\mathbf{x}$ for some $\mathbf{x} \in S_p$. Since

$$\dot{\alpha}(1) \cdot \mathbf{w} = \dot{\alpha}(1) \cdot \dot{\beta}(0) = \frac{d\alpha}{dt}(1) \cdot \frac{d\beta}{dt}(0) = \mathbf{v} \cdot \mathbf{x}$$

the condition that $\mathbf{w}$ be orthogonal to the ray α says that $\mathbf{v} \cdot \mathbf{x} = 0$.

We must show that $(\exp \dot{\circ} \alpha)(1) \cdot (d\exp)(\mathbf{w}) = 0$. But

$$(d\exp)(\mathbf{w}) = (d\exp)(\dot{\beta}(0)) = (\exp \dot{\circ} \beta)(0)$$

so

$$\begin{aligned}(\exp \dot{\circ} \alpha)(1) \cdot (d\exp)(\mathbf{w}) &= (\exp \dot{\circ} \alpha)(1) \cdot (\exp \dot{\circ} \beta)(0) \\ &= \mathbf{E}_1(1, 0) \cdot \mathbf{E}_2(1, 0)\end{aligned}$$

where $\mathbf{E}_1$ and $\mathbf{E}_2$ are the coordinate vector fields along the map $\psi\colon [0, 1] \times (-\varepsilon, \varepsilon) \to S$ defined by

$$\psi(t, s) = \exp(t(\mathbf{v} + s\mathbf{x})),$$

$\varepsilon > 0$ being chosen small enough that $t(\mathbf{v} + s\mathbf{x}) \in U$ whenever $0 \le t \le 1$ and $|s| < \varepsilon$ (see Figure 19.7). So we must show that $(\mathbf{E}_1 \cdot \mathbf{E}_2)(1, 0) =$

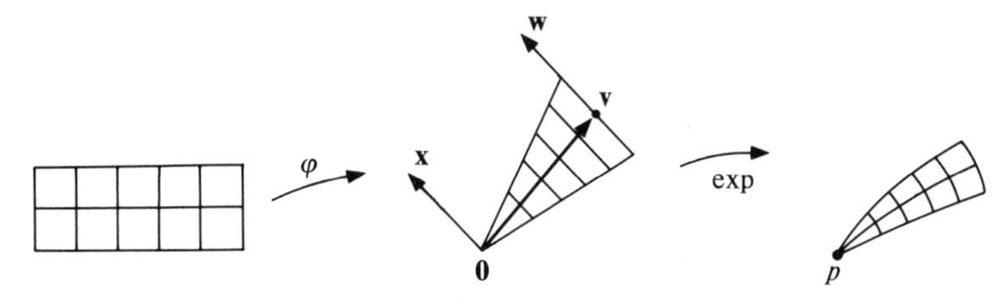

Figure 19.7 $\varphi(t, s) = t(\mathbf{v} + s\mathbf{x})$ maps $[0, 1] \times (-\varepsilon, \varepsilon)$ onto a triangle in S_p. The map $\psi = \exp \circ \varphi$ is a variation of the geodesic α.

$\mathbf{E}_1(1, 0) \cdot \mathbf{E}_2(1, 0) = 0$. We shall do this by showing that $(\mathbf{E}_1 \cdot \mathbf{E}_2)(t, 0) = 0$ for all $t \in [0, 1]$. Since $(\mathbf{E}_1 \cdot \mathbf{E}_2)(0, 0) = 0$ (because $\mathbf{E}_2(0, 0) = \mathbf{0}$), it suffices to check that $(\mathbf{E}_1 \cdot \mathbf{E}_2)(t, 0)$ is constant.

Note first that for each $s \in (-\varepsilon, \varepsilon)$ the coordinate curve $\alpha_s\colon [0, 1] \to S$ defined by

$$\alpha_s(t) = \exp(t(\mathbf{v} + s\mathbf{x}))$$

is a geodesic in S with initial velocity $\mathbf{v} + s\mathbf{x}$. Since geodesics have constant speed and $\mathbf{v} \cdot \mathbf{x} = 0$,

$$\|\mathbf{E}_1(s, t)\|^2 = \|\dot{\alpha}_s(t)\|^2 = \|\dot{\alpha}_s(0)\|^2 = \|\mathbf{v}\|^2 + s^2\|\mathbf{x}\|^2$$

for all $(t, s) \in [0, 1] \times (-\varepsilon, \varepsilon)$.

Now

$$\frac{\partial}{\partial t}(\mathbf{E}_1 \cdot \mathbf{E}_2) = (\nabla_{\mathbf{E}_1}\mathbf{E}_1) \cdot \mathbf{E}_2 + \mathbf{E}_1 \cdot (\nabla_{\mathbf{E}_1}\mathbf{E}_2)$$

where $(\nabla_{\mathbf{E}_1}\mathbf{E}_j)(t, s) = \nabla_{(t, s, 1, 0)}\mathbf{E}_j$ for $j \in \{1, 2\}$. Since each coordinate curve α_s is a geodesic, $(\nabla_{\mathbf{E}_1}\mathbf{E}_1)(s, t) = \ddot{\alpha}_s(t)$ is orthogonal to S and hence $(\nabla_{\mathbf{E}_1}\mathbf{E}_1) \cdot \mathbf{E}_2 = 0$. Since, in addition,

$$\begin{aligned}(\nabla_{\mathbf{E}_1}\mathbf{E}_2)(t, s) &= \left(\psi(t, s), \frac{\partial^2\psi}{\partial t\, \partial s}(t, s)\right) \\ &= \left(\psi(t, s), \frac{\partial^2\psi}{\partial s\, \partial t}(t, s)\right) = (\nabla_{\mathbf{E}_2}\mathbf{E}_1)(t, s),\end{aligned}$$

we find that

$$\begin{aligned}\frac{\partial}{\partial t}(\mathbf{E}_1 \cdot \mathbf{E}_2) &= \mathbf{E}_1 \cdot (\nabla_{\mathbf{E}_2}\mathbf{E}_1) = \frac{1}{2}\frac{\partial}{\partial s}(\mathbf{E}_1 \cdot \mathbf{E}_1) \\ &= \frac{1}{2}\frac{\partial}{\partial s}(\|\mathbf{v}\|^2 + s^2\|\mathbf{x}\|^2) = s\|\mathbf{x}\|^2\end{aligned}$$

so

$$\left.\frac{\partial}{\partial t}(\mathbf{E}_1 \cdot \mathbf{E}_2)\right|_{s=0} = 0$$

and hence $(\mathbf{E}_1 \cdot \mathbf{E}_2)(t, 0)$ is constant, as required. □

Theorem 3. *Let S be an n-surface in $\mathbb{R}^{n+1}$, let $p \in S$, and let $\varepsilon > 0$ be such that the exponential map of S maps the ball $B_\varepsilon = \{\mathbf{v} \in S_p : \|\mathbf{v}\| < \varepsilon\}$ diffeomorphically onto an open set U_ε in S. Then, for each $q \in U_\varepsilon$, the parametrized curve $\alpha(t) = \exp(t\mathbf{v})$, $0 \le t \le 1$, where $\mathbf{v} \in B_\varepsilon$ is such that $\exp(\mathbf{v}) = q$, is a geodesic in S joining p to q, and if $\beta: [a, b] \to S$ is any other parametrized curve in S joining p to q then $l(\beta) \ge l(\alpha)$.*

PROOF. Let $r: S_p \to \mathbb{R}$ be defined by $r(\mathbf{x}) = \|\mathbf{x}\|$. We shall use the following facts about the 1-form dr on $S_p - \{\mathbf{0}\}$:

(a) If $\mathbf{w} \in (S_p)_\mathbf{v}$ is tangent to the ray in S_p through $\mathbf{v} \in S_p$ then $|dr(\mathbf{w})| = \|\mathbf{w}\|$
(b) If $\mathbf{w} \in (S_p)_\mathbf{v}$ is orthogonal to the ray in S_p through $\mathbf{v} \in S_p$ then $dr(\mathbf{w}) = 0$.

To verify these facts, note that each $\mathbf{w} \in (S_p)_\mathbf{v}$ is of the form $\mathbf{w} = \dot{\gamma}(0)$ where $\gamma(s) = \mathbf{v} + s\mathbf{x}$ for some $\mathbf{x} \in S_p$. If $\mathbf{w}$ is tangent to the ray through $\mathbf{v}$ then $\mathbf{x} = \lambda\mathbf{v}$ for some $\lambda \in \mathbb{R}$ and so $\gamma(s) = (1 + \lambda s)\mathbf{v}$ and

$$\begin{aligned}|dr(\mathbf{w})| &= |dr(\dot{\gamma}(0))| = |(r \circ \gamma)'(0)| \\ &= \left|\left.\frac{d}{ds}\right|_0 \|(1 + \lambda s)\mathbf{v}\|\right| = |\lambda|\,\|\mathbf{v}\| = \|\mathbf{x}\| = \|\mathbf{w}\|.\end{aligned}$$

If $\mathbf{w}$ is orthogonal to the ray through $\mathbf{v}$ then $\mathbf{x} \perp \mathbf{v}$, $r \circ \gamma(s) = \|\mathbf{v} + s\mathbf{x}\| = (\|\mathbf{v}\|^2 + s^2\|\mathbf{x}\|^2)^{1/2}$, and $dr(\mathbf{w}) = dr(\dot{\gamma}(0)) = (r \circ \gamma)'(0) = 0$.

Now the fact that $\alpha(t) = \exp(t\mathbf{v})$ is a geodesic joining p to q is already clear from Theorem 2. So suppose β: $[a, b] \to S$ has $\beta(a) = p$ and $\beta(b) = q$. Let c denote the least upper bound of the set

$$\{t \in [a, b]: \beta([a, t]) \subset U_\varepsilon\}$$

so that $\beta(I) \subset U_\varepsilon$ where $I = [a, b]$ if $c = b$ and $I = [a, c]$ otherwise. Letting $\bar{r}$: $U_\varepsilon \to \mathbb{R}$ be defined by $\bar{r} = r \circ (\exp|_{B_\varepsilon})^{-1}$ (see Figure 19.8), we see that

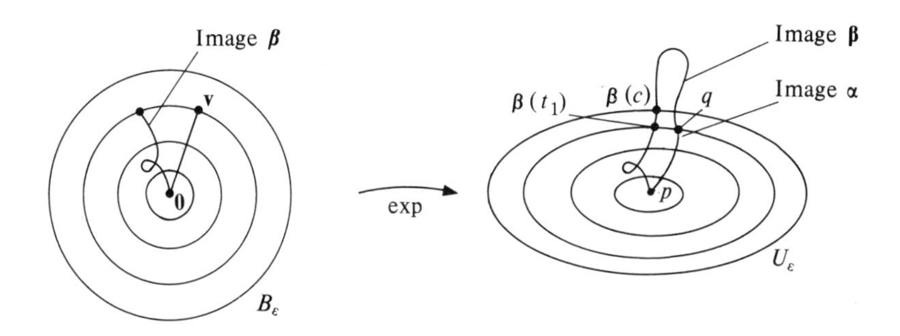

Figure 19.8 The concentric spheres in the ball B_ε are level sets of r: $S_p \to \mathbb{R}$. The images of these sets under exp are level sets of $\bar{r}$: $U_\varepsilon \to \mathbb{R}$.

$\bar{r}(\beta(a)) = \bar{r}(p) = 0$ and $\lim_{t \to c} \bar{r}(\beta(t)) = \varepsilon > \bar{r}(q)$ if $c \neq b$, $\bar{r}(\beta(b)) = \bar{r}(q)$ if $c = b$. In either case, by the intermediate value theorem, $\bar{r}(\beta(t)) = \bar{r}(q)$ for some $t \in I$; let t_1 be the smallest such t. Let $\boldsymbol{\beta}$: $[a, t_1] \to B_\varepsilon$ be defined by $\boldsymbol{\beta}(t) = (\exp|_{B_\varepsilon})^{-1}(\beta(t))$. Then $\dot{\boldsymbol{\beta}}(t) = \dot{\boldsymbol{\beta}}_T(t) + \dot{\boldsymbol{\beta}}_\perp(t)$ where $\dot{\boldsymbol{\beta}}_T(t)$ is tangent to the ray in S_p through $\boldsymbol{\beta}(t)$ and $\boldsymbol{\beta}_\perp(t)$ is orthogonal to this ray. Using the above facts about dr, we find

$$\begin{aligned}
l(\alpha) &= \int_0^1 \|\dot{\alpha}\| = \int_0^1 \|\mathbf{v}\| = \|\mathbf{v}\| = r(\mathbf{v}) = \bar{r}(q) = \bar{r}(\beta(t_1)) - \bar{r}(\beta(a)) \\
&= \int_a^{t_1} (\bar{r} \circ \beta)' = \int_a^{t_1} (r \circ \boldsymbol{\beta})' = \int_a^{t_1} dr(\dot{\boldsymbol{\beta}}) = \int_a^{t_1} dr(\dot{\boldsymbol{\beta}}_T) \\
&\leq \int_a^{t_1} |dr(\dot{\boldsymbol{\beta}}_T)| = \int_a^{t_1} \|\dot{\boldsymbol{\beta}}_T\| \overset{(1)}{=} \int_a^{t_1} \|(d\exp)(\dot{\boldsymbol{\beta}}_T)\| \\
&\overset{(2)}{\leq} \int_0^{t_1} \|(d\exp)(\dot{\boldsymbol{\beta}}_T) + (d\exp)(\dot{\boldsymbol{\beta}}_\perp)\| \\
&= \int_a^{t_1} (d\exp)(\dot{\boldsymbol{\beta}}) = \int_a^{t_1} \|\exp \dot{\circ} \boldsymbol{\beta}\| = \int_a^{t_1} \|\dot{\beta}\| \leq \int_a^{b} \|\dot{\beta}\| = l(\beta)
\end{aligned}$$

when the equality (1) and the inequality (2) are valid by the lemma. ☐

Remark 1. $l(\beta)$ can equal $l(\alpha)$ only if each of the three inequalities above is an equality. This can happen only if (working backwards through the inequalities)

(i) $\beta(t) = \beta(t_1)$ for all $t \geq t_1$.
(ii) $\dot{\boldsymbol{\beta}}(t)$ has no component orthogonal to the ray in S_p through $\boldsymbol{\beta}(t)$, for all $t \leq t_1$, and
(iii) $r \circ \boldsymbol{\beta}$ is monotone on $[a, t_1]$.

These three conditions imply that, *under the hypotheses of the theorem, if $l(\beta) = l(\alpha)$ then $\beta = \alpha \circ h$ where h: $[a, b] \to [0, \|\mathbf{v}\|]$ is monotone; in particular, α and β have the same image.*

Remark 2. A review of the proof of Theorem 3 will verify that if V is any open set in $U \cap S_p$ (U the domain of the exponential map) such that exp maps V diffeomorphically onto an open set W in S and if $\exp(t\mathbf{v}) \in W$ for $0 \leq t \leq t_0$ then $\alpha_{\mathbf{v}}(t) = \exp(t\mathbf{v})$ $(0 \leq t \leq t_0)$ is shortest among all parametrized curves in W joining p to $\exp(t_0\mathbf{v})$. However, $\alpha_{\mathbf{v}}(t) = \exp(t\mathbf{v})$ $(0 \leq t \leq t_0)$ need not be shortest among all curves in S joining p to $\exp(t_0\mathbf{v})$ (see Exercise 19.4).

Remark 3. A point $q = \alpha_{\mathbf{v}}(\tau)$ is said to be *conjugate* to $p = \alpha_{\mathbf{v}}(0)$ along the geodesic $\alpha_{\mathbf{v}}(t) = \exp(t\mathbf{v})$ if $(d \exp)(\mathbf{w}) = \mathbf{0}$ for some non-zero $\mathbf{w} \in (S_p)_{\tau\mathbf{v}}$. By the lemma of this chapter, each $\mathbf{w} \in (S_p)_{\tau\mathbf{v}}$ such that $(d \exp)(\mathbf{w}) = \mathbf{0}$ must be orthogonal to the ray $\alpha(t) = t\mathbf{v}$ in S_p, so $\mathbf{w} = \dot{\beta}(0)$ where $\beta(s) = \mathbf{v} + s\mathbf{x}$ for some $\mathbf{x} \in S_p$ with $\mathbf{x} \perp \mathbf{v}$. Defining ψ: $[0, \tau] \times (-\varepsilon, \varepsilon) \to S$ (ε sufficiently small) by

$$\psi(t, s) = \exp(t(\mathbf{v} + s\mathbf{x})),$$

we obtain a variation of the geodesic $\alpha_{\mathbf{v}}|_{[0, \tau]}$ such that each of the coordinate curves $\alpha_s(t) = \psi(t, s)$ is a geodesic starting at p, and these geodesics tend to focus at q (see Figure 19.9). Thus conjugate points along geodesics from p are analogous to focal points along normal lines to an n-surface in $\mathbb{R}^{n+1}$. This analogy is made more complete by the observation that the geodesics radiating from p in S are the same as the geodesics normal to the

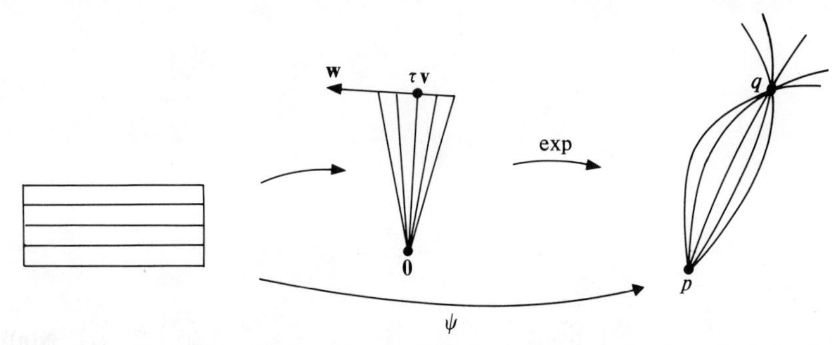

Figure 19.9 There is a 1-parameter family of geodesics through p which tends to focus at the conjugate point q.

$(n-1)$-surface $\exp\{\mathbf{v} \in S_p: \|\mathbf{v}\| = \delta\}$, $\delta > 0$ being chosen small enough that exp is a diffeomorphism on some ball B_ε about the origin in S_p of radius $\varepsilon > \delta$. The proof of Theorem 3 can be modified slightly to show that up to the first conjugate point the geodesic $\alpha_\mathbf{v}$ locally minimizes the length integral in the sense that if χ is any fixed endpoint variation of $\alpha_\mathbf{v}|_{[0,t_1]}$ and if there are no conjugate points $\alpha_\mathbf{v}(\tau)$ for $0 < \tau < t_1$ then $l(\gamma_s) \geq l(\alpha_\mathbf{v}|_{[0,t_1]})$ for all sufficiently small s, where $\gamma_s(t) = \chi(t, s)$. It can be shown that $\alpha_\mathbf{v}$ does not minimize the length integral, even locally, beyond the first conjugate point (see Figure 19.4).

The set of points $q \in S$ such that q is conjugate to p along some geodesic through p is called the *conjugate locus* of p in S (see Figure 19.10).

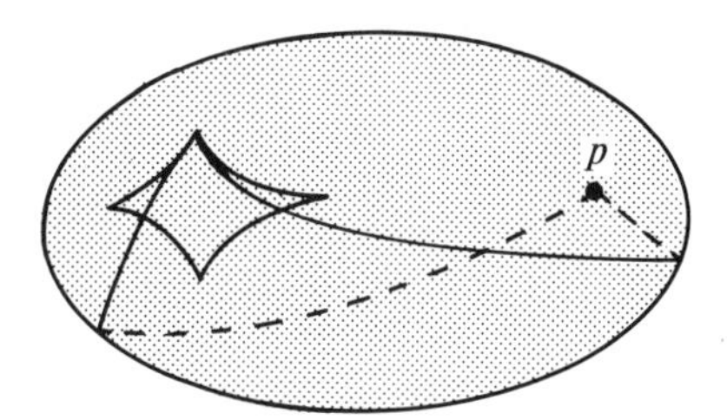

Figure 19.10 The conjugate locus of a point p on an ellipsoid. Two geodesics from p are also shown.

Exercises

19.1. Let S be an n-surface in $\mathbb{R}^{n+1}$. For $\alpha: [a, b] \to S$ a parametrized curve in S, define the *energy* of α to be the integral $\int_a^b \|\dot{\alpha}(t)\|^2\, dt$. Show that α is a geodesic in S if and only if the energy integral is stationary at α with respect to fixed endpoint variations.

19.2. (a) Show that each vector tangent to the unit circle $\mathbf{S}^1 \subset \mathbb{R}^2$ is of the form

$$\mathbf{v}(\varphi, \theta) = (\cos\varphi, \sin\varphi, -\theta\sin\varphi, \theta\cos\varphi)$$

for some $\varphi, \theta \in \mathbb{R}$.

(b) Show that the exponential map on $\mathbf{S}^1$ is given by

$$\exp(\mathbf{v}(\varphi, \theta)) = e^{i(\varphi+\theta)}$$

where $\mathbb{R}^2$ is viewed as the set of complex numbers by identifying (a, b) with $a + bi$.

19.3. Let S be an oriented n-surface in $\mathbb{R}^{n+1}$, let $T(S) = \bigcup_{p \in S} S_p \subset \mathbb{R}^{2(n+1)}$, and let $\mathbf{v} = (p, v) \in T(S)$.

(a) Show that the tangent space $(T(S))_\mathbf{v}$ to $T(S)$ at $\mathbf{v}$ is

$$(T(S))_\mathbf{v} = \{(x_1, x_2, x_3, x_4) \in \mathbb{R}^{4(n+1)}: \\ x_1 = p,\ x_2 = v,\ (p, x_3) \in S_p,\ (p, x_3) \cdot L_p(\mathbf{v}) = (p, x_4) \cdot \mathbf{N}(p)\}$$

where L_p is the Weingarten map of S at p.

(b) Show that the tangent space $(S_p)_{\mathbf{v}}$ to S_p at $\mathbf{v}$ is

$$(S_p)_{\mathbf{v}} = \{(p, v, 0, x): (p, x) \in S_p\}.$$

(c) Show that $(p, v, 0, x) \in (S_p)_{\mathbf{v}}$ is tangent to the ray $\alpha(t) = t\mathbf{v}$ in S_p if and only if $x = \lambda v$ for some $\lambda \in \mathbb{R}$, and that $(p, v, 0, x)$ is orthogonal to this ray if and only if $v \cdot x = 0$ (orthogonality in $(S_p)_{\mathbf{v}}$!).

19.4. Let S be the cylinder $x_1^2 + x_2^2 = 1$ in $\mathbb{R}^3$ and let $p = (1, 0, 0) \in S$.

(a) Show that $S_p = \{(p, 0, a, b): a, b \in \mathbb{R}\}$.
(b) Compute $\exp(\mathbf{v})$ for $\mathbf{v} = (p, 0, a, b) \in S_p$.
(c) Show that the conjugate locus of p in S is empty.
(d) Show that there is an open set in S_p containing the ray $\alpha(t) = t\mathbf{v}$, $\mathbf{v} = (p, 0, 1, 1)$, which is mapped diffeomorphically by exp onto an open set in S containing the geodesic $\alpha_{\mathbf{v}}(t) = \exp(t\mathbf{v})$.
(e) Show that, nevertheless, there is a $t_0 \in \mathbb{R}$ such that $\alpha_{\mathbf{v}}(t) = \exp(t\mathbf{v})$ $(0 \le t \le t_0)$ is not a shortest curve in S joining p to $\exp(t_0 \mathbf{v})$.

19.5. Let $\mathbf{S}^2$ be the unit 2-sphere in $\mathbb{R}^3$ and let $p = (0, 0, 1) \in \mathbf{S}^2$.

(a) Show that $\mathbf{S}_p^2 = \{(p, a, b, 0): a, b \in \mathbb{R}\}$.
(b) Compute $\exp(\mathbf{v})$ for $\mathbf{v} = (p, a, b, 0) \in \mathbf{S}_p^2$.
(c) Show that the conjugate locus of p consists of the single point $q = (0, 0, -1)$.
(d) Show that exp maps the ball $\{\mathbf{v} \in \mathbf{S}_p^2: \|\mathbf{v}\| < \pi\}$ diffeomorphically onto $\mathbf{S}^2 - \{q\}$.

19.6. Let S be a connected n-surface in $\mathbb{R}^{n+1}$. For p_1 and $p_2 \in S$, define the *intrinsic distance* $d(p_1, p_2)$ from p_1 to p_2 to be the greatest lower bound of the set

$$\{l(\alpha): \alpha \text{ is a piecewise smooth parametrized curve in } S \text{ joining } p_1 \text{ to } p_2\}.$$

Show that, for all p_1, p_2, and $p_3 \in S$,

(a) $d(p_1, p_2) = d(p_2, p_1)$
(b) $d(p_1, p_2) + d(p_2, p_3) \ge d(p_1, p_3)$
(c) $d(p_1, p_2) \ge 0$ and $d(p_1, p_2) = 0$ if and only if $p_1 = p_2$.

[*Hint*: for (c), take $p = p_1$ and choose ε as in Theorem 3 but small enough so that $p_2 \notin U_\varepsilon$. Then argue that $d(p_1, p_2) \ge \varepsilon$.]

19.7. Let S be an n-surface in $\mathbb{R}^{n+1}$ and let $T_1(S)$ denote the unit sphere bundle of S (Exercise 15.6).

(a) Show that the restriction to $T_1(S)$ of the geodesic spray is a tangent vector field on $T_1(S)$.
(b) Using the fact that $T_1(S)$ is compact if S is compact, show that each compact n-surface in $\mathbb{R}^{n+1}$ is geodesically complete.
(c) Conclude that if S is compact then the domain of the exponential map of S is all of $T(S)$.

20 Surfaces with Boundary

In this chapter we shall develop some machinery which we shall need in the next chapter to prove one of the most celebrated theorems in differential geometry, the Gauss–Bonnet theorem. We shall first discuss n-surfaces-with-boundary. Then we shall develop a little bit of the differential calculus of forms.

An *n-surface-with-boundary* in $\mathbb{R}^{n+1}$ is a non-empty subset S of $\mathbb{R}^{n+1}$ of the form

$$\begin{aligned} S &= f^{-1}(c) \cap g_1^{-1}((-\infty, c_1]) \cap \cdots \cap g_k^{-1}((-\infty, c_k]) \\ &= \{p \in U : f(p) = c, g_1(p) \le c_1, \ldots, g_k(p) \le c_k\} \end{aligned}$$

where k is a positive integer, $\{c, c_1, \ldots, c_k\} \subset \mathbb{R}$, and $f: U \to \mathbb{R}$ and $g_i: U_i \to \mathbb{R}$ for $i \in \{1, \ldots, k\}$ are smooth functions defined on open subsets of $\mathbb{R}^{n+1}$ and satisfying the conditions

(i) $\nabla f(p) \neq \mathbf{0}$ for all $p \in S$
(ii) $g_i^{-1}(c_i) \cap g_j^{-1}(c_j) \cap S$ is empty whenever $i \neq j$
(iii) For each $i \in \{1, \ldots, k\}$, $\{\nabla f(p), \nabla g_i(p)\}$ is linearly independent for all $p \in g_i^{-1}(c_i) \cap S$.

The *boundary* ∂S of S is the set

$$\partial S = \{p \in S : g_i(p) = c_i \text{ for some } i\} = \bigcup_{i=1}^{k} g_i^{-1}(c_i) \cap S.$$

The *interior* of S is the set $S - \partial S$.

Condition (i) guarantees that the interior of S is an n-surface in $\mathbb{R}^{n+1}$ and in fact that S itself is part of an n-surface ($f^{-1}(c)$) in $\mathbb{R}^{n+1}$. Condition (ii) guarantees that the parts of the boundary ∂S defined by the various functions g_i are disjoint. Condition (iii) then guarantees that ∂S is an $(n-1)$-surface in $\mathbb{R}^{n+1}$.

Remark. Equivalently, an n-surface-with-boundary in $\mathbb{R}^{n+1}$ may be described as

$$S = \{p \in \tilde{S}: g_1(p) \le c_1, \ldots, g_k(p) \le c_k\}$$

where $\tilde{S}$ is an n-surface in $\mathbb{R}^{n+1}$ and $g_1, \ldots, g_k: \tilde{S} \to \mathbb{R}$ are smooth functions on $\tilde{S}$ such that $g_i^{-1}(c_i) \cap g_j^{-1}(c_j)$ is empty whenever $i \ne j$ and such that $(\text{grad } g_i)(p) \ne \mathbf{0}$ whenever $p \in g_i^{-1}(c_i)$.

EXAMPLE 1. The hemisphere

$$S = \{(x_1, x_2, x_3) \in \mathbb{R}^3: x_1^2 + x_2^2 + x_3^2 = 1, x_3 \ge 0\}$$

is a 2-surface-with-boundary in $\mathbb{R}^3$ (take $f(x_1, x_2, x_3) = x_1^2 + x_2^2 + x_3^2$, $c = 1$, $g(x_1, x_2, x_3) = -x_3$, and $c_1 = 0$). Its boundary is the equator:

$$\partial S = \{(x_1, x_2, x_3) \in S: x_3 = 0\}$$

(see Figure 20.1).

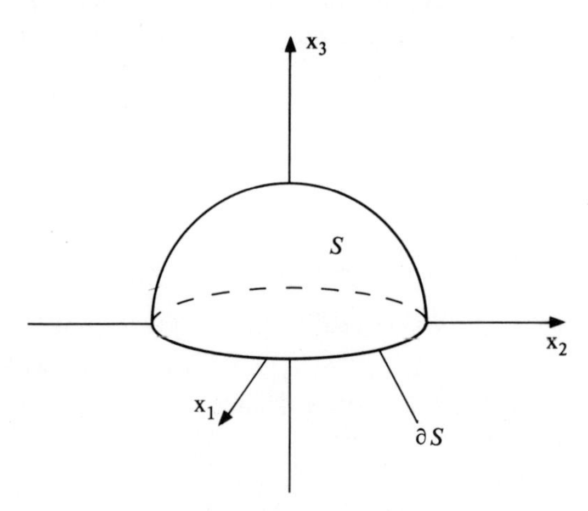

Figure 20.1 The hemisphere $x_1^2 + x_2^2 + x_3^2 = 1$, $x_3 \ge 0$.

EXAMPLE 2. For $S = f^{-1}(c)$ an $(n-1)$-surface (without boundary) in $\mathbb{R}^n$, consider the set

$$S \times I = \{(x_1, \ldots, x_{n+1}) \in \mathbb{R}^{n+1}: f(x_1, \ldots, x_n) = c, 0 \le x_{n+1} \le 1\}.$$

Thus $S \times I$ is a portion of the cylinder over S (see Figure 20.2). $S \times I$ is an n-surface-with-boundary in $\mathbb{R}^{n+1}$. Its boundary consists of two copies of S: $g_1^{-1}(0)$ where $g_1(x_1, \ldots, x_{n+1}) = -x_{n+1}$ and $g_2^{-1}(1)$ where $g_2(x_1, \ldots, x_{n+1}) = x_{n+1}$.

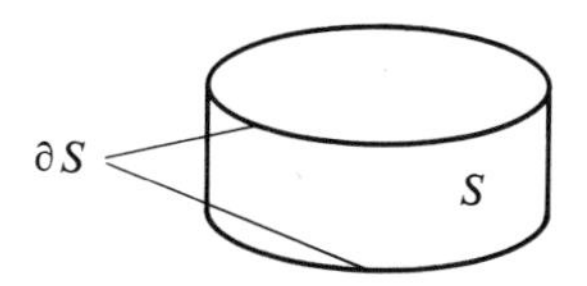

Figure 20.2 The cylinder-with-boundary $S = \mathbf{S}^1 \times I$.

The *tangent space* at a point $p \in S$, $S = f^{-1}(c) \cap \bigcap_{i=1}^{k} g_i^{-1}((-\infty, c_i])$ an n-surface-with-boundary in $\mathbb{R}^{n+1}$, is the n-dimensional vector space

$$S_p = \{\mathbf{v} \in \mathbb{R}_p^{n+1} : \mathbf{v} \cdot \nabla f(p) = 0\}.$$

A vector $\mathbf{v} \in S_p$, $p \in \partial S$ ($p \in g_i^{-1}(c_i)$ for some i), is (see Figure 20.3)

(i) *outward-pointing* if $\mathbf{v} \cdot \nabla g_i(p) > 0$
(ii) *inward-pointing* if $\mathbf{v} \cdot \nabla g_i(p) < 0$
(iii) *tangent to the boundary* if $\mathbf{v} \cdot \nabla g_i(p) = 0$
(iv) *normal to the boundary* if $\mathbf{v} \cdot \mathbf{w} = 0$ for all $\mathbf{w} \in S_p$ which are tangent to the boundary.

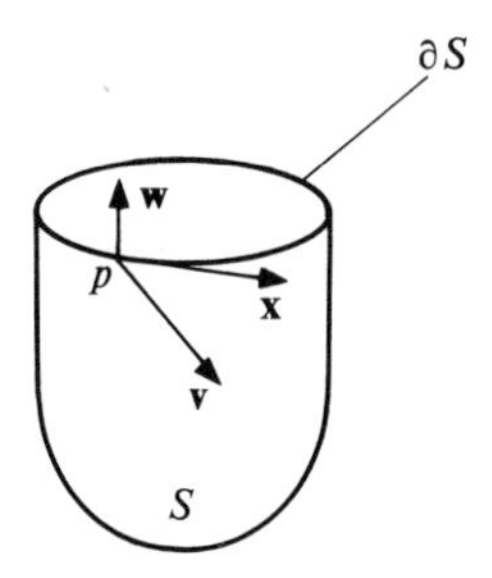

Figure 20.3 Three vectors in the tangent space S_p, $p \in \partial S$. $\mathbf{v}$ is inward pointing, $\mathbf{w}$ is outward pointing and normal to the boundary, and $\mathbf{x}$ is tangent to the boundary.

Note, for each $p \in \partial S$, that the set $(\partial S)_p$ of all vectors in S_p which are tangent to the boundary is an $(n-1)$-dimensional subspace of S_p and that there is exactly one outward-pointing unit vector in S_p which is normal to the boundary.

Note also that all the conditions above can be reformulated without reference to the functions $f, g_1, \ldots, g_k$ and hence depend only on the surface and not on the functions defining it. Thus, for $p \in S$, the tangent space S_p can be described as the set of all vectors $\mathbf{v} \in \mathbb{R}_p^{n+1}$ of the form $\mathbf{v} = \dot{\alpha}(t_0)$ where $\alpha: I \to \mathbb{R}^{n+1}$ (I an open interval) is a parametrized curve such that $\alpha(t_0) = p$ for some $t_0 \in I$ and either $\alpha(t) \in S$ for all $t \in I$ with $t \le t_0$, or $\alpha(t) \in S$ for all $t \in I$ with $t \ge t_0$, or both. A vector $\mathbf{v} \in S_p$, $p \in \partial S$, is tangent to ∂S if $\mathbf{v} = \dot{\alpha}(t_0)$ for some $\alpha: I \to \mathbb{R}^{n+1}$ with $\alpha(t) \in \partial S$ for all $t \in I$. A vector $\mathbf{v} \in S_p$, $p \in \partial S$, which is not tangent to ∂S is outward-pointing if $\mathbf{v} = \dot{\alpha}(t_0)$ for some $\alpha: I \to \mathbb{R}^{n+1}$ with $\alpha(t) \in S$ for all $t \in I$ with $t \le t_0$, and $\mathbf{v}$ is inward-pointing if $\mathbf{v} = \dot{\alpha}(t_0)$ for some α with $\alpha(t) \in S$ for all $t \in I$ with $t \ge t_0$.

An *orientation* on S is a choice of smooth unit vector field $\mathbf{N}$ on S with $\mathbf{N}(p) \perp S_p$ for all $p \in S$, where smoothness is defined exactly as in the case of n-surfaces without boundary. Note that each orientation $\mathbf{N}$ on S defines a volume form on S, i.e., a smooth n-form ζ on S with $\zeta(\mathbf{v}_1, \ldots, \mathbf{v}_n) = \pm 1$ whenever $\{\mathbf{v}_1, \ldots, \mathbf{v}_n\}$ is an orthonormal basis for S_p, by the formula

$$\zeta(\mathbf{v}_1, \ldots, \mathbf{v}_n) = \det \begin{pmatrix} \mathbf{v}_1 \\ \vdots \\ \mathbf{v}_n \\ \mathbf{N}(p) \end{pmatrix}.$$

Conversely, each volume form ζ on S uniquely determines an orientation $\mathbf{N}$ on S by the requirement that $\mathbf{N}(p)$ for $p \in S$ be the unique unit vector in $S_p^\perp$ such that

$$\det \begin{pmatrix} \mathbf{v}_1 \\ \vdots \\ \mathbf{v}_n \\ \mathbf{N}(p) \end{pmatrix} = \zeta(\mathbf{v}_1, \ldots, \mathbf{v}_n)$$

for $\{\mathbf{v}_1, \ldots, \mathbf{v}_n\}$ any orthonormal basis for S_p. Thus an orientation on S determines, and is determined by, a choice of volume form on S; we could therefore *define* an orientation on S to be a choice of volume form. This definition makes sense also for n-surfaces in $\mathbb{R}^{n+m}$ ($m \geq 0$) so we shall reformulate the concept of orientation in this more general setting.

Let S be either an n-surface in $\mathbb{R}^{n+m}$ or an n-surface-with-boundary in $\mathbb{R}^{n+1}$. A *volume form* on S is a smooth n-form ζ on S such that $\zeta(\mathbf{v}_1, \ldots, \mathbf{v}_n) = \pm 1$ whenever $\{\mathbf{v}_1, \ldots, \mathbf{v}_n\}$ is an orthonormal basis for S_p, $p \in S$. An *orientation* on S is a choice of volume form ζ on S. An ordered basis $\{\mathbf{v}_1, \ldots, \mathbf{v}_n\}$ (not necessarily orthonormal) for S_p, $p \in S$, is said to be *consistent with the orientation* ζ if (and only if) $\zeta(\mathbf{v}_1, \ldots, \mathbf{v}_n) > 0$. S is said to be *oriented* if there is given an orientation ζ on S.

Remark. These definitions extend in an obvious way to n-surfaces-with-boundary in $\mathbb{R}^{n+m}$. We leave it to the interested reader to formulate a definition of "n-surface-with-boundary in $\mathbb{R}^{n+m}$". For n-surfaces or n-surfaces-with-boundary in $\mathbb{R}^{n+1}$ we shall, whenever convenient, continue to view an orientation as a choice of smooth unit normal vector field.

For S an n-surface-with-boundary in $\mathbb{R}^{n+1}$, an orientation ζ on S defines an orientation $\zeta_{\partial S}$ on the $(n-1)$-surface ∂S by the formula $\zeta_{\partial S} = \mathbf{V} \lrcorner \zeta$ where $\mathbf{V}$ is the smooth vector field on ∂S defined by $\mathbf{V}(p) =$ the outward pointing unit vector in S_p which is normal to the boundary. This orientation $\zeta_{\partial S}$ is called the *induced orientation* on ∂S.

Integration of differential n-forms over compact oriented n-surfaces in $\mathbb{R}^{n+m}$ or over compact oriented n-surfaces-with-boundary in $\mathbb{R}^{n+1}$ can now be defined exactly as for n-surfaces in $\mathbb{R}^{n+1}$. We first define *local parametrizations*. For S an oriented n-surface in $\mathbb{R}^{n+m}$, a local parametrization of S is a parametrized n-surface $\varphi: U \to \mathbb{R}^{n+m}$ such that $\varphi(U) \subset S$ and such that φ is *consistent with the orientation* ζ on S in the sense that $\zeta(\mathbf{E}_1, \ldots, \mathbf{E}_n) > 0$ where the $\mathbf{E}_1, \ldots, \mathbf{E}_n$ are the coordinate vector fields along φ. For S an oriented n-surface-with-boundary in $\mathbb{R}^{n+1}$, a local parametrization is a smooth map φ of one of the following two types:

(i) $\varphi: U \to \mathbb{R}^{n+1}$ is a parametrized n-surface such that $\varphi(U)$ is an open set in S (i.e., $\varphi(U)$ is the intersection with S of an open set in $\mathbb{R}^{n+1}$) and such that φ is consistent with the orientation ζ in the sense described above (these are the local parametrizations whose images are contained in the interior of S);

(ii) $\varphi: U \to \mathbb{R}^{n+1}$ is the restriction to $U = V \cap \mathbb{R}^n_-$, where $\mathbb{R}^n_- = \{(x_1, \ldots, x_n) \in \mathbb{R}^n: x_n \leq 0\}$, of a parametrized n-surface $\tilde{\varphi}: V \to \mathbb{R}^{n+1}$ such that $\varphi(U)$ is an open set in S and such that φ is consistent with the orientation ζ on S in the sense described above (these are the local parametrizations whose images contains points of ∂S; see Figure 20.4).

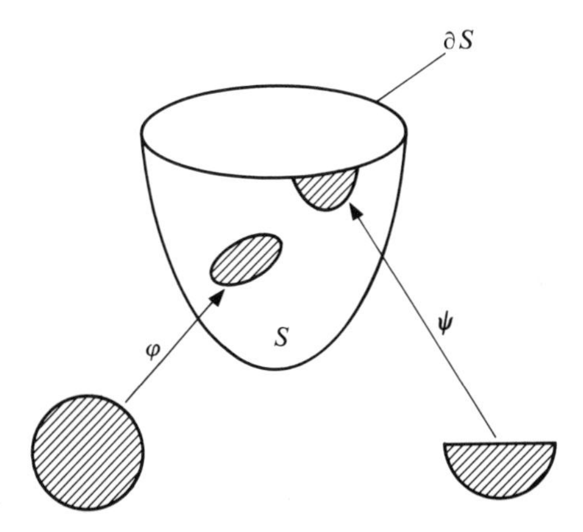

Figure 20.4 Local parametrizations of a 2-surface-with-boundary.

The existence of one to one local parametrizations whose images cover the given n-surface (or n-surface-with-boundary) is guaranteed by Theorem 1 of Chapter 15 and its generalizations (see Exercises 15.10 and 20.1). We may even insist, if we wish, that each of the sets U be either an open ball in $\mathbb{R}^n$ or the intersection with $\mathbb{R}^n_-$ of an open ball centered on the $(n-1)$-plane $x_n = 0$ (Figure 20.4).

For ω an n-form on the compact oriented n-surface $S \subset \mathbb{R}^{n+m}$ or on the compact oriented n-surface-with-boundary $S \subset \mathbb{R}^{n+1}$, the *integral* $\int_S \omega$ is defined to be the real number

$$\int_S \omega = \sum_i \int_{\varphi_i} (f_i \omega)$$

where $\{f_i\}$ is any partition of unity on S subordinate to a finite collection $\{\varphi_i\}$ of one to one local parametrizations of S. The existence of a partition of unity on S and the fact that $\int_S \omega$ is independent of the particular partition of unity used are proved exactly as in the case of oriented n-surfaces in $\mathbb{R}^{n+1}$ (see Chapter 17).

Having defined the integral over S of an arbitrary smooth n-form ω, we can define the *volume* of a compact oriented n-surface S in $\mathbb{R}^{n+m}$ or of a compact oriented n-surface-with-boundary S in $\mathbb{R}^{n+1}$ to be the integral over S of the orientation volume form:

$$V(S) = \int_S \zeta,$$

and we can define the *integral* over such an S *of any smooth function* $f\colon S \to \mathbb{R}$ by the formula

$$\int_S f = \int_S f\zeta.$$

The constructions above are part of the integral calculus of forms. We shall also need to use some of the differential calculus of forms.

Let S be an n-surface or an n-surface-with-boundary in $\mathbb{R}^{n+1}$. The *differential* of a smooth function $f\colon S \to \mathbb{R}$ is the smooth 1-form df on S defined by $df(\mathbf{v}) = \nabla_{\mathbf{v}} f$ for $\mathbf{v} \in S_p$, $p \in S$. The *exterior derivative* of a smooth 1-form ω on S is the smooth 2-form $d\omega$ on S defined by

$$d\omega(\mathbf{v}_1, \mathbf{v}_2) = \nabla_{\mathbf{v}_1}\omega(\mathbf{V}_2) - \nabla_{\mathbf{v}_2}\omega(\mathbf{V}_1) - \omega([\mathbf{V}_1, \mathbf{V}_2](p))$$

where, for $\mathbf{v}_1, \mathbf{v}_2 \in S_p$, $p \in S$, $\mathbf{V}_1$ and $\mathbf{V}_2$ are arbitrarily chosen smooth tangent vector fields, defined on an open set U of S containing p, such that $\mathbf{V}_1(p) = \mathbf{v}_1$ and $\mathbf{V}_2(p) = \mathbf{v}_2$, and where $[\mathbf{V}_1, \mathbf{V}_2]$, the *Lie bracket* of the vector fields $\mathbf{V}_1$ and $\mathbf{V}_2$, is the smooth tangent vector field on S defined by

$$[\mathbf{V}_1, \mathbf{V}_2](q) = \nabla_{\mathbf{V}_1(q)} \mathbf{V}_2 - \nabla_{\mathbf{V}_2(q)} \mathbf{V}_1$$

(see Exercise 9.12). The verification that the right hand side of the formula defining $d\omega$ is independent of the choice of vector fields $\mathbf{V}_1$ and $\mathbf{V}_2$ is left as an exercise (Exercise 20.2). Note that the multilinearity, skewsymmetry, and smoothness of $d\omega$ are evident from the definition.

Remark. The formula defining $d\omega$ often appears in the literature with a factor of $\frac{1}{2}$ on the right hand side. This is to compensate for a factor of $\frac{1}{2}$ which is also introduced, in these sources, into the definition of exterior product of 1-forms.

Lemma 1. *Let* $f\colon S \to \mathbb{R}$ *be a smooth function on* S *and let* ω *be a smooth* 1-*form on* S. *Then*

(i) $d(df) = 0$
(ii) $d(f\omega) = df \wedge \omega + f\, d\omega$.

PROOF. (i) Since $d(df)$ is bilinear, it suffices to check that $d(df)(\mathbf{v}_i, \mathbf{v}_j) = 0$ for all $i, j \in \{1, \ldots, n\}$ where $\{\mathbf{v}_1, \ldots, \mathbf{v}_n\}$ is any basis for S_p ($p \in S$ arbitrary). We shall take $\mathbf{v}_i = \mathbf{E}_i(q)$ where the $\mathbf{E}_i$ are the coordinate vector fields of some one to one local parametrization $\varphi\colon U \to S$ with $\varphi(q) = p$. Setting $\mathbf{V}_i = \mathbf{E}_i \circ \varphi^{-1}$, we see that $\mathbf{V}_j$ has vector part $(\partial\varphi/\partial x_j) \circ \varphi^{-1}$ so $\nabla_{\mathbf{V}_i(p)} \mathbf{V}_j = \nabla_{\mathbf{E}_i(q)} \mathbf{V}_j$ has vector part $(\partial^2\varphi/\partial x_i\, \partial x_j)(q)$ for all i and j, and $[\mathbf{V}_i, \mathbf{V}_j](p) = \nabla_{\mathbf{V}_i(p)} \mathbf{V}_j - \nabla_{\mathbf{V}_j(p)} \mathbf{V}_i = \mathbf{0}$. Since $\mathbf{V}_i(p) = \mathbf{E}_i(q) = \mathbf{v}_i$ for all i, it follows that

$$\begin{aligned}
d(df)(\mathbf{v}_i, \mathbf{v}_j) &= \nabla_{\mathbf{E}_i(q)}\, df(\mathbf{V}_j) - \nabla_{\mathbf{E}_j(q)}\, df(\mathbf{V}_i) \\
&= \nabla_{\mathbf{E}_i(q)} \nabla_{\mathbf{E}_j \circ \varphi^{-1}} f - \nabla_{\mathbf{E}_j(q)} \nabla_{\mathbf{E}_i \circ \varphi^{-1}} f \\
&= \nabla_{\mathbf{E}_i(q)} \nabla_{\mathbf{E}_j}(f \circ \varphi) - \nabla_{\mathbf{E}_j(q)} \nabla_{\mathbf{E}_i}(f \circ \varphi) \\
&= \frac{\partial^2(f \circ \varphi)}{\partial x_i\, \partial x_j}(q) - \frac{\partial^2(f \circ \varphi)}{\partial x_j\, \partial x_i}(q) \\
&= 0.
\end{aligned}$$

(ii) Adopting the same notation used in defining $d\omega$, we have

$$\begin{aligned}
d(f\omega)(\mathbf{v}_1, \mathbf{v}_2) &= \nabla_{\mathbf{v}_1}(f\omega(\mathbf{V}_2)) - \nabla_{\mathbf{v}_2}(f\omega(\mathbf{V}_1)) - f\omega([\mathbf{V}_1, \mathbf{V}_2](p)) \\
&= (\nabla_{\mathbf{v}_1} f)\omega(\mathbf{V}_2(p)) + f(p)\nabla_{\mathbf{v}_1}\omega(\mathbf{V}_2) \\
&\quad - (\nabla_{\mathbf{v}_2} f)\omega(\mathbf{V}_1(p)) - f(p)\nabla_{\mathbf{v}_2}\omega(\mathbf{V}_1) - f(p)\omega([\mathbf{V}_1, \mathbf{V}_2](p)) \\
&= df(\mathbf{v}_1)\omega(\mathbf{v}_2) - df(\mathbf{v}_2)\omega(\mathbf{v}_1) + f(p)\, d\omega(\mathbf{v}_1, \mathbf{v}_2) \\
&= (df \wedge \omega)(\mathbf{v}_1, \mathbf{v}_2) + (f\, d\omega)(\mathbf{v}_1, \mathbf{v}_2). \qquad \square
\end{aligned}$$

Lemma 2. *Let ω be a smooth 1-form on S and let $\varphi\colon U \to S$ be a singular 2-surface in S. Then*

$$d\omega(\mathbf{E}_1, \mathbf{E}_2) = \frac{\partial \omega_2}{\partial x_1} - \frac{\partial \omega_1}{\partial x_2}$$

where $\mathbf{E}_1$, $\mathbf{E}_2$ are the coordinate vector fields along φ and ω_i for $i \in \{1, 2\}$ is the smooth function along φ defined by $\omega_i = \omega(\mathbf{E}_i)$.

PROOF. First note that if $\psi\colon V \to S$ is a one to one local parametrization of S and ω is any smooth 1-form on $\psi(V)$ then $\omega = \sum_{i=1}^n f_i\, dg_i$ for some choice of smooth functions f_i and g_i on $\psi(V)$. Indeed, if we define $f_i = \omega(\mathbf{E}_i^\psi \circ \psi^{-1})$ and $g_i = x_i \circ \psi^{-1}$ where the $x_1, \ldots, x_n$ are the coordinate functions on $\mathbb{R}^n$ ($x_i(a_1, \ldots, a_n) = a_i$) then, for each $p \in V$ and $j \in \{1, \ldots, n\}$,

$$\begin{aligned}
\left(\sum f_i\, dg_i\right)(\mathbf{E}_j^\psi(p)) &= \sum f_i(\psi(p)) \nabla_{\mathbf{E}_j^\psi(p)}\, g_i = \sum f_i(\psi(p)) \frac{\partial}{\partial x_j} g_i \circ \psi \\
&= f_j(\psi(p)) = \omega(E_j^\psi(p))
\end{aligned}$$

so the linear functions $\omega_{\psi(p)}$ and $\left(\sum f_i\, dg_i\right)_{\psi(p)}$ agree on a basis for $S_{\psi(p)}$, and hence are equal, for each $p \in V$.

Thus for ω a 1-form on S and $\varphi\colon U \to S$ a singular 2-surface we may, in some open set W about any given point of Image φ, express ω as $\omega = \sum f_i\, dg_i$. Then, on $\varphi^{-1}(W)$, by Lemma 1,

$$\begin{aligned} d\omega(\mathbf{E}_1, \mathbf{E}_2) &= \sum df_i \wedge dg_i(\mathbf{E}_1, \mathbf{E}_2) \\ &= \sum (df_i(\mathbf{E}_1)\, dg_i(\mathbf{E}_2) - df_i(\mathbf{E}_2)\, dg_i(\mathbf{E}_1)) \\ &= \sum \left(\frac{\partial f_i \circ \varphi}{\partial x_1} \frac{\partial g_i \circ \varphi}{\partial x_2} - \frac{\partial f_i \circ \varphi}{\partial x_2} \frac{\partial g_i \circ \varphi}{\partial x_1} \right) \end{aligned}$$

whereas

$$\begin{aligned} \frac{\partial \omega_2}{\partial x_1} = \frac{\partial}{\partial x_1} \omega(\mathbf{E}_2) &= \frac{\partial}{\partial x_1} \sum (f_i \circ \varphi)\, dg_i(\mathbf{E}_2) \\ &= \sum \left(\frac{\partial f_i \circ \varphi}{\partial x_1} \frac{\partial g_i \circ \varphi}{\partial x_2} + (f_i \circ \varphi) \frac{\partial^2 g_i \circ \varphi}{\partial x_1\, \partial x_2} \right) \\ \frac{\partial \omega_1}{\partial x_2} = \frac{\partial}{\partial x_2} \omega(\mathbf{E}_1) &= \frac{\partial}{\partial x_2} \sum (f_i \circ \varphi)\, dg_i(\mathbf{E}_1) \\ &= \sum \left(\frac{\partial f_i \circ \varphi}{\partial x_2} \frac{\partial g_i \circ \varphi}{\partial x_1} + (f_i \circ \varphi) \frac{\partial^2 g_i \circ \varphi}{\partial x_2\, \partial x_1} \right) \end{aligned}$$

so $d\omega(\mathbf{E}_1, \mathbf{E}_2) = (\partial \omega_2 / \partial x_1) - (\partial \omega_1 / \partial x_2)$. □

We shall need a formula (Stokes' formula) relating differentiation to integration for forms on 2-surfaces. This formula is the natural generalization to 2-surfaces of the fundamental theorem of calculus as applied to line integrals ($\int_\alpha df = f(\alpha(b)) - f(\alpha(a))$. We shall integrate first over special "singular 2-surfaces-with-boundary".

Let S be an n-surface or an n-surface-with-boundary in $\mathbb{R}^{n+1}$.

A *singular disc* in S is a smooth map $\varphi\colon \mathsf{D} \to S$, where $\mathsf{D} = \{(x_1, x_2) \in \mathbb{R}^2\colon x_1^2 + x_2^2 \le 1\}$. Smoothness here means, as usual, that φ can be extended to a smooth map defined on some open set containing D. The *boundary* of the singular disc $\varphi\colon \mathsf{D} \to S$ is the parametrized curve $\partial\varphi = \varphi \circ \alpha$ where $\alpha\colon [0, 2\pi] \to \mathsf{D}$ is defined by $\alpha(t) = (\cos t, \sin t)$ (see Figure 20.5).

A *singular half-disc* in S is a smooth map $\varphi\colon \mathsf{D} \cap \mathbb{R}^2_- \to S$ where $\mathbb{R}^2_- = \{(x_1, x_2) \in \mathbb{R}^2\colon x_2 \le 0\}$; its *boundary* is the piecewise smooth parametrized curve $\partial\varphi = \varphi \circ \alpha$ where $\alpha\colon [0, 2 + \pi] \to S$ is defined by

$$\alpha(t) = \begin{cases} (1 - t, 0) & \text{if } \ 0 \le t \le 2 \\ (\cos(t - 2 + \pi), \sin(t - 2 + \pi)) & \text{if } \ 2 \le t \le \pi + 2 \end{cases}$$

(Figure 20.5).

A *singular triangle* in S is a smooth map $\varphi\colon \Delta \to S$ where

$$\Delta = \{(x_1, x_2) \in \mathbb{R}^2\colon x_1 \ge 0, x_2 \ge 0, x_1 + x_2 \le 1\};$$

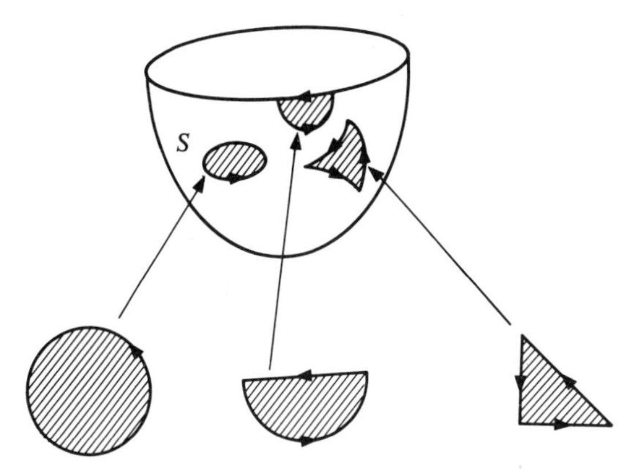

Figure 20.5 A singular disc, a singular half-disc, and a singular triangle in the 2-surface-with-boundary S.

its *boundary* is the piecewise smooth parametrized curve $\partial\varphi = \varphi \circ \alpha$ where α: $[0, 3] \to \Delta$ is defined by

$$\alpha(t) = \begin{cases} (t, 0) & \text{if } 0 \le t \le 1 \\ (2 - t, t - 1) & \text{if } 1 \le t \le 2 \\ (0, 3 - t) & \text{if } 2 \le t \le 3 \end{cases}$$

(Figure 20.5).

The *integral* of a smooth 2-form ω on S over one of these singular 2-surfaces-with-boundary φ is defined in the same way as the integral of ω over a singular 2-surface:

$$\int_\varphi \omega = \int_{\mathscr{D}(\varphi)} \omega(\mathbf{E}_1, \mathbf{E}_2)$$

where $\mathscr{D}(\varphi) \subset \mathbb{R}^2$ is the domain of φ and $\mathbf{E}_1$, $\mathbf{E}_2$ are the coordinate vector fields along φ.

Theorem 1 (Local Stokes' Theorem). *Let $S \subset \mathbb{R}^{n+1}$ be an n-surface or an n-surface-with-boundary, let ω be a smooth 1-form on S, and let φ be either a singular disc, a singular half-disc, or a singular triangle in S. Then*

$$\int_\varphi d\omega = \int_{\partial\varphi} \omega.$$

PROOF. By Lemma 2,

$$d\omega(\mathbf{E}_1, \mathbf{E}_2) = \frac{\partial\omega_2}{\partial x_1} - \frac{\partial\omega_1}{\partial x_2}$$

where $\omega_1 = \omega(\mathbf{E}_1)$ and $\omega_2 = \omega(\mathbf{E}_2)$ are smooth functions along φ. By

Green's Theorem (see Exercise 20.5)

$$\int_{\mathscr{D}(\varphi)} \left(\frac{\partial \omega_2}{\partial x_1} - \frac{\partial \omega_1}{\partial x_2} \right) = \int_{\alpha} (\omega_1 \, dx_1 + \omega_2 \, dx_2)$$

where α is the piecewise smooth parametrized curve used in the definition of $\partial\varphi$ and $x_1, x_2 \colon \mathbb{R}^2 \to \mathbb{R}$ are the coordinate functions on $\mathbb{R}^2$ $(x_i(a_1, a_2) = a_i)$. Hence, letting $[a, b]$ denote the domain of α and letting α_1, α_2 denote the coordinate functions of α $(\alpha(t) = (\alpha_1(t), \alpha_2(t))$ for $t \in [a, b])$ we have

$$\begin{aligned}
\int_{\varphi} d\omega &= \int_{\mathscr{D}(\varphi)} d\omega(\mathbf{E}_1, \mathbf{E}_2) = \int_{\mathscr{D}(\varphi)} \left(\frac{\partial \omega_2}{\partial x_1} - \frac{\partial \omega_1}{\partial x_2} \right) = \int_{\alpha} (\omega_1 \, dx_1 + \omega_2 \, dx_2) \\
&= \int_a^b ((\omega_1 \circ \alpha) \, dx_1(\dot{\alpha}) + (\omega_2 \circ \alpha) \, dx_2(\dot{\alpha})) \\
&= \int_a^b \left(\omega(\mathbf{E}_1 \circ \alpha) \frac{d\alpha_1}{dt} + \omega(\mathbf{E}_2 \circ \alpha) \frac{d\alpha_2}{dt} \right) \\
&= \int_a^b \omega \left(\frac{d\alpha_1}{dt} \mathbf{E}_1 \circ \alpha + \frac{d\alpha_2}{dt} \mathbf{E}_2 \circ \alpha \right) \\
&= \int_a^b \omega(d\varphi(\dot{\alpha})) = \int_a^b \omega(\varphi \,\dot{\circ}\, \alpha) = \int_{\varphi \circ \alpha} \omega = \int_{\partial\varphi} \omega.
\end{aligned}$$

□

Theorem 2 (Global Stokes' Theorem). (i) *Let S be a compact oriented 2-surface-with-boundary in $\mathbb{R}^3$, let its boundary ∂S be oriented by its induced orientation, and let ω be a smooth 1-form on S. Then*

$$\int_S d\omega = \int_{\partial S} \omega.$$

(ii) *Let S be a compact oriented 2-surface (without boundary) in $\mathbb{R}^3$ and let ω be a smooth 1-form on S. Then*

$$\int_S d\omega = 0.$$

PROOF. For each $p \in S$ we can find a one to one local parametrization φ_p of S with $p \in \text{Image } \varphi_p$. We may assume that the domain of each φ_p, for p in the interior of S, is an open ball and, in fact, by composing with a diffeomorphism of $\mathbb{R}^2$ if necessary, that the domain of φ_p is the ball of radius 2 centered at the origin in $\mathbb{R}^2$. For $p \in \partial S$ we may similarly assume that the domain of φ_p is the intersection with $\mathbb{R}^2_-$ of the ball of radius 2 centered at the origin in $\mathbb{R}^2$ and that $\varphi_p \circ \beta$, where $\beta(t) = (1 - t, 0)$ for $-1 \leq t \leq 3$, is a local parametrization of ∂S (see Exercise 20.1). Note that the parametrized 1-surface $\varphi_p \circ \beta$ is consistent with the orientation on ∂S; indeed, the induced orientation on ∂S was constructed precisely so that this would be true. Proceeding as in the proof of Theorem 4, Chapter 17, we can construct a partition of unity $\{f_i\}$ on S subordinate to a finite collection $\{\varphi_i = \varphi_{p_i}\}$ of these local parametrizations and in fact such that each f_i is identically zero outside $\varphi_i(\mathsf{D} \cap \mathscr{D}(\varphi_i))$, where $\mathsf{D} \cap \mathscr{D}(\varphi_i) = \mathsf{D}$ if $p_i \in S - \partial S$ and $\mathsf{D} \cap \mathscr{D}(\varphi_i) = \mathsf{D} \cap R^2_-$ if $p_i \in \partial S$ (take $r_p = 1$ in the partition of unity construction). Then, by Lemma 1,

$$d\omega = \sum f_i \, d\omega = \sum (d(f_i \omega) - df_i \wedge \omega).$$

Since

$$\sum df_i \wedge \omega = d(\sum f_i) \wedge \omega = d(1) \wedge \omega = 0,$$

we have

$$d\omega = \sum d(f_i \omega)$$

and hence

$$\int_S d\omega = \sum \int_S d(f_i \omega) = \sum \int_{\varphi_i} d(f_i \omega)$$

where the last equality holds because $f_i \omega$, and hence $d(f_i \omega)$, is identically zero outside $\varphi_i(\mathsf{D} \cap \mathscr{D}(\varphi_i))$ (see Exercise 20.6). Letting $\psi_i = \varphi_i|_{\mathsf{D} \cap \mathscr{D}(\varphi_i)}$ we see further that

$$\int_S d\omega = \sum \int_{\psi_i} d(f_i \omega) = \sum \int_{\partial \psi_i} f_i \omega$$

where the last equality is a consequence of Theorem 1. Letting $\mathscr{I} = \{i \colon p_i \in \partial S\}$ we find that, for $i \notin \mathscr{I}, f_i \circ \varphi_i$ is zero outside D, hence on the boundary of D, and so $f_i \circ \partial\psi_i = 0$ which implies that $\int_{\partial\psi_i} f_i \omega = 0$. It follows that if S is without boundary then $\int_S d\omega = 0$. On the other hand, if S has a boundary then $\{f_i|_{\partial S} \colon i \in \mathscr{I}\}$ is a partition of unity on ∂S subordinate to the local parametrizations $\{\varphi_i \circ \beta \colon i \in \mathscr{I}\}$, and $f_i \circ \partial\psi_i = 0$ for $2 \le t \le \pi + 2$ so $\int_{\partial\psi_i} f_i \omega = \int_{\varphi_i \circ \beta} f_i \omega$ for $i \in \mathscr{I}$ $(f_i \circ \varphi_i \circ \beta = 0$ for $t \le 0$ and for $t \ge 2)$ and

$$\int_S d\omega = \sum_{i \in \mathscr{I}} \int_{\partial\psi_i} f_i \omega = \sum_{i \in \mathscr{I}} \int_{\varphi_i \circ \beta} f_i \omega = \int_{\partial S} \omega. \qquad \square$$

Exercises

20.1. Let $S = f^{-1}(c) \cap g_1^{-1}((-\infty, c_1]) \cap \cdots \cap g_k^{-1}((-\infty, c_k])$ be a n-surface-with-boundary in $\mathbb{R}^{n+1}$ and suppose $p \in g_i^{-1}(c_i)$. Show that there exists a parametrized n-surface $\varphi \colon B_\varepsilon \to \mathbb{R}^{n+1}$, where B_ε is a ball of radius ε about 0 in $\mathbb{R}^n$, such that $\varphi(0) = p$ and $\varphi|_{B_\varepsilon \cap \mathbb{R}^n_-}$ maps $B_\varepsilon \cap \mathbb{R}^n_-$ one to one onto an open set W about p in S. [*Hint:* First find a local parametrization $\psi \colon U \to f^{-1}(c)$ with image containing p. Then apply the inverse function theorem to the map $\phi \colon U \to \mathbb{R}^n$ defined by $\phi(x_1, \ldots, x_n) = (x_1, \ldots, x_{j-1}, x_{j+1}, \ldots, x_n, g_i(\psi(x_1, \ldots, x_n)) - c_i)$ where j is such that $(\partial/\partial x_j)(g_i \circ \psi)(\psi^{-1}(p)) \neq 0$.

20.2. Let S be an n-surface or n-surface-with-boundary in $\mathbb{R}^{n+1}$ and let ω be a smooth 1-form on S. For $\mathbf{V}_1$, $\mathbf{V}_2$ any two smooth tangent vector fields on an open set $U \subset S$, define $\mu(\mathbf{V}_1, \mathbf{V}_2)\colon U \to \mathbb{R}$ by

$$(\mu(\mathbf{V}_1, \mathbf{V}_2))(p) = \nabla_{\mathbf{V}_1(p)}\omega(\mathbf{V}_2) - \nabla_{\mathbf{V}_2(p)}\omega(\mathbf{V}_1) - \omega([\mathbf{V}_1, \mathbf{V}_2](p)).$$

(a) Show that

$$\mu(f\mathbf{V}_1, \mathbf{V}_2) = f\mu(\mathbf{V}_1, \mathbf{V}_2) = \mu(\mathbf{V}_1, f\mathbf{V}_2)$$

for all smooth functions $f\colon U \to \mathbb{R}$.

(b) Show that if $\mathbf{W}_1$ and $\mathbf{W}_2$ are smooth tangent vector fields on U such that $\mathbf{W}_1(p) = \mathbf{V}_1(p)$ and $\mathbf{W}_2(p) = \mathbf{V}_2(p)$ for some $p \in U$ then $\mu(\mathbf{V}_1, \mathbf{V}_2)(p) = \mu(\mathbf{W}_1, \mathbf{W}_2)(p)$. [*Hint:* Take $\mathbf{X}_1, \ldots, \mathbf{X}_n$ smooth tangent vector fields, defined on some open set $V \subset U$ with $p \in V$, such that $\{\mathbf{X}_1(q), \ldots, \mathbf{X}_n(q)\}$ is a basis for S_q for all $q \in V$. Express the given vector fields as linear combinations of the $\mathbf{X}_i$'s and apply part a).]

(c) Conclude that the value of the right hand side of the formula used in this chapter to define $d\omega$ is independent of the choice of vector fields $\mathbf{V}_1$ and $\mathbf{V}_2$.

20.3. Let S be an n-surface in $\mathbb{R}^{n+1}$, let $\tilde{S}$ be an m-surface in $\mathbb{R}^{m+1}$, and let $f\colon S \to \tilde{S}$ be a smooth map.

(a) Show that if ω_1 and ω_2 are 1-forms on $\tilde{S}$ then $f^*(\omega_1 \wedge \omega_2) = f^*\omega_1 \wedge f^*\omega_2$.

(b) Show that if $g\colon \tilde{S} \to \mathbb{R}$ is smooth then $f^*(dg) = d(g \circ f)$.

(c) Show that if ω is a smooth 1-form on $\tilde{S}$ then $f^*(d\omega) = d(f^*\omega)$.

[*Hint:* Use the fact that, for U a suitably small open set in $\tilde{S}$, $\omega|_U = \sum_{i=1}^{n} f_i\, dg_i$ where f_i and $g_i\colon U \to \mathbb{R}$ are smooth functions (see the proof of Lemma 2).]

20.4. Let S be an n-surface in $\mathbb{R}^{n+1}$ and let ω be a smooth 2-form on S. Show that if $\varphi\colon U \to S$ is a one to one local parametrization of S then there exist smooth real valued functions f_{ij} $(1 \le i < j \le n)$ and g_i $(1 \le i \le n)$ on $\varphi(U)$ such that $\omega|_{\varphi(U)} = \sum_{1 \le i < j \le n} f_{ij}\, dg_i \wedge dg_j$. [*Hint:* See the proof of Lemma 2.]

20.5. Let U be an open set in $\mathbb{R}^2$ containing $\mathscr{D}$ where $\mathscr{D}$ is Δ, D, or D $\cap\, \mathbb{R}^2_-$, and suppose ω_1 and ω_2 are smooth real valued functions on U. Prove *Green's theorem*:

$$\int_{\mathscr{D}} \left(\frac{\partial \omega_2}{\partial x_1} - \frac{\partial \omega_1}{\partial x_2}\right) = \int_{\alpha} (\omega_1\, dx_1 + \omega_2\, dx_2)$$

where α is the piecewise smooth parametrization of the boundary of $\mathscr{D}$ described in this chapter. [*Hint:* Break the left hand side into a difference of two integrals, evaluate these by iterated integration, and reparametrize the curve which appears in the resulting line integral.]

20.6. Let S be an n-surface in $\mathbb{R}^{n+1}$ and let ω be a smooth n-form on S. Suppose ω is identically zero outside $\varphi(C)$ where $\varphi\colon U \to S$ is a local parametrization of S and C is a compact subset of U. Show that $\int_S \omega = \int_\varphi \omega$. [*Hint:* Construct a partition of unity $\{f_i\}$ on S with the property that for each i either (i) f_i is identically zero outside $\varphi(U)$, or (ii) f_i is identically zero on $\varphi(C)$.]

20.7. Let ω be a smooth k-form on an n-surface S. For $p \in S$ and $\mathbf{v}_1, \ldots, \mathbf{v}_{k+1} \in S_p$, let

$$d\omega(\mathbf{v}_1, \ldots, \mathbf{v}_{k+1}) = \sum_{1 \le i \le k+1} (-1)^{i-1} \nabla_{\mathbf{v}_i} \omega\, (\mathbf{V}_1, \ldots, \mathbf{V}_{i-1}, \mathbf{V}_{i+1}, \ldots, \mathbf{V}_{k+1})$$

$$+ \sum_{1 \le i < j \le k+1} (-1)^{i+j} \omega\, ([\mathbf{V}_i, \mathbf{V}_j], \mathbf{V}_1, \ldots, \mathbf{V}_{i-1}, \mathbf{V}_{i+1}, \ldots,$$

$$\mathbf{V}_{j-1}, \mathbf{V}_{j+1}, \ldots, \mathbf{V}_{k+1})(p)$$

where $\mathbf{V}_1, \ldots, \mathbf{V}_{k+1}$ are smooth tangent vector fields, defined on an open set in S, such that $\mathbf{V}_i(p) = \mathbf{v}_i$ for each i. Show that the value of the right hand side of this formula is independent of the choice of vector fields $\mathbf{V}_1, \ldots, \mathbf{V}_{k+1}$, and that $d\omega$ is a smooth $(k+1)$-form on S. [*Hint:* See Exercise 20.2.] $d\omega$ is the *exterior derivative* of the k-form ω.

20.8. Show that exterior differentiation of smooth k-forms (Exercise 20.7) has the following properties:

(a) If ω and η are smooth k-forms on S then $d(\omega + \eta) = d\omega + d\eta$.
(b) If $f\colon S \to \mathbb{R}$ is a smooth function and ω is a smooth k-form on S then $d(f\omega) = df \wedge \omega + f\, d\omega$.
(c) If ω is a smooth k-form on S and η is a smooth l-form on S then $d(\omega \wedge \eta) = d\omega \wedge \eta + (-1)^k \omega \wedge d\eta$.
(d) $d^2 = 0$.

20.9. Let $\mathbf{X}$ be a smooth vector field on an n-surface S and let $\omega_{\mathbf{X}}$ be its dual 1-form.
(a) Show that, for $\mathbf{v}, \mathbf{w} \in S_p$, $p \in S$,

$$d\omega_{\mathbf{X}}(\mathbf{v}, \mathbf{w}) = (\nabla_{\mathbf{v}} \mathbf{X}) \cdot \mathbf{w} - (\nabla_{\mathbf{w}} \mathbf{X}) \cdot \mathbf{v}.$$

(b) Show that if $S = \mathbb{R}^3$ then

$$d\omega_{\mathbf{X}}(\mathbf{v}, \mathbf{w}) = (\operatorname{curl} \mathbf{X}) \cdot (\mathbf{v} \times \mathbf{w})$$

where

$$(\operatorname{curl} \mathbf{X})(p) = \left(p, \frac{\partial X_3}{\partial x_2} - \frac{\partial X_2}{\partial x_3}, \frac{\partial X_1}{\partial x_3} - \frac{\partial X_3}{\partial x_1}, \frac{\partial X_2}{\partial x_1} - \frac{\partial X_1}{\partial x_2} \right)\bigg|_p,$$

X_1, X_2, and X_3 being the component functions of $\mathbf{X}$.

20.10. Let S be a compact oriented 2-surface-with-boundary in $\mathbb{R}^3$ and let $\mathbf{X}$ be a smooth vector field defined on an open set U in $\mathbb{R}^3$ containing S. Prove the *classical Stokes' formula*

$$\int_S (\operatorname{curl} \mathbf{X}) \cdot \mathbf{N} = \int_{\partial S} \mathbf{X} \cdot \mathbf{T}$$

where curl $\mathbf{X}$ is as in Exercise 20.9, $\mathbf{N}$ is the orientation vector field on S, and $\mathbf{T}(p)$ is, for each $p \in \partial S$, the unique unit vector tangent to ∂S at p such that $\{T(p)\}$ is consistent with the induced orientation on ∂S. [*Hint:* Apply Theorem 2 to the 1-form $i^*\omega_{\mathbf{X}}$ where $i\colon S \to \mathbb{R}^3$ is defined by $i(q) = q$ for all $q \in S$.]

21 The Gauss-Bonnet Theorem

In this chapter we shall study the integral $\int_S K$ of the Gaussian curvature over a compact oriented 2-surface S. We shall see that $(1/2\pi) \int_S K$ always turns out to be an integer, the Euler characteristic of S. This is the 2-dimensional version of the Gauss–Bonnet theorem. A similar result is valid in all higher even dimensions but the computations are less transparent so we shall be content with a few comments about this more general case at the end of the chapter.

The Gauss–Bonnet theorem is obtained by applying Stokes' Theorem to a 1-form constructed with the aid of a unit tangent vector field. Let S be an oriented 2-surface or 2-surface-with-boundary in $\mathbb{R}^3$. Suppose $\mathbf{X}$ is a smooth unit tangent vector field defined on an open set U in S. We use the vector field $\mathbf{X}$ to construct a 1-form ω on U as follows. For any $\mathbf{v} \in S_p$, $p \in S$, let $J\mathbf{v} \in S_p$ be the vector obtained from $\mathbf{v}$ by a positive rotation in S_p through the angle $\pi/2$. Thus $J\mathbf{v} = \mathbf{N}(p) \times \mathbf{v}$ where $\mathbf{N}$ is the orientation vector field on S. Note that $\{\mathbf{v}, J\mathbf{v}\}$ is an ordered orthonormal basis for S_p consistent with the orientation of S. We define the 1-form ω on U by

$$\omega(\mathbf{v}) = (D_{\mathbf{v}}\mathbf{X}) \cdot J\mathbf{X}(p) = (\nabla_{\mathbf{v}}\mathbf{X}) \cdot J\mathbf{X}(p)$$

where D denotes covariant differentiation ($D_{\mathbf{v}}\mathbf{X}$ is the tangential component of $\nabla_{\mathbf{v}}\mathbf{X}$). This 1-form ω is called the *connection form* on U associated with $\mathbf{X}$. Note that $J\mathbf{X}$, defined by $(J\mathbf{X})(p) = J\mathbf{X}(p)$, is a smooth unit vector field on U which is everywhere orthogonal to $\mathbf{X}$ and that

$$\boxed{\begin{aligned} D_{\mathbf{v}}\mathbf{X} &= \omega(\mathbf{v})J\mathbf{X}(p) \\ D_{\mathbf{v}}(J\mathbf{X}) &= -\omega(\mathbf{v})\mathbf{X}(p) \end{aligned}}$$

Indeed, $\mathbf{X}$, being a unit vector field on S, has derivative $D_{\mathbf{v}}\mathbf{X}$ orthogonal to $\mathbf{X}(p)$. Hence $D_{\mathbf{v}}\mathbf{X} = aJ\mathbf{X}(p)$ for some $a \in \mathbb{R}$, and $a = D_{\mathbf{v}}\mathbf{X} \cdot J\mathbf{X}(p) = \omega(\mathbf{v})$. Similarly, $D_{\mathbf{v}}(J\mathbf{X}) = b\mathbf{X}(p)$ where

$$b = D_{\mathbf{v}}(J\mathbf{X}) \cdot \mathbf{X}(p) = \nabla_{\mathbf{v}}(J\mathbf{X} \cdot \mathbf{X}) - J\mathbf{X}(p) \cdot D_{\mathbf{v}}\mathbf{X} = -\omega(\mathbf{v}).$$

The connection form ω measures, up to sign, the rotation rate, relative to $\mathbf{X}$, of parallel vector fields along parametrized curves in U. In order to see this, we must first formulate precisely the meaning of "rotation rate". Suppose $\alpha: [a, b] \to S$ is a parametrized curve in S and $\mathbf{Y}$ and $\mathbf{Z}$ are smooth unit vector fields tangent to S along α. Then

$$\mathbf{Z}(t) = \cos\theta(t)\mathbf{Y}(t) + \sin\theta(t)J\mathbf{Y}(t)$$

for some smooth function $\theta: [a, b] \to \mathbb{R}$ (see Figure 21.1).

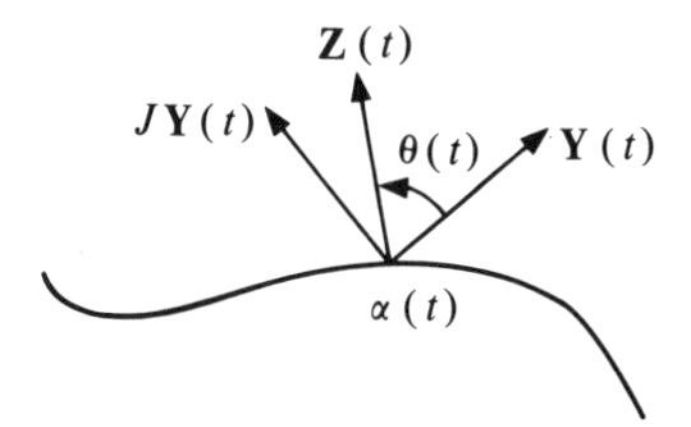

Figure 21.1. $\theta(t)$ measures the angle of rotation from $\mathbf{Y}(t)$ to $\mathbf{Z}(t)$.

An explicit formula for such a function θ may be obtained as follows. Let $\beta: I \to \mathbb{R}^2$ be defined by

$$\beta(t) = (\mathbf{Z}(t) \cdot \mathbf{Y}(t), \mathbf{Z}(t) \cdot J\mathbf{Y}(t)).$$

Then $\|\beta(t)\| = 1$ for all $t \in I$ so in particular we can find a $\theta_0 \in \mathbb{R}$ such that $\beta(a) = (\cos\theta_0, \sin\theta_0)$. Set $\theta(t) = \theta_0 + \int_{\beta_t} \eta$ where β_t is the restriction of β to the interval $[a, t]$ and η is the 1-form on $\mathbb{R}^2$ given by

$$\eta = -[x_2/(x_1^2 + x_2^2)]\, dx_1 + [x_1/(x_1^2 + x_2^2)]\, dx_2.$$

Then $\beta(t) = (\cos\theta(t), \sin\theta(t))$ for all $t \in I$, as required (see the proof of Theorem 3, Chapter 11).

The function θ measures the angle of rotation from $\mathbf{Y}$ to $\mathbf{Z}$ along α. It is not uniquely determined, but any two such functions must differ by a multiple of 2π. Hence, the derivative $\theta'(t)$ *is* uniquely determined. θ' is called the *rotation rate* of $\mathbf{Z}$, relative to $\mathbf{Y}$, along α. The real number $\theta(b) - \theta(a)$ is also uniquely determined; it is called the *total angle of rotation* of $\mathbf{Z}$ along α, relative to $\mathbf{Y}$.

Lemma 1. *Let S be an oriented 2-surface in $\mathbb{R}^3$, let $\mathbf{X}$ be a smooth unit tangent vector field on an open set U in S, and let ω be the connection form on U associated with $\mathbf{X}$. Suppose $\alpha: [a, b] \to U$ is any parametrized curve in U and $\mathbf{Z}$*

is any parallel unit vector field along α. *Then*

(i) $\omega(\dot{\alpha})$ *is equal to the negative of the rotation rate of* $\mathbf{Z}$, *relative to* $\mathbf{X}$ (*or, more precisely, relative to* $\mathbf{X} \circ \alpha$), *along* α.

(ii) $\int_\alpha \omega$ *is equal to the negative of the total angle of rotation of* $\mathbf{Z}$, *relative to* $\mathbf{X}$, *along* α.

PROOF. (i) Let $\theta\colon [a, b] \to \mathbb{R}$ measure the angle of rotation from $\mathbf{X}$ to $\mathbf{Z}$ along α. Since $\mathbf{Z}$ is parallel along α,

$$\begin{aligned} \mathbf{0} = \mathbf{Z}' &= (\cos\theta\, \mathbf{X} \circ \alpha + \sin\theta\, J\mathbf{X} \circ \alpha)' \\ &= -\theta' \sin\theta\, \mathbf{X} \circ \alpha + \theta' \cos\theta\, J\mathbf{X} \circ \alpha + \cos\theta\, D_{\dot{\alpha}}\mathbf{X} + \sin\theta\, D_{\dot{\alpha}} J\mathbf{X} \\ &= (\theta' + \omega(\dot{\alpha}))(-\sin\theta\, \mathbf{X} \circ \alpha + \cos\theta\, J\mathbf{X} \circ \alpha) \end{aligned}$$

(we have used here the above boxed formulas) and hence $\theta' + \omega(\dot{\alpha}) = 0$; i.e., $\omega(\dot{\alpha}) = -\theta'$.

(ii) $\int_\alpha \omega = \int_a^b \omega(\dot{\alpha}) = -\int_a^b \theta' = -(\theta(b) - \theta(a))$. □

If $\alpha\colon [a, b] \to U$ is a unit speed geodesic in S then the velocity field $\dot{\alpha}$ is parallel along α and may be used as the vector field $\mathbf{Z}$ in Lemma 1. Then Lemma 1 says that $\int_\alpha \omega$ measures the negative of the total angle of rotation of $\dot{\alpha}$ with respect to the vector field $\mathbf{X}$. The 1-form ω can also be used to measure the angle of rotation of $\dot{\alpha}$ with respect to the vector field $\mathbf{X}$ for α any smooth unit speed curve in U. The relevant formula contains also the *geodesic curvature* $\kappa_g\colon [a, b] \to \mathbb{R}$ of α, defined by

$$\kappa_g = (\dot{\alpha})' \cdot J\dot{\alpha}.$$

The geodesic curvature measures how much α deviates from being a geodesic. Its magnitude $|\kappa_g|$ is just the magnitude $\|\dot{\alpha}'\|$ of the covariant acceleration $\dot{\alpha}'$ of α since $\dot{\alpha}$, being a unit vector field along α, has covariant derivative orthogonal to itself and hence a multiple of $J\dot{\alpha}$. Note that α is a geodesic if and only if κ_g is identically zero.

Lemma 2. *Let* S, $\mathbf{X}$, U *and* ω *be as in Lemma 1 and let* $\alpha\colon [a, b] \to S$ *be a smooth unit speed curve in* U. *Then the total angle of rotation of* $\dot{\alpha}$ *with respect to the vector field* $\mathbf{X}$ *is equal to* $\int_a^b \kappa_g - \int_\alpha \omega$.

PROOF. Let $\mathbf{Z}$ be a parallel unit vector field along α, let $\theta\colon [a, b] \to \mathbb{R}$ measure the angle of rotation from $\mathbf{X}$ to $\mathbf{Z}$ along α, and let $\phi\colon [a, b] \to \mathbb{R}$ measure the angle of rotation from $\mathbf{X}$ to $\dot{\alpha}$. Then $\phi - \theta$ measures the angle of rotation from $\mathbf{Z}$ to $\dot{\alpha}$; that is,

$$\dot{\alpha} = \cos(\phi - \theta)\mathbf{Z} + \sin(\phi - \theta)J\mathbf{Z},$$

and

$$J\dot{\alpha} = -\sin(\phi - \theta)\mathbf{Z} + \cos(\phi - \theta)J\mathbf{Z}.$$

Taking the covariant derivative of $\dot{\alpha}$ and using the fact that $\mathbf{Z}$ and $J\mathbf{Z}$ are both parallel along α (JZ is parallel because by the existence and uniqueness theorem for parallel vector fields there is a unique parallel vector field along α with initial value $J\mathbf{Z}(a)$; this vector field must be smooth, of unit length, and orthogonal to $\mathbf{Z}$ along α; and JZ is the only such vector field) we find

$$\dot{\alpha}' = (\phi' - \theta')(-\sin(\phi - \theta)\mathbf{Z} + \cos(\phi - \theta)J\mathbf{Z})$$

so

$$\kappa_g = \dot{\alpha}' \cdot J\dot{\alpha} = \phi' - \theta'.$$

Hence the total angle of rotation of $\dot{\alpha}$ relative to $\mathbf{X}$ is

$$\int_a^b \phi' = \int_a^b \kappa_g + \int_a^b \theta' = \int_a^b \kappa_g - \int_\alpha \omega. \qquad \square$$

The relation between the 1-form ω and the Gaussian curvature is as follows.

Lemma 3. *Let S, U, $\mathbf{X}$, and ω be as in Lemma 1. Then, on U,*

$$d\omega = -K\zeta$$

where K is the Gaussian curvature of S and ζ is the volume form on S.

Proof. By Exercise 17.13, $d\omega = f\zeta$ for some $f\colon U \to \mathbb{R}$. To find $f(p)$ ($p \in U$) we need only evaluate $d\omega$ and ζ on a basis for S_p. The basis we shall use is the coordinate basis $\{\mathbf{E}_1(p), \mathbf{E}_2(p)\}$ attached to a local parametrization φ of S whose image contains p and is contained in U. Then, by Lemma 2, Chapter 20,

$$\begin{aligned} d\omega(\mathbf{E}_1, \mathbf{E}_2) &= \frac{\partial \omega_2}{\partial x_1} - \frac{\partial \omega_1}{\partial x_2} = \frac{\partial}{\partial x_1}\omega(\mathbf{E}_2) - \frac{\partial}{\partial x_2}\omega(\mathbf{E}_1) \\ &= \frac{\partial}{\partial x_1}(\nabla_{\mathbf{E}_2}\mathbf{X} \cdot J\mathbf{X} \circ \varphi) - \frac{\partial}{\partial x_2}(\nabla_{\mathbf{E}_1}\mathbf{X} \cdot J\mathbf{X} \circ \varphi) \\ &= (\nabla_{\mathbf{E}_1}\nabla_{\mathbf{E}_2}\mathbf{X} - \nabla_{\mathbf{E}_2}\nabla_{\mathbf{E}_1}\mathbf{X}) \cdot J\mathbf{X} \circ \varphi \\ &\quad + \nabla_{\mathbf{E}_2}\mathbf{X} \cdot \nabla_{\mathbf{E}_1}J\mathbf{X} - \nabla_{\mathbf{E}_1}\mathbf{X} \cdot \nabla_{\mathbf{E}_2}J\mathbf{X} \end{aligned}$$

where $\nabla_{\mathbf{E}_i}\mathbf{Z}$ for $\mathbf{Z}$ a smooth vector field on U is the smooth vector field along φ defined by $(\nabla_{\mathbf{E}_i}\mathbf{Z})(p) = \nabla_{\mathbf{E}_i(p)}\mathbf{Z}$, and $\nabla_{\mathbf{E}_i}\mathbf{Z}$ for $\mathbf{Z}$ a smooth vector field along φ is the smooth vector field along φ defined by $(\nabla_{\mathbf{E}_i}\mathbf{Z})(p) = \nabla_{\mathbf{e}_i}\mathbf{Z}$. Here $\mathbf{e}_1 = (p, 1, 0)$ and $\mathbf{e}_2 = (p, 0, 1)$. The first term in the above expression vanishes by the equality of mixed partial derivatives. Furthermore,

$$\begin{aligned} \nabla_{\mathbf{E}_i}\mathbf{X} &= D_{\mathbf{E}_i}\mathbf{X} + ((\nabla_{\mathbf{E}_i}\mathbf{X}) \cdot \mathbf{N} \circ \varphi)\mathbf{N} \circ \varphi \\ &= D_{\mathbf{E}_i}\mathbf{X} + (L(\mathbf{E}_i) \cdot \mathbf{X} \circ \varphi)\mathbf{N} \circ \varphi \end{aligned}$$

where $L(\mathbf{E}_i)$ is the vector field along φ defined by $L(\mathbf{E}_i)(p) = L_p(\mathbf{E}_i(p)) = -\nabla_{\mathbf{E}_i}\mathbf{N}$, L_p being the Weingarten map of φ at p. Using this and the corresponding formula for $\nabla_{\mathbf{E}_i} J\mathbf{X}$ we find

$$\begin{aligned} d\omega(\mathbf{E}_1, \mathbf{E}_2) = D_{\mathbf{E}_2}\mathbf{X} \cdot D_{\mathbf{E}_1} J\mathbf{X} - D_{\mathbf{E}_1}\mathbf{X} \cdot D_{\mathbf{E}_2} J\mathbf{X} \\ + (L(\mathbf{E}_2) \cdot \mathbf{X} \circ \varphi)(L(\mathbf{E}_1) \cdot J\mathbf{X} \circ \varphi) \\ - (L(\mathbf{E}_1) \cdot \mathbf{X} \circ \varphi)(L(\mathbf{E}_2) \cdot J\mathbf{X} \circ \varphi). \end{aligned}$$

The first two terms on the right hand side of this last formula vanish because, for each i and j, the derivative $D_{\mathbf{E}_i}\mathbf{X}$ of the unit vector field $\mathbf{X}$ must be orthogonal to $\mathbf{X}$, hence a multiple of $J\mathbf{X}$, and hence orthogonal to $D_{\mathbf{E}_j} J\mathbf{X}$. Applying the vector identity

$$(\mathbf{v}_1 \times \mathbf{v}_2) \cdot (\mathbf{v}_3 \times \mathbf{v}_4) = (\mathbf{v}_1 \cdot \mathbf{v}_3)(\mathbf{v}_2 \cdot \mathbf{v}_4) - (\mathbf{v}_1 \cdot \mathbf{v}_4)(\mathbf{v}_2 \cdot \mathbf{v}_3)$$

to the remaining two terms yields

$$\begin{aligned} d\omega(\mathbf{E}_1, \mathbf{E}_2) &= (L(\mathbf{E}_2) \times L(\mathbf{E}_1)) \cdot (\mathbf{X} \circ \varphi \times J\mathbf{X} \circ \varphi) \\ &= -L(\mathbf{E}_1) \times L(\mathbf{E}_2) \cdot \mathbf{N} \circ \varphi \\ &= -(\det L)\, \mathbf{E}_1 \times \mathbf{E}_2 \cdot \mathbf{N} \circ \varphi \\ &= -(K \circ \varphi) \det \begin{pmatrix} \mathbf{E}_1 \\ \mathbf{E}_2 \\ \mathbf{N} \circ \varphi \end{pmatrix} \\ &= -(K \circ \varphi)\zeta(\mathbf{E}_1, \mathbf{E}_2) \\ &= -(K\zeta)(\mathbf{E}_1, \mathbf{E}_2) \end{aligned}$$

from which we can conclude that $d\omega = -K\zeta$. □

Theorem 1. *Let S be an oriented 2-surface in $\mathbb{R}^3$ and let U be an open subset of S on which there is defined a smooth unit tangent vector field $\mathbf{X}$. Then, for $\varphi\colon \mathsf{D} \to U$ any singular disc in U and $\mathbf{Z}$ any parallel vector field along $\partial\varphi$, $\int_\varphi K\zeta$ is equal to the total angle of rotation of $\mathbf{Z}$, relative to $\mathbf{X}$, along $\partial\varphi$.*

PROOF. Let ω be the connection form associated with $\mathbf{X}$. By Lemma 3 and Stokes' theorem,

$$\int_\varphi K\zeta = -\int_\varphi d\omega = -\int_{\partial\varphi} \omega$$

which, by Lemma 1, is equal to the total angle of rotation of $\mathbf{Z}$, relative to $\mathbf{X}$, along $\partial\varphi$. □

Remark. Theorem 1 shows, in particular, that the total angle of rotation of $\mathbf{Z}$ relative to $\mathbf{X}$ along $\partial\varphi$ is independent of both $\mathbf{X}$ and $\mathbf{Z}$ and in fact depends only on φ; this angle is called the *holonomy angle* of φ. Note that this angle depends crucially on φ and not just on $\partial\varphi$ (see Figure 21.2).

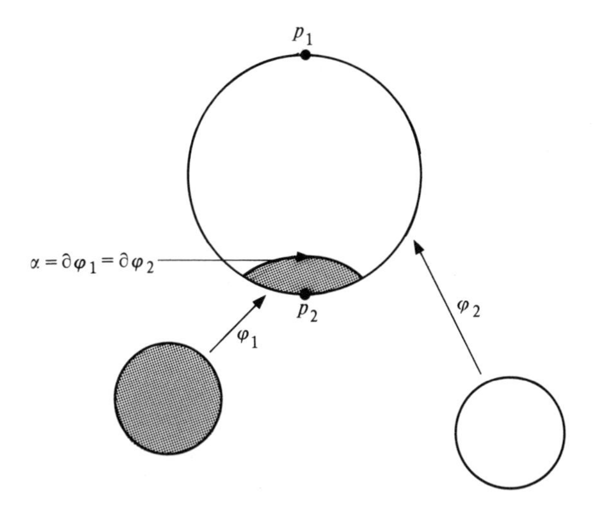

Figure 21.2. The total angle of rotation of a unit vector field $\mathbf{Z}$ parallel along $\alpha = \partial\varphi_1 = \partial\varphi_2$, with respect to a unit tangent vector field $\mathbf{X}_1$ on $U_1 = \mathsf{S}^2 - \{p_1\}$, will be considerably different from the total angle of rotation of $\mathbf{Z}$ along α relative to a unit tangent vector field $\mathbf{X}_2$ on $U_2 = \mathsf{S}^2 - \{p_2\}$.

Theorem 1 also yields an interesting interpretation of the Gaussian curvature: $K(p)$, $p \in S$, is the limit, as a disc φ about p shrinks to zero, of the ratio: holonomy angle of φ/area of φ. More precisely, we have the following.

Corollary. *Let S be an oriented 2-surface in $\mathbb{R}^3$, let $p \in S$, and let φ: $\mathsf{D} \to S$ be a singular disc in S with $\varphi(0) = p$ and $d\varphi_0$: $\mathbb{R}^2_0 \to S_p$ non-singular. Then*

$$K(p) = \lim_{\varepsilon \to 0} \theta(\varphi_\varepsilon)/A(\varphi_\varepsilon)$$

where φ_ε: $\mathsf{D} \to S$ is defined by $\varphi_\varepsilon(q) = \varphi(\varepsilon q)$, $\theta(\varphi_\varepsilon)$ is the holonomy angle of φ_ε, and $A(\varphi_\varepsilon)$ is the area of φ_ε (see Figure 21.3).

Proof. For ε sufficiently small, Image φ_ε is contained in the image of a one to one local parametrization of S and hence there exists a smooth unit vector field $\mathbf{X}$ (e.g., a normalized coordinate vector field) on an open set containing Image φ_ε. The regularity of φ at 0 guarantees that $A(\varphi_\varepsilon) \neq 0$ for all $\varepsilon > 0$. Using Theorem 1, the mean value theorem for integrals, and the fact that the coordinate vector fields $\mathbf{E}_i^\varepsilon$ along φ_ε are related to the coordinate vector fields $\mathbf{E}_i$ along φ by $\mathbf{E}_i^\varepsilon(q) = \varepsilon\mathbf{E}_i(\varepsilon q)$, we find

$$\frac{\theta(\varphi_\varepsilon)}{A(\varphi_\varepsilon)} = \frac{\int_{\varphi_\varepsilon} K\zeta}{\int_{\varphi_\varepsilon} \zeta} = \frac{\int_{\mathsf{D}} (K \circ \varphi_\varepsilon)\zeta(\mathbf{E}_1^\varepsilon, \mathbf{E}_2^\varepsilon)}{\int_{\mathsf{D}} \zeta(\mathbf{E}_1^\varepsilon, \mathbf{E}_2^\varepsilon)} = \frac{K(\varphi_\varepsilon(q_1))\zeta(\mathbf{E}_1^\varepsilon(q_1), \mathbf{E}_2^\varepsilon(q_1)) \int_{\mathsf{D}} 1}{\zeta(\mathbf{E}_1^\varepsilon(q_2), \mathbf{E}_2^\varepsilon(q_2)) \int_{\mathsf{D}} 1}$$

$$= \frac{K(\varphi(\varepsilon q_1))\zeta(\mathbf{E}_1(\varepsilon q_1), \mathbf{E}_2(\varepsilon q_1))}{\zeta(\mathbf{E}_1(\varepsilon q_2), \mathbf{E}_2(\varepsilon q_2))}$$

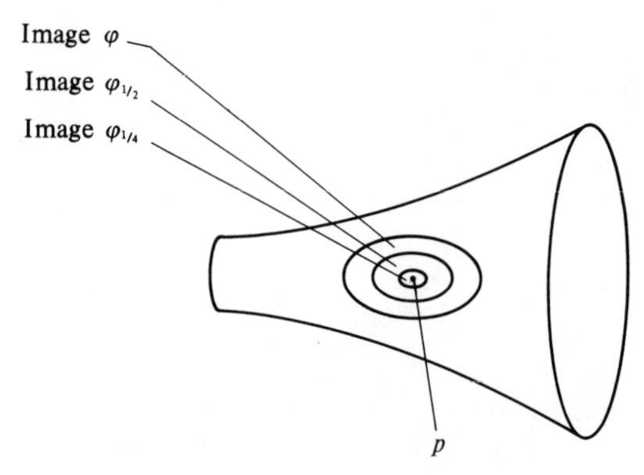

Figure 21.3. The Gaussian curvature $K(p)$ is equal to the limit, as $\varepsilon \to 0$, of the ratio: holonomy angle of φ_ε/area of φ_ε.

for some $q_1, q_2 \in \mathsf{D}$ (depending on ε). Taking the limit as $\varepsilon \to 0$ completes the proof. □

The Gauss–Bonnet theorem in its local form relates the integral of the Gaussian curvature over a regular triangle to the integral of the geodesic curvature over its boundary. By a *regular triangle* in an oriented 2-surface S we mean a singular triangle $\varphi: \Delta \to S$ which is the restriction to Δ of a one to one local parametrization of S defined on some open set in $\mathbb{R}^2$ containing Δ. The boundary $\partial\varphi: [0, 3] \to S$ is then a piecewise smooth curve in S with the property that $\alpha_i = \partial\varphi|_{[i-1, i]}$ is a regular parametrized curve ($\dot{\alpha}_i \neq \mathbf{0}$) for $i \in \{1, 2, 3\}$. The *exterior angles* of a regular triangle φ are the unique real numbers $\theta_1, \theta_2, \theta_3 \in (-\pi, \pi]$ such that

$$\mathbf{v}_i = (\cos \theta_i)\mathbf{u}_i + (\sin \theta_i)J\mathbf{u}_i$$

where $\mathbf{u}_i = \dot{\alpha}_i(i)/\|\dot{\alpha}_i(i)\|$ for $i \in \{1, 2, 3\}$, $\mathbf{v}_i = \dot{\alpha}_{i+1}(i)/\|\dot{\alpha}_{i+1}(i)\|$ for $i \in \{1, 2\}$, and $\mathbf{v}_3 = \dot{\alpha}_1(0)/\|\dot{\alpha}_1(0)\|$ (see Figure 21.4). In fact, $0 < \theta_i < \pi$ for each i since φ is orientation preserving.

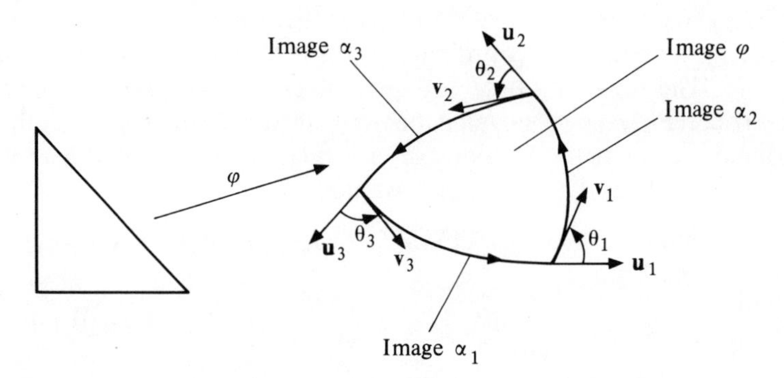

Figure 21.4. The exterior angles of a regular triangle.

Theorem 2 (Local Gauss–Bonnet Theorem). *Let S be an oriented 2-surface in $\mathbb{R}^3$ and let $\varphi\colon \Delta \to S$ be a regular triangle in S. Then*

$$\int_\varphi K\zeta + \int_a^b \kappa_g = 2\pi - \sum_{i=1}^3 \theta_i$$

where K is the Gaussian curvature of S, ζ is the volume form on S, $\beta\colon [a, b] \to S$ is a unit speed reparametrization of $\partial\varphi$, κ_g is the geodesic curvature of β, and $\theta_1, \theta_2, \theta_3$ are the exterior angles of φ.

Proof. Since φ is the restriction to Δ of a one to one local parametrization $\tilde{\varphi}$, there exists a smooth unit tangent vector field $\mathbf{X}$ defined on an open set U in S containing Image φ. Indeed, we may take $U =$ Image $\tilde{\varphi}$ and $\mathbf{X} = \mathbf{E}_1 \circ \tilde{\varphi}^{-1}/\|\mathbf{E}_1 \circ \tilde{\varphi}^{-1}\|$ where $\mathbf{E}_1$ is the first coordinate vector field of $\tilde{\varphi}$. Let ω be the connection form on U associated with $\mathbf{X}$. Then, by Lemma 3 and the local version of Stokes' theorem,

$$\int_\varphi K\zeta = -\int_\varphi d\omega = -\int_{\partial\varphi} \omega = -\int_\beta \omega = -\int_{\beta_1} \omega - \int_{\beta_2} \omega - \int_{\beta_3} \omega$$

where the $\beta_i\colon [a_i, b_i] \to S$ are the three smooth segments of β. By Lemma 2,

$$\int_{\beta_i} \omega = \int_{a_i}^{b_i} \kappa_g - \phi_i$$

where ϕ_i is the total angle of rotation of $\dot{\beta}_i$ with respect to $\mathbf{X}$. But, choosing θ_0 so that

$$\dot{\beta}_1(a_1) = \cos\theta_0 \mathbf{X}(\beta(a_1)) + \sin\theta_0 J\mathbf{X}(\beta(a_1)),$$

we see that (see Figure 21.5).

$$\dot{\beta}_1(b_1) = \cos(\theta_0 + \phi_1)\mathbf{X}(\beta(a_2)) + \sin(\theta_0 + \phi_1)J\mathbf{X}(\beta(a_2))$$

$$\dot{\beta}_2(a_2) = \cos(\theta_0 + \phi_1 + \theta_1)\mathbf{X}(\beta(a_2)) + \sin(\theta_0 + \phi_1 + \theta_1)J\mathbf{X}(\beta(a_2))$$

$$\dot{\beta}_2(b_2) = \cos(\theta_0 + \phi_1 + \theta_1 + \phi_2)\mathbf{X}(\beta(a_3)) + \sin(\theta_0 + \phi_1 + \theta_2 + \phi_2)J\mathbf{X}(\beta(a_3))$$

$$\vdots$$

$$\dot{\beta}_1(a_1) = \cos(\theta_0 + \textstyle\sum \phi_i + \sum \theta_i)\mathbf{X}(\beta(a_1)) + \sin(\theta_0 + \sum \phi_i + \sum \theta_i)J\mathbf{X}(\beta(a_1)).$$

Comparison of the two formulas above for $\dot{\beta}_1(a_1)$ shows that

$$\sum \phi_i + \sum \theta_i = 2\pi k \text{ for some integer } k.$$

Hence

$$\int_\varphi K\zeta = -\sum \int_{a_i}^{b_i} \kappa_g + \sum \phi_i$$

$$= -\int_a^b \kappa_g + 2\pi k - \sum \theta_i$$

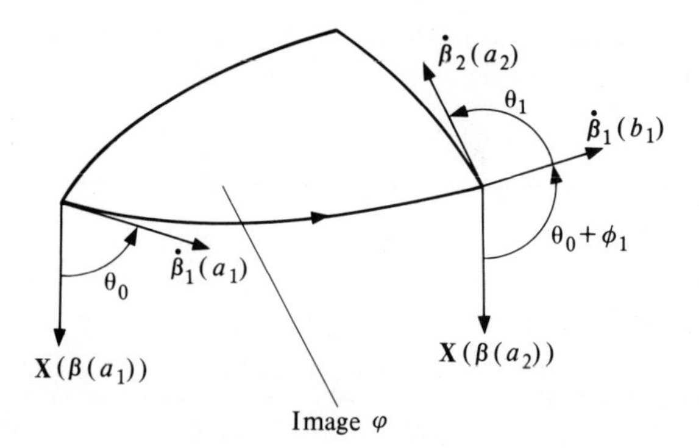

Figure 21.5. The angle of rotation of $\dot{\beta}$ with respect to $\mathbf{X}$ increases by ϕ_i along β_i and increases by θ_i at the ith vertex of φ.

so

$$\frac{1}{2\pi}\left(\int_{\varphi} K\zeta + \int_a^b \kappa_g + \sum \theta_i\right) = k$$

for some integer k.

To see that $k = 1$, consider the singular triangle $\varphi_t: \Delta \to \mathbb{R}^3$ defined for each $t \in [0, 1]$ by

$$\varphi_t(x_1, x_2) = \begin{cases} \varphi(0, 0) + \dfrac{\varphi(tx_1, tx_2) - \varphi(0, 0)}{t} & \text{if } 0 < t \le 1 \\ \varphi(0, 0) + x_1 \dfrac{\partial \varphi}{\partial x_1}(0, 0) + x_2 \dfrac{\partial \varphi}{\partial x_2}(0, 0) & \text{if } t = 0. \end{cases}$$

Then φ_t is in fact a regular triangle in the oriented 2-surface $S_t = \text{Image } \tilde{\varphi}_t$ where $\tilde{\varphi}_t: W \to \mathbb{R}^3$ is the parametrized 2-surface obtained by replacing φ everywhere in the above formula by a local parametrization $\tilde{\varphi}: W \to S$ of S with $\tilde{\varphi}|_\Delta = \varphi$, the open set W being chosen so that $tp \in W$ whenever $p \in W$ and $0 \le t \le 1$. The 2-surfaces S_t $(0 \le t \le 1)$ describe a continuous deformation of the 2-surface $S_1 = \text{Image } \tilde{\varphi}$ onto the 2-surface S_0 which is a portion of a 2-plane (see Figure 21.6). Letting K^t and ζ^t denote the Gaussian curvature and volume form of S_t and $\kappa_g^t: [a_t, b_t] \to \mathbb{R}$ and θ_i^t denote the geodesic

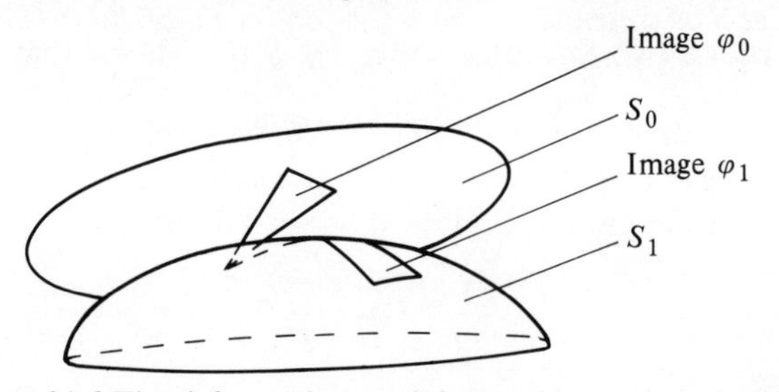

Figure 21.6 The deformation φ_t deforms the regular triangle φ to the plane triangle φ_0.

curvature and exterior angles associated with the regular triangle φ_t, the argument above shows that

$$\frac{1}{2\pi}\left(\int_{\varphi_t} K^t\zeta^t + \int_{a_t}^{b_t} \kappa_g^t + \sum \theta_i^t\right) = k^t$$

where k^t is an integer. But the left hand side of this equation varies continuously with t, hence so must the right hand side. Since k^t is always an integer, k^t must therefore have the same value for all $t \in [0, 1]$. But when $t = 0$, the regular triangle φ_t is just an ordinary plane triangle bounded by straight line segments, so $K^0 = 0$, $\kappa_g^0 = 0$, $\sum \theta_i^0 = 2\pi$, and hence $1 = k^0 = k^1 = k$. □

Remark. The formula in this theorem can be rephrased in terms of the *interior angles* $\delta_i = \pi - \theta_i$ of φ as follows:

$$\int_\varphi K\zeta + \int_a^b \kappa_g = \left(\sum_{i=1}^3 \delta_i\right) - \pi.$$

This formula has an interesting interpretation for geodesic triangles. A *geodesic triangle* in S is a regular triangle $\varphi: \Delta \to S$ such that each smooth segment of $\partial\varphi$ is a reparametrization of a geodesic. For such triangles, $\int_a^b \kappa_g = 0$ so the local Gauss–Bonnet formula becomes

$$\int_\varphi K\zeta = \left(\sum_{i=1}^3 \delta_i\right) - \pi.$$

Since $(\sum \delta_i) - \pi = 0$ when S is a 2-plane (and, in fact, whenever $K = 0$), this formula says that $\int_\varphi K\zeta$ *measures the excess (in comparison with the geodesic triangles of plane geometry) angular content of the geodesic triangle* φ. In particular, if $K > 0$ everywhere then geodesic triangles in S have angle sum $> \pi$, and if $K < 0$ everywhere the geodesic triangles in S have angle sum $< \pi$.

The Gauss–Bonnet theorem in its global form expresses the integral $\int_S K$ of the Gaussian curvature of a compact oriented 2-surface S as 2π times a certain integer associated with S. If there is a smooth nowhere zero tangent vector field on S, this integer must be zero:

Theorem 3. *Let S be a compact oriented 2-surface in $\mathbb{R}^3$. Suppose there exists a smooth nowhere zero tangent vector field on S. Then $\int_S K = 0$.*

Proof. If $\mathbf{X}$ is such a vector field then $\mathbf{X}/\|\mathbf{X}\|$ is a smooth unit tangent vector field on S. Letting ω be the connection form associated with $\mathbf{X}/\|\mathbf{X}\|$ we have, by Lemma 3 and the global version of Stokes' theorem,

$$\int_S K = \int_S K\zeta = -\int_S d\omega = 0.$$ □

Corollary. *Let S be a compact oriented 2-surface in $\mathbb{R}^3$ whose Gaussian curvature is everywhere ≥ 0. Then there can be no smooth nowhere zero tangent* *vector field on S. In particular, there is no smooth nowhere zero tangent vector field on* $\mathbf{S}^2$.

PROOF. By Theorem 4 of Chapter 12 there must be a point $p \in S$ with $K(p) > 0$. K must therefore be > 0 on an open set in S about p and since $K \geq 0$ everywhere it follows that $\int_S K > 0$. $\square$

To get some insight into what happens when there is no smooth unit tangent vector field on S, consider the case where S is the sphere $\mathbf{S}^2$ oriented by its outward normal. Although there can be no smooth unit tangent vector field on $\mathbf{S}^2$, there is one on $\mathbf{S}^2 - \{p\}$ where $p = (0, 0, -1)$ and there is another on $\mathbf{S}^2 - \{q\}$ where $q = (0, 0, 1)$. For example, we can define $\mathbf{X}_1$ on $\mathbf{S}^2 - \{p\}$ by $\mathbf{X}_1 = \mathbf{E}_1^{\varphi_1} \circ \varphi_1^{-1}/\|\mathbf{E}_1^{\varphi_1} \circ \varphi_1^{-1}\|$ where $\varphi_1 \colon \mathbb{R}^2 \to \mathbf{S}^2 - \{p\}$ is the inverse of stereographic projection from the south pole p of $\mathbf{S}^2$ and $\mathbf{E}_1^{\varphi_1}$ is its first coordinate vector field, and we can similarly define $\mathbf{X}_2$ on $\mathbf{S}^2 - \{q\}$ by $\mathbf{X}_2 = \mathbf{E}_1^{\varphi_2} \circ \varphi_2^{-1}/\|\mathbf{E}_1^{\varphi_2} \circ \varphi_2^{-1}\|$ where $\varphi_2 \colon \mathbb{R}^2 \to \mathbf{S}^2 - \{q\}$ is the inverse of stereographic projection from the north pole q. Letting ω_1 and ω_2 be the connection forms associated with $\mathbf{X}_1$ and $\mathbf{X}_2$ respectively and letting

$$\mathbf{S}_+^2 = \{(x_1, x_2, x_3) \in \mathbf{S}^2 \colon x_3 \geq 0\} \quad \text{and} \quad \mathbf{S}_-^2 = \{(x_1, x_2, x_3) \in \mathbf{S}^2 \colon x_3 \leq 0\}$$

we see that $\mathbf{S}_+^2$ and $\mathbf{S}_-^2$ are 2-surfaces-with-boundary whose union is $\mathbf{S}^2$ and whose intersection is the equator in $\mathbf{S}^2$. Furthermore, ω_1 is defined on $\mathbf{S}_+^2$ and ω_2 on $\mathbf{S}_-^2$. Applying Lemma 3 and the global version of Stokes' theorem yields

$$\int_{\mathbf{S}^2} K = \int_{\mathbf{S}^2} K\zeta = \int_{\mathbf{S}_+^2} K\zeta + \int_{\mathbf{S}_-^2} K\zeta = -\int_{\mathbf{S}_+^2} d\omega_1 - \int_{\mathbf{S}_-^2} d\omega_2$$

$$= -\int_{\partial \mathbf{S}_+^2} \omega_1 - \int_{\partial \mathbf{S}_-^2} \omega_2 = -\int_{\partial \mathbf{S}_+^2} \omega_1 + \int_{\partial \mathbf{S}_+^2} \omega_2$$

where the last equality is due to the fact that $\partial \mathbf{S}_-^2$ and $\partial \mathbf{S}_+^2$ are the same 1-surface in $\mathbb{R}^2$ but provided with opposite orientations. Letting $\alpha(t) = (\cos t, \sin t, 0)$ for $0 \leq t \leq 2\pi$, we see that $\alpha|_{(0, 2\pi)}$ is a local parametrization of $\partial \mathbf{S}_+^2$ whose image misses only one point of $\partial \mathbf{S}_+^2$ and hence

$$\int_{\mathbf{S}^2} K = -\int_{\partial \mathbf{S}_+^2} \omega_1 + \int_{\partial \mathbf{S}_+^2} \omega_2 = -\int_\alpha \omega_1 + \int_\alpha \omega_2.$$

But, by Lemma 1, $-\int_\alpha \omega_1$ is equal to the total angle of rotation of $\mathbf{Z}$ relative to $\mathbf{X}_1$ along α, where $\mathbf{Z}$ is any parallel vector field along α (we could take $\mathbf{Z} = \dot{\alpha}$), and $\int_\alpha \omega_2$ is equal to the total angle of rotation of $\mathbf{X}_2$ relative to $\mathbf{Z}$ along α; i.e., $\int_{\mathbf{S}^2} K = -\int_\alpha \omega_1 + \int_\alpha \omega_2$ is equal to the total angle of rotation of $\mathbf{X}_2$ relative to $\mathbf{X}_1$ along α. Since $\mathbf{X}_1(\alpha(2\pi)) = \mathbf{X}_1(\alpha(0))$ and $\mathbf{X}_2(\alpha(2\pi)) = \mathbf{X}_2(\alpha(0))$, this total angle of rotation must be an integer multiple of 2π; i.e., $(1/2\pi) \int_{\mathbf{S}^2} K$ must be an integer. From Figure 21.7 it is easy to see that this integer is 2 (as expected since $\int_{\mathbf{S}^2} K = \int_{\mathbf{S}^2} 1 = V(\mathbf{S}^2) = 4\pi$).

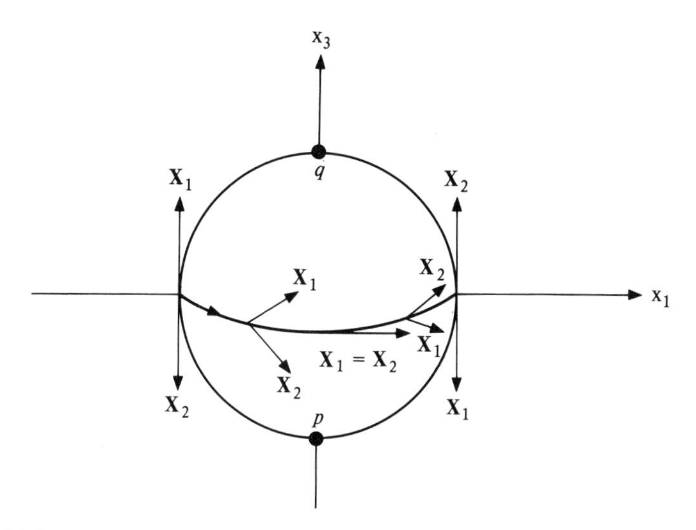

Figure 21.7. The rotation angle of $\mathbf{X}_2$ with respect to $\mathbf{X}_1$ increases by 2π along each half of the equator.

The global Gauss–Bonnet theorem is obtained by generalizing the above construction to an arbitrary compact oriented 2-surface $S \subset \mathbb{R}^3$. We shall first establish an integral formula for the total angle of rotation, along a parametrized curve, of one unit tangent vector field relative to another. Then we shall study this total rotation angle along closed curves which encircle "singularities" of one of the vector fields.

Lemma 4. *Let* $\mathbf{X}$ *and* $\mathbf{Y}$ *be smooth unit tangent vector fields defined on an open set* U *in an oriented* 2-*surface* S. *Let* ω_{XY} *be the smooth* 1-*form on* U *defined by*

$$\omega_{\mathbf{XY}} = f\,dg - g\,df$$

where $f = \mathbf{X} \cdot \mathbf{Y}$ *and* $g = \mathbf{X} \cdot J\mathbf{Y}$. *Then*

(i) $d\omega_{\mathbf{XY}} = 0$

(ii) $\int_\alpha \omega_{\mathbf{XY}}$, *where* α *is any parametrized curve in* U, *is equal to the total angle of rotation of* $\mathbf{X}$ *relative to* $\mathbf{Y}$ *along* α.

Remark. Some insight into why this lemma is correct may be gleaned from the observation that $\omega_{\mathbf{XY}} = d\tan^{-1}(g/f)$ wherever f is not zero.

Proof of Lemma 4.

(i) By Lemma 1 of Chapter 20,

$$d\omega_{\mathbf{XY}} = df \wedge dg - dg \wedge df = 2df \wedge dg.$$

But $f^2 + g^2 = 1$ and hence

$$0 = d(f^2 + g^2) = 2f\,df + 2g\,dg.$$

Taking exterior products of this equation with dg and with df yields

$$0 = 2f df \wedge dg \quad \text{and} \quad 0 = 2g df \wedge dg.$$

Since f and g are never simultaneously zero, this implies that

$$0 = 2df \wedge dg = d\omega_{\mathbf{XY}}.$$

(ii) Let $\theta\colon [a, b] \to \mathbb{R}$ measure the angle of rotation from $\mathbf{Y}$ to $\mathbf{X}$ along α so that

$$\mathbf{X} \circ \alpha = \cos\theta\, \mathbf{Y} \circ \alpha + \sin\theta\, J\mathbf{Y} \circ \alpha.$$

Then $f \circ \alpha = \cos\theta$ and $g \circ \alpha = \sin\theta$ and hence

$$\begin{aligned}\omega_{\mathbf{XY}}(\dot{\alpha}) &= (f \circ \alpha) dg(\dot{\alpha}) - (g \circ \alpha) df(\dot{\alpha}) \\ &= (f \circ \alpha)(g \circ \alpha)' - (g \circ \alpha)(f \circ \alpha)' = \theta'.\end{aligned}$$

Integrating yields

$$\int_\alpha \omega_{\mathbf{XY}} = \int_a^b \omega_{\mathbf{XY}}(\dot{\alpha}) = \int_a^b \theta' = \theta(b) - \theta(a). \qquad \square$$

Let $\mathbf{X}$ be a smooth unit tangent vector field defined on an open set U in an oriented 2-surface S. An *isolated singularity* of $\mathbf{X}$ is a point $p \in S$ such that $p \notin U$ but $V - \{p\} \subset U$ for some open set V in S containing p. Given an isolated singularity p of $\mathbf{X}$, we may choose

(i) $\varepsilon > 0$ so that the exponential map exp of S maps the open ball B_ε of radius ε about $\mathbf{0}$ in S_p diffeomorphically onto an open set $U_\varepsilon \subset U \cup \{p\}$,
(ii) $\mathbf{u} \in S_p$ with $\|\mathbf{u}\| = 1$,
(iii) $r \in \mathbb{R}$ with $0 < r < \varepsilon$, and
(iv) $\mathbf{Y}$ a smooth unit tangent vector field on U_ε.

The existence of such an ε is guaranteed by Theorem 2 of Chapter 19; the vector field $\mathbf{Y}$ can be obtained, for example, by applying d exp to any smooth non-zero vector field on B_ε and normalizing the result. Having chosen ε, $\mathbf{u}$, r and $\mathbf{Y}$, we define the *index* $\iota(\mathbf{X}, p)$ of $\mathbf{X}$ at the isolated singularity p to be $1/2\pi$ times the total angle of rotation of $\mathbf{X}$ relative to $\mathbf{Y}$ along the closed curve α_r, where $\alpha_r\colon [0, 2\pi] \to U_\varepsilon$ is defined by

$$\alpha_r(t) = \exp(r \cos t\, \mathbf{u} + r \sin t\, J\mathbf{u})$$

(see Figure 21.8).

Lemma 5. *The index $\iota(\mathbf{X}, p)$ is an integer which depends only on $\mathbf{X}$ and p (and not on the choices of ε, $\mathbf{u}$, r, and $\mathbf{Y}$).*

Proof. $\iota(\mathbf{X}, p)$ is an integer because if $\theta_r\colon [0, 2\pi] \to \mathbb{R}$ measures the angle of rotation from $\mathbf{Y}$ to $\mathbf{X}$ along α_r then the equation $\mathbf{X} \circ \alpha_r(2\pi) = \mathbf{X} \circ \alpha_r(0)$ im-

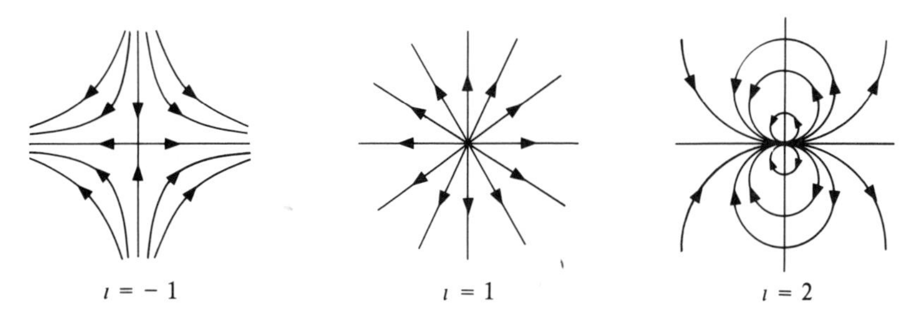

Figure 21.8. Isolated singularities of vector fields. In each case the integral curves are shown and the index is indicated.

plies that

$$\begin{aligned}\cos \theta_r(2\pi)\mathbf{Y} \circ \alpha_r(2\pi) + \sin \theta_r(2\pi)J\mathbf{Y} \circ \alpha_r(2\pi)\\ = \mathbf{X} \circ \alpha_r(2\pi)\\ = \mathbf{X} \circ \alpha_r(0)\\ = \cos \theta_r(0)\mathbf{Y} \circ \alpha_r(0) + \sin \theta_r(0)J\mathbf{Y} \circ \alpha_r(0)\end{aligned}$$

and, since $\mathbf{Y} \circ \alpha_r(2\pi) = \mathbf{Y} \circ \alpha_r(0)$, this can happen only if $\theta_r(2\pi) - \theta_r(0)$ is an integer multiple of 2π.

Independence of $\mathbf{Y}$. Suppose $\mathbf{Z}$ were another smooth unit tangent vector field on U_ε. We must show that the total angle of rotation of $\mathbf{X}$ relative to $\mathbf{Z}$ along α_r is equal to the total angle of rotation of $\mathbf{X}$ relative to $\mathbf{Y}$ along α_r. The difference of these angles is just the total angle of rotation of $\mathbf{Y}$ relative to $\mathbf{Z}$ which, by Lemma 4, is equal to $\int_{\alpha_r} \omega_{\mathbf{YZ}}$, so we must show that $\int_{\alpha_r} \omega_{\mathbf{YZ}} = 0$. But since $\omega_{\mathbf{YZ}}$ is defined on all of U_ε, and α_r is the boundary of the singular disc φ_r: $\mathsf{D} \to U_\varepsilon$ defined by $\varphi_r(x_1, x_2) = \exp(r(x_1\mathbf{u} + x_2 J\mathbf{u}))$, Stokes' theorem and Lemma 4 imply that

$$\int_{\alpha_r} \omega_{\mathbf{YZ}} = \int_{\partial\varphi_r} \omega_{\mathbf{YZ}} = \int_{\varphi_r} d\omega_{\mathbf{YZ}} = 0,$$

as required.

Independence of r: Let $\omega_{\mathbf{XY}}$ be the 1-form on $U_\varepsilon - \{p\}$ defined as in Lemma 4. Then $\iota(\mathbf{X}, p) = (1/2\pi) \int_{\alpha_r} \omega_{\mathbf{XY}}$. This formula shows that $\iota(\mathbf{X}, p)$ varies continuously with r. But since $\iota(\mathbf{X}, p)$ is always an integer, this can happen only if $\iota(\mathbf{X}, p)$ is constant as a function of r, i.e. $\iota(\mathbf{X}, p)$ is independent of r.

Independence of $\mathbf{u}$: For $0 < r < \varepsilon$, let

$$S_r = \left\{q \in U_\varepsilon : \left\| \left(\exp\Big|_{B_\varepsilon}\right)^{-1}(q)\right\|^2 \le r^2\right\}.$$

Then S_r is an oriented 2-surface-with-boundary, oriented by the restriction to S_r of the orientation on S. Moreover, α_r: $(0, 2\pi) \to \partial S_r$ is a local parametrization of ∂S_r whose image misses just one point of ∂S_r. It follows that

$$\iota(\mathbf{X}, p) = \int_{\alpha_r} \omega_{\mathbf{XY}} = \int_{\partial S_r} \omega_{\mathbf{XY}}.$$

This last integral does not depend on the choice of **u**.

Independence of ε: If exp maps both B_{ε_1} and B_{ε_2} diffeomorphically onto open sets in $U \cup \{p\}$ then we can choose r less than both ε_1 and ε_2 and use this r to compute $\iota(\mathbf{X}, p)$; the choice of $\varepsilon \in \{\varepsilon_1, \varepsilon_2\}$ is then clearly irrelevant. □

Theorem 4 (Global Gauss–Bonnet Theorem). *Let S be a compact oriented 2-surface in $\mathbb{R}^3$ and let $\mathbf{X}$ be any smooth unit tangent vector field defined on S except at isolated singularities $\{p_1, \ldots, p_k\}$. Then*

$$(1/2\pi) \int_S K = \sum_{i=1}^{k} \iota(\mathbf{X}, p_i).$$

In particular $(1/2\pi) \int_S K$ is always an integer.

Remarks. This theorem can be read in two ways. On the one hand it says that $(1/2\pi) \int_S K$ is always an integer, a remarkable result. On the other hand, it says that *the sum of the indices of any smooth unit tangent vector field defined on S except at isolated singularities is the same as the sum of the indices of any other such vector field*, another remarkable result. This common integer is called the *Euler characteristic* χ of S. It can be shown that χ is equal to 2-2g where g is the genus (the "number of holes") of S (see Figure 21.9).

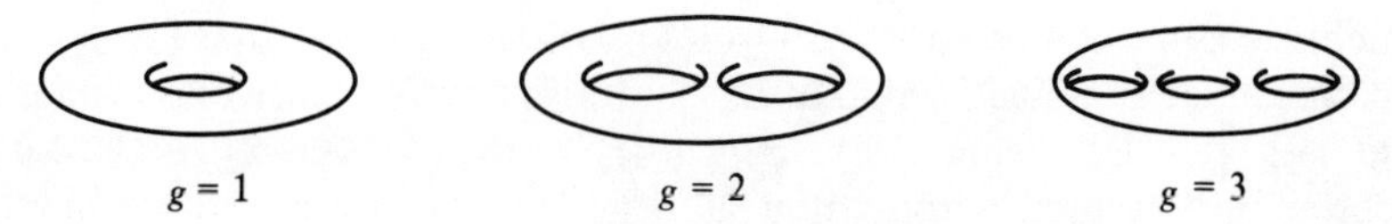

Figure 21.9. 2-surfaces with genus $g \in \{1, 2, 3\}$.

(The theorem that $(1/2\pi) \int_S K = \chi$ is actually the Gauss–Bonnet theorem. The theorem that $\sum \iota(X, p_i) = \chi$ is called the *Poincaré–Hopf theorem.*) For another interpretation of χ see Exercise 21.4.

We have implicitly assumed here that there is at least one smooth unit tangent vector field defined on S except at isolated singularities. That this is so is left as an exercise (Exercise 21.5). Note that compactness of S guarantees that there can be only finitely many isolated singularities for any given vector field.

PROOF OF THEOREM 4. For each $i \in \{1, \ldots, k\}$ choose $\varepsilon_i > 0$ so that the exponential map exp maps the ball B_{ε_i} of radius ε_i about $\mathbf{0}$ in S_{p_i} diffeomorphically onto an open set U_i about p_i in S. We may also insist that the ε_i's are chosen small enough so that $U_i \cap U_j$ is empty whenever $i \neq j$. Choose $r > 0$

such that $r < \varepsilon_i$ for all i. Let

$$S_+ = \bigcup_{i=1}^{k} S_i, \quad \text{where } S_i = \left\{ q \in U_i : \left\| \left(\exp \Big|_{B_{\varepsilon_i}} \right)^{-1} (q) \right\| \le \varepsilon \right\}$$

and let

$$S_- = \left\{ q \in S : \left\| \left(\exp \Big|_{B_{\varepsilon_i}} \right)^{-1} (q) \right\| \ge r \text{ whenever } q \in U_i \text{ for some } i \right\}$$

(see Figure 21.10). Then S_+ and S_- are compact oriented 2-surfaces-with-boundary in $\mathbb{R}^3$, oriented by restricting the orientation on S, and

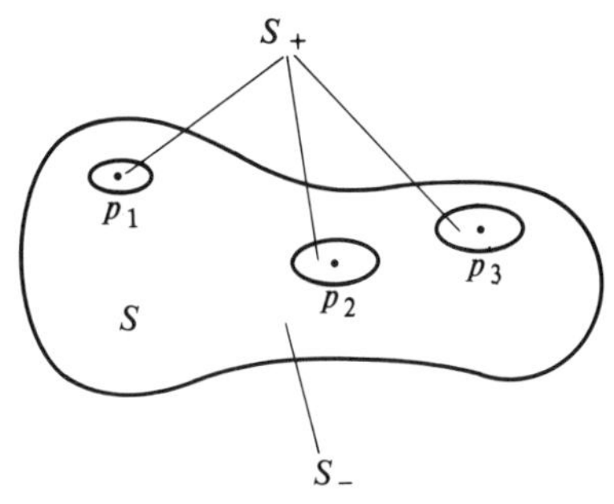

Figure 21.10. S is the union of two 2-surfaces-with-boundary, one (S_+) a union of discs about the singularities of $\mathbf{X}$, and the other (S_-) consisting of the complement of the interiors of these discs.

$\partial S_+ = \partial S_- = S_+ \cap S_-$. Note however that the induced orientation on ∂S_+ is opposite to the induced orientation on ∂S_-.

Now let $\mathbf{Y}$ be a smooth unit tangent vector field on $\bigcup_{i=1}^{k} U_i$. ($\mathbf{Y}$ can be obtained, for example, by applying $d \exp$ to smooth non-zero vector fields on each B_{ε_i} and normalizing the result.) Letting ω_1 be the connection form associated with $\mathbf{Y}$ and ω_2 be the connection form associated with $\mathbf{X}$, Lemma 3 and Stokes' Theorem imply that

$$\begin{aligned}
\int_S K &= \int_S K\zeta = \int_{S_+} K\zeta + \int_{S_-} K\zeta = -\int_{S_+} d\omega_1 - \int_{S_-} d\omega_2 \\
&= -\int_{\partial S_+} \omega_1 - \int_{\partial S_-} \omega_2 = -\int_{\partial S_+} \omega_1 + \int_{\partial S_+} \omega_2 \\
&= \sum_{i=1}^{k} \left(-\int_{\partial S_i} \omega_1 + \int_{\partial S_i} \omega_2 \right) = \sum_{i=1}^{k} \left(-\int_{\alpha_i} \omega_2 + \int_{\alpha_i} \omega_2 \right),
\end{aligned}$$

where $\alpha_i \colon [0, 2\pi] \to \partial S_i$ is defined by $\alpha_i(t) = \exp(r \cos t\, \mathbf{u}_i + r \sin t\, J\mathbf{u}_i)$, $\mathbf{u}_i$ a unit vector in S_{p_i}. But if $\mathbf{Z}_i$ is any parallel unit vector field along α_i then, by Lemma 1, $-\int_{\alpha_i} \omega_1 + \int_{\alpha_i} \omega_2$ equals the total angle of rotation of $\mathbf{Z}$ relative to

$\mathbf{Y}$ along α_i plus the total angle of rotation of $\mathbf{X}$ relative to $\mathbf{Z}$ along α_i which is the same as the total angle of rotation of $\mathbf{X}$ relative to $\mathbf{Y}$ along α_i. This last angle is just the index of $\mathbf{X}$ at p_i, so

$$\int_S K = \sum_{i=1}^{k} \iota(\mathbf{X}, p_i). \qquad \square$$

The Gauss–Bonnet formula generalizes to compact oriented n-surfaces S of any even dimension as follows: $(2/V(\mathbf{S}^n)) \int_S K = \chi$ where $V(\mathbf{S}^n)$ is the volume of the unit n-sphere $\mathbf{S}^n$, K is the Gauss–Kronecker curvature of S, and χ is the Euler characteristic of S, an integer. One proof (see S. S. Chern, A simple intrinsic proof of the Gauss–Bonnet formula for closed Riemannian manifolds, *Annals of Mathematics* **45** (1944) 747–752) is by a argument which directly generalizes the argument used in the proof of Theorem 4. Another proof is based on the following fact.

Lemma 6. *Let S be an oriented n-surface in $\mathbb{R}^{n+1}$, let ζ be the volume form on S, and let ξ be the volume form on the unit sphere $\mathbf{S}^n$ with its standard orientation. Then*

$$N^*\xi = K\zeta$$

where $N: S \to \mathbf{S}^n$ is the Gauss map and K is the Gauss–Kronecker curvature of S.

Proof. For $\mathbf{v}_1, \ldots, \mathbf{v}_n \in S_p$, $p \in S$, we have, using Theorem 5 of Chapter 12 together with the fact that $dN(\mathbf{v})$ and $\nabla_{\mathbf{v}} \mathbf{N}$ have the same vector part for all $\mathbf{v} \in S_p$,

$$\begin{aligned}(N^*\xi)(\mathbf{v}_1, \ldots, \mathbf{v}_n) &= \xi(dN(\mathbf{v}_1), \ldots, dN(\mathbf{v}_n)) \\ &= \det\begin{pmatrix} dN(\mathbf{v}_1) \\ \vdots \\ dN(\mathbf{v}_n) \\ \mathbf{N}^{\mathbf{S}^n}(\mathbf{N}(p)) \end{pmatrix} = (-1)^n \det\begin{pmatrix} \nabla_{\mathbf{v}_1}\mathbf{N} \\ \vdots \\ \nabla_{\mathbf{v}_n}\mathbf{N} \\ \mathbf{N}(p) \end{pmatrix} = K(p)\det\begin{pmatrix} \mathbf{v}_1 \\ \vdots \\ \mathbf{v}_n \\ \mathbf{N}(p) \end{pmatrix} \\ &= K(p)\zeta(\mathbf{v}_1, \ldots, \mathbf{v}_n). \qquad \square\end{aligned}$$

For $S \subset \mathbb{R}^{n+1}$ a compact connected oriented n-surface with $K > 0$ everywhere, the fact that $(2/V(\mathbf{S}^n)) \int_S K$ is an integer is now immediate because, in this case, N is an orientation preserving diffeomorphism and hence, using Exercise 17.15,

$$\int_S K = \int_S K\zeta = \int_S N^*\xi = \int_{\mathbf{S}^n} \xi = V(\mathbf{S}^n)$$

so $(2/V(\mathbf{S}^n)) \int_S K = 2$. In the general case, if $p \in S$ is such that $K(p) \neq 0$ then dN_p will be non-singular so there will be an open set U_p about p which is mapped diffeomorphically by N onto an open set in $\mathbf{S}^n$. If $K(p) > 0$ this diffeomorphism will be orientation preserving, so

$$\int_{U_p} K = \int_{U_p} N^*\xi = \int_{N(U_p)} \xi = V(N(U_p)).$$

If $K(p) < 0$ the diffeomorphism $N|_{U_p}$ will be orientation reversing so

$$\int_{U_p} K = \int_{U_p} N^*\xi = -\int_{N(U_p)} \xi = -V(N(U_p)).$$

It can be shown (see J. Milnor, *Topology from the Differentiable Viewpoint*, The University Press of Virginia, 1965) that "most" points q of $\mathbf{S}^n$ are regular values of N in that dN_p is non-singular ($K(p) \neq 0$) for all $p \in N^{-1}(q)$ and, furthermore, that the integer

$$d = \#\{p \in N^{-1}(q)\colon K(p) > 0\} - \#\{p \in N^{-1}(q)\colon K(p) < 0\}$$

is independent of the regular value q, where $\#\{-\}$ denotes the number of points in the (finite) set $\{-\}$. The number d is called the *degree* of the Gauss map $N\colon S \to \mathbf{S}^n$. For a sufficiently small open set U about a regular value q of N it follows that $N^{-1}(U)$ consists of $\#(N^{-1}(q))$ disjoint open sets in S, each mapped diffeomorphically by N onto U, and that

$$\int_{N^{-1}(U)} K = \int_{N^{-1}(U)} N^*\xi = d\int_U \xi = dV(U).$$

Since the regions where $K = 0$ contribute nothing to the integral, a careful choice of partition of unity will yield $\int_S K = dV(\mathbf{S}^n)$, or $(1/V(\mathbf{S}^n)) \int_S K = d$. For n even, $d = \chi/2$ where χ is the Euler characteristic of S.

Exercises

21.1. Let S be an oriented 2-surface in $\mathbb{R}^3$. Suppose $\varphi\colon \mathsf{D} \to S$ and $\psi\colon \mathsf{D} \to S$ are singular discs in S such that $\partial\varphi = \partial\psi$. Show that the holonomy angle of φ differs from the holonomy angle of ψ by an integer multiple of 2π.

21.2. A *singular rectangle* in an oriented 2-surface S is a smooth map $\varphi\colon \square \to S$ where

$$\square = \{(x_1, x_2) \in \mathbb{R}^2\colon 0 \le x_1 \le 1, 0 \le x_2 \le 1\}.$$

Its *boundary* is the piecewise smooth parametrized curve $\alpha\colon [0, 4] \to \mathbb{R}^2$ defined by

$$\alpha(t) = \begin{cases} (t, 0) & \text{if } 0 \le t \le 1 \\ (1, t-1) & \text{if } 1 \le t \le 2 \\ (3-t, 1) & \text{if } 2 \le t \le 3 \\ (0, 4-t) & \text{if } 3 \le t \le 4. \end{cases}$$

φ is *regular* if it is the restriction to $\square$ of a one to one local parametrization of S. Prove the Gauss–Bonnet theorem for regular rectangles:

$$\int_{\varphi} K\zeta + \int_a^b \kappa_g = 2\pi - \sum_{i=1}^{4} \theta_i$$

where κ_g: $[a, b] \to \mathbb{R}$ and θ_i $(i \in \{1, 2, 3, 4\})$ are defined in the same way as for regular triangles (see Theorem 2).

21.3. Let **X** and **Y** be smooth unit tangent vector fields defined on an open set U in an oriented 2-surface S and let h: $U \to \mathbb{R}^2$ be defined by $h(p) = (\mathbf{X}(p) \cdot \mathbf{Y}(p), \mathbf{X}(p) \cdot J\mathbf{Y}(p))$. Show that $\omega_{\mathbf{XY}} = h^*\eta$ where $\omega_{\mathbf{XY}}$ is the 1-form on U defined in Lemma 4 and η is the 1-form on $\mathbb{R}^2 - \{0\}$ defined in Theorem 3 of Chapter 11.

21.4. Let S be a compact oriented 2-surface in $\mathbb{R}^3$. Suppose there exists a finite collection of regular triangles φ_i: $\Delta \to S (i \in \{1, \dots, m\})$ such that

(i) $\bigcup_{i=1}^m$ Image $\varphi_i = S$

(ii) If $p \in S$ is in the image of more than 2 of the φ_i then p is, for each such i, the image under φ_i of a vertex of Δ.

(iii) If p is in the image of exactly two of the φ_i then the intersection of the images of these φ_i is equal to the image of a smooth segment of the boundary of each.

Such a collection $\{\varphi_1, \dots, \varphi_m\}$ is called a *triangulation* T of S (see Figure 21.11). Points of type (ii) are called vertices of T and subsets of the form Image $\varphi_i \cap$ Image φ_j as in (iii) are called *edges* of T. The φ_i themselves are called the *faces* of T.

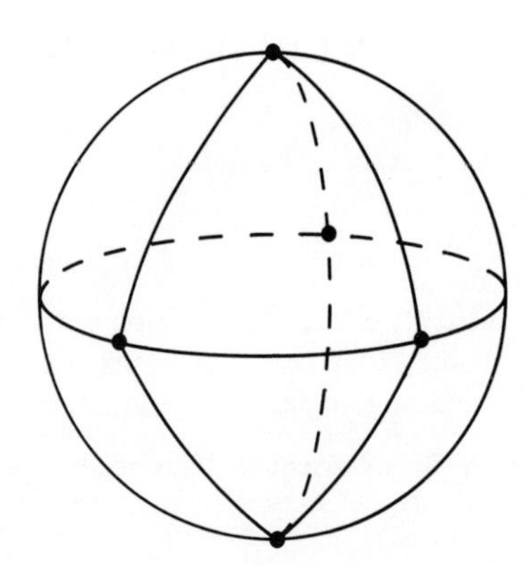

Figure 21.11. A triangulation of the 2-sphere.

Show that

$$(1/2\pi) \int_S K = v - e + f$$

where v is the number of vertices, e the number of edges, and f the number of faces, of T. [*Hint*: Apply the local Gauss–Bonnet theorem to each triangle φ_i.]

21.5. Let S be an oriented n-surface in $\mathbb{R}^{n+1}$. For $q \notin S$, let f_q: $S \to \mathbb{R}$ be the smooth function defined by $f_q(p) = \|q - p\|^2$.

(a) Show that $p \in S$ is a critical point of f_q if and only if $q - p = \lambda N(p)$ for some $\lambda \in \mathbb{R}$, where N is the Gauss map of S.

(b) Show that if $p \in S$ is a critical point of f_q and $\mathbf{v} \in S_p$, $\mathbf{v} \neq 0$, then $\nabla_{\mathbf{v}}$ grad $f_q = \mathbf{0}$ if and only if $L_p(\mathbf{v}) = (1/\lambda)\mathbf{v}$ where L_p is the Weingarten map of S at p and λ is as in part (a).

(c) Conclude that if q does not lie on the focal locus of S then all critical points of f_q are non-degenerate and hence isolated. (It follows then that grad $f_q/\|\text{grad } f_q\|$ is a smooth unit tangent vector field defined on S except at isolated singularities.)

21.6. Let S be an oriented 2-surface in $\mathbb{R}^3$ and let $\mathbf{X}$ be a smooth unit tangent vector field defined on an open set U in S. Let θ_1 be the 1-form on U dual to $\mathbf{X}$, let θ_2 be the 1-form on U dual to $J\mathbf{X}$, and let ω be the connection 1-form on U associated with $\mathbf{X}$. Show that

$$d\theta_1 = \omega \wedge \theta_2$$

$$d\theta_2 = -\omega \wedge \theta_1$$

$$d\omega = -K\theta_1 \wedge \theta_2$$

where K is the Gaussian curvature of S.

(These equations are called the *Cartan structural equations.*)

22 Rigid Motions and Congruence

A *rigid motion* of $\mathbb{R}^{n+1}$ is a map $\psi: \mathbb{R}^{n+1} \to \mathbb{R}^{n+1}$ such that $\|\psi(p) - \psi(q)\| = \|p - q\|$ for all $p, q \in \mathbb{R}^{n+1}$. Thus a rigid motion is a map which preserves distances between points.

EXAMPLE 1. For $a \in \mathbb{R}^{n+1}$, define $\psi: \mathbb{R}^{n+1} \to \mathbb{R}^{n+1}$ by $\psi(p) = p + a$. Then ψ is a rigid motion of $\mathbb{R}^{n+1}$, called *translation* by a.

EXAMPLE 2. For $\theta \in \mathbb{R}$, define $\psi: \mathbb{R}^2 \to \mathbb{R}^2$ by

$$\psi(x_1, x_2) = (x_1 \cos \theta - x_2 \sin \theta, x_1 \sin \theta + x_2 \cos \theta).$$

Then ψ is a rigid motion of $\mathbb{R}^2$, called *rotation* through θ (see Figure 22.1).

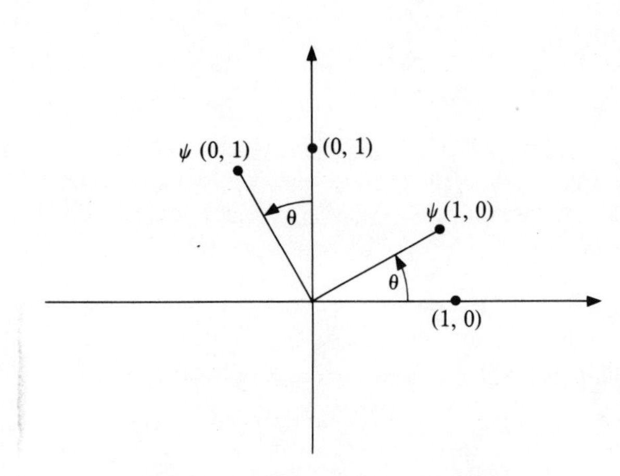

Figure 22.1 Rotation of $\mathbb{R}^2$ through θ.

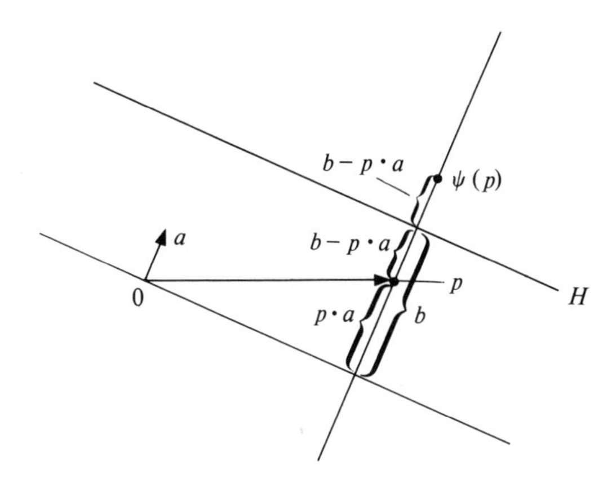

Figure 22.2 Reflection through the n-plane H.

EXAMPLE 3. For $a \in \mathbb{R}^{n+1}$, $\|a\| = 1$, and $b \in \mathbb{R}$, define $\psi: \mathbb{R}^{n+1} \to \mathbb{R}^{n+1}$ by (see Figure 22.2)

$$\psi(p) = p + 2(b - p \cdot a)a.$$

Then ψ is a rigid motion of $\mathbb{R}^{n+1}$, called *reflection* through the n-plane $H = \{x \in \mathbb{R}^{n+1}: a \cdot x = b\}$.

EXAMPLE 4. Let $\psi: \mathbb{R}^{n+1} \to \mathbb{R}^{n+1}$ be a linear transformation such that $\|\psi(v)\| = \|v\|$ for all $v \in \mathbb{R}^{n+1}$. Then ψ is a rigid motion because $\|\psi(p) - \psi(q)\| = \|\psi(p - q)\| = \|p - q\|$ for all $p, q \in \mathbb{R}^{n+1}$. ψ is called an *orthogonal transformation* of $\mathbb{R}^{n+1}$. Note that a linear transformation of $\mathbb{R}^{n+1}$ is a rigid motion if and only if it is an orthogonal transformation.

The composition $\psi_2 \circ \psi_1$ of two rigid motions of $\mathbb{R}^{n+1}$ is a rigid motion. In particular, an orthogonal transformation followed by a translation is a rigid motion. It turns out that every rigid motion can be obtained this way.

Theorem 1. *Let ψ be a rigid motion of $\mathbb{R}^{n+1}$. Then there exists a unique orthogonal transformation ψ_1 and a unique translation ψ_2 such that $\psi = \psi_2 \circ \psi_1$.*

PROOF. Let $a = \psi(0)$, let ψ_2 be translation by a, and let $\psi_1 = \psi_2^{-1} \circ \psi$. We shall show that ψ_1 is an orthogonal transformation. Clearly ψ_1 is a rigid motion with $\psi_1(0) = 0$. ψ_1 preserves norms because

$$\|\psi_1(v)\| = \|\psi_1(v) - \psi_1(0)\| = \|v - 0\| = \|v\|$$

for all $v \in \mathbb{R}^{n+1}$. Thus we need only show that ψ_1 is linear. First observe that

ψ_1 preserves dot products:

$$\begin{aligned}\psi_1(v) \cdot \psi_1(w) &= \tfrac{1}{2}(\|\psi_1(v)\|^2 + \|\psi_1(w)\|^2 - \|\psi_1(v) - \psi_1(w)\|^2) \\ &= \tfrac{1}{2}(\|v\|^2 + \|w\|^2 - \|v - w\|^2) \\ &= v \cdot w,\end{aligned}$$

for all $v, w \in \mathbb{R}^{n+1}$. Finally to establish linearity, we must show that

$$\psi_1(c_1 v_1 + c_2 v_2) = c_1 \psi_1(v_1) + c_2 \psi_1(v_2)$$

or, equivalently, that

$$\psi_1(c_1 v_1 + c_2 v_2) - c_1 \psi_1(v_1) - c_2 \psi_1(v_2) = 0$$

for all $v_1, v_2 \in \mathbb{R}^{n+1}$ and $c_1, c_2 \in \mathbb{R}$. For this it suffices to show that the vector $\psi_1(c_1 v_1 + c_2 v_2) - c_1 \psi_1(v_1) - c_2 \psi_1(v_2)$ is orthogonal to every vector in some basis for $\mathbb{R}^{n+1}$. But if $\{e_1, \ldots, e_{n+1}\}$ is any orthonormal basis for $\mathbb{R}^{n+1}$ then $\{\psi(e_1), \ldots, \psi(e_{n+1})\}$ is also an orthonormal basis for $\mathbb{R}^{n+1}$, since ψ_1 preserves dot products, and

$$\begin{aligned}&[\psi_1(c_1 v_1 + c_2 v_2) - c_1 \psi_1(v_1) - c_2 \psi_1(v_2)] \cdot \psi_1(e_i) \\ &\qquad = (c_1 v_1 + c_2 v_2) \cdot e_i - c_1(v_1 \cdot e_i) - c_2(v_2 \cdot e_i) = 0\end{aligned}$$

for $i \in \{1, \ldots, n+1\}$. Thus ψ_1 is linear.

Uniqueness of ψ_1 and ψ_2 follows from the requirements that the translation ψ_2 must satisfy $\psi_2(0) = \psi_2 \circ \psi_1(0) = \psi(0)$ and that ψ_1 must equal $\psi_2^{-1} \circ \psi$. □

Corollary. *Let $\psi: \mathbb{R}^{n+1} \to \mathbb{R}^{n+1}$ be a rigid motion. Then*

(i) *ψ is smooth.*
(ii) *ψ maps $\mathbb{R}^{n+1}$ onto $\mathbb{R}^{n+1}$, and*
(iii) $d\psi(\mathbf{v}) \cdot d\psi(\mathbf{w}) = \mathbf{v} \cdot \mathbf{w}$ for all $\mathbf{v}, \mathbf{w} \in \mathbb{R}^{n+1}_p$, $p \in \mathbb{R}^{n+1}$.

PROOF. (i) Linear transformations and translations are smooth, hence so are rigid motions.

(ii) Translations are onto, as is any orthogonal transformation of $\mathbb{R}^{n+1}$ (the kernel must be zero, since norms are preserved). Hence rigid motions are onto.

(iii) First note that if $\psi = \psi_2 \circ \psi_1$ is the unique decomposition of the rigid motion ψ into an orthogonal transformation ψ_1 followed by a translation ψ_2 then

$$(*) \qquad d\psi(p, v) = (\psi(p), \psi_1(v))$$

for all $(p, v) \in \mathbb{R}^{n+1}_p$, $p \in \mathbb{R}^{n+1}$. Indeed, setting $\alpha(t) = p + tv$, so that

$\dot{\alpha}(0) = (p, v)$, we have (where ψ_2 is translation by a)

$$d\psi(p, v) = \psi \dot{\circ} \alpha(0) = \left(\psi(p), \left.\frac{d}{dt}\right|_0 \psi(p + tv)\right)$$

$$= \left(\psi(p), \left.\frac{d}{dt}\right|_0 \psi_2 \circ \psi_1(p + tv)\right)$$

$$= \left(\psi(p), \left.\frac{d}{dt}\right|_0 (\psi_1(p) + t\psi_1(v) + a)\right)$$

$$= (\psi(p), \psi_1(v)).$$

Hence, for $\mathbf{v} = (p, v)$ and $\mathbf{w} = (p, w) \in \mathbb{R}_p^{n+1}$,

$$d\psi(\mathbf{v}) \cdot d\psi(\mathbf{w}) = (\psi(p), \psi_1(v)) \cdot (\psi(p), \psi_1(w))$$
$$= \psi_1(v) \cdot \psi_1(w) = v \cdot w = \mathbf{v} \cdot \mathbf{w}. \qquad \square$$

Two n-surfaces S and $\tilde{S}$ in $\mathbb{R}^{n+1}$ are *congruent* if there exists a rigid motion $\psi: \mathbb{R}^{n+1} \to \mathbb{R}^{n+1}$ such that $\psi(S) = \tilde{S}$. The differential $d\psi$ of such a rigid motion maps the tangent space S_p to S at each $p \in S$ onto the tangent space $\tilde{S}_{\psi(p)}$ to $\tilde{S}$ at $\psi(p)$ so, since $d\psi$ preserves dot products, $d\psi(\mathbf{N}(p)) = \pm\tilde{\mathbf{N}}(\psi(p))$ where $\mathbf{N}$ and $\tilde{\mathbf{N}}$ are any given orientation vector fields on S and $\tilde{S}$ respectively. Note that an orientation $\tilde{\mathbf{N}}$ can always be chosen on $\tilde{S}$ so that $d\psi(\mathbf{N}(p)) = +\tilde{\mathbf{N}}(\psi(p))$ for all $p \in S$ (i.e., so that $d\psi \circ \mathbf{N} = \tilde{\mathbf{N}} \circ \psi$).

Theorem 2. *Let S and $\tilde{S}$ be congruent n-surface in $\mathbb{R}^{n+1}$, let $\psi: \mathbb{R}^{n+1} \to \mathbb{R}^{n+1}$ be a rigid motion such that $\psi(S) = \tilde{S}$, and assume that S and $\tilde{S}$ are oriented so that $d\psi \circ \mathbf{N} = \tilde{\mathbf{N}} \circ \psi$. Then*

(i) $d\psi(\mathbf{v}) \cdot d\psi(\mathbf{w}) = \mathbf{v} \cdot \mathbf{w}$ *for all* $\mathbf{v}, \mathbf{w} \in S_p$, $p \in S$, *and*
(ii) *the second fundamental forms $\mathscr{S}_p$ of S at $p \in S$ and $\tilde{\mathscr{S}}_{\psi(p)}$ of $\tilde{S}$ at $\psi(p)$ are related by $\mathscr{S}_p = \tilde{\mathscr{S}}_{\psi(p)} \circ d\psi$.*

Proof. (i) is immediate from the corollary above.

(ii) Let ψ_1 be the orthogonal part of ψ as in Theorem 1. Given $p \in S$ and $\mathbf{v} \in S_p$, let $\alpha: I \to S$ be such that $\dot{\alpha}(t_0) = \mathbf{v}$. Then $d\psi(\mathbf{v}) = \psi \circ \alpha(t_0)$ so the value of the Weingarten map $\tilde{L}_{\psi(p)}$ of $\tilde{S}$ at $\psi(p)$ on $d\psi(\mathbf{v})$ is

$$\tilde{L}_{\psi(p)}(d\psi(\mathbf{v})) = -\nabla_{d\psi(\mathbf{v})}\tilde{\mathbf{N}}$$
$$= -(\tilde{\mathbf{N}} \circ \dot{\psi} \circ \alpha)(t_0)$$
$$= -(d\psi \circ \dot{\mathbf{N}} \circ \alpha)(t_0)$$
$$= -(\psi(p), (\psi_1 \circ N \circ \alpha)'(t_0)) \quad \text{(by equation (*), above)}$$
$$= -(\psi(p), \psi_1((N \circ \alpha)'(t_0)) \quad \text{(by linearity of } \psi_1)$$
$$= -d\psi(p, (N \circ \alpha)'(t_0)) \quad \text{(again, by equation (*))}$$
$$= -d\psi(\nabla_{\mathbf{v}}\mathbf{N})$$
$$= d\psi(L_p(\mathbf{v})),$$

where L_p is the Weingarten map of S at p. Thus

$$\begin{aligned}\mathscr{S}_{\psi(p)}(d\psi(\mathbf{v})) &= \tilde{L}_{\psi(p)}(d\psi(\mathbf{v})) \cdot d\psi(\mathbf{v}) = d\psi(L_p(\mathbf{v})) \cdot d\psi(\mathbf{v}) \\ &= L_p(\mathbf{v}) \cdot \mathbf{v} = \mathscr{S}_p(\mathbf{v}). \end{aligned}$$

□

Corollary. *Let S and $\tilde{S}$ be congruent oriented n-surfaces in $\mathbb{R}^{n+1}$. Then S and $\tilde{S}$ have the same geometry in that if ψ is a rigid motion mapping S onto $\tilde{S}$ with $\tilde{\mathbf{N}} \circ \psi = d\psi \circ \mathbf{N}$ and ψ_1 is the orthogonal part of ψ then*

(i) *the length of $\alpha: I \to S$ is the same as the length of $\psi \circ \alpha: I \to \tilde{S}$,*
(ii) *the Gauss map N of S is related to the Gauss map $\tilde{N}$ of $\tilde{S}$ by $\tilde{N} \circ \psi = \psi_1 \circ N$, and in particular the spherical image of $\tilde{S}$ is the image under ψ_1 of the spherical image of S,*
(iii) *$\alpha: I \to S$ is a geodesic in S if and only if $\psi \circ \alpha$ is a geodesic in $\tilde{S}$,*
(iv) *a vector field $\mathbf{X}$ along $\alpha: I \to S$ is parallel if and only if $d\psi \circ \mathbf{X}$ is parallel along $\psi \circ \alpha$,*
(v) *the Weingarten maps L_p of S at $p \in S$ and $L_{\psi(p)}$ of $\tilde{S}$ at $\psi(p)$ are related by*

$$\tilde{L}_{\psi(p)} \circ d\psi_p = d\psi_p \circ L_p,$$

(vi) *the normal curvatures k of S and $\tilde{k}$ of $\tilde{S}$ are related by $k = \tilde{k} \circ d\psi$ and in particular the principal curvatures of S at $p \in S$ are the same as the principal curvatures of $\tilde{S}$ at $\psi(p)$,*
(vii) *the Gauss-Kronecker and mean curvatures of S at $p \in S$ are equal to the Gauss-Kronecker and mean curvatures of $\tilde{S}$ at $\psi(p)$,*
(viii) *S is convex at $p \in S$ if and only if $\tilde{S}$ is convex at $\psi(p)$,*
(ix) *the focal locus of $\tilde{S}$ is the image under ψ of the focal locus of S,*
(x) *the volume of S equals the volume of $\tilde{S}$,*
(xi) *S is a minimal surface if and only if $\tilde{S}$ is a minimal surface, and*
(xii) *the conjugate locus of $\psi(p)$ in $\tilde{S}$ is the image under ψ of the conjugate locus of p in S.*

PROOF.

(i) $l(\psi \circ \alpha) = \int_I \|\psi \dot{\circ} \alpha\| = \int_I \|d\psi \circ \dot{\alpha}\| = \int_I \|\dot{\alpha}\| = l(\alpha)$.
(ii) Immediate from the equation $\tilde{\mathbf{N}} \circ \psi = d\psi \circ \mathbf{N}$.
(iii) Follows from (iv) since α is a geodesic if and only if $\dot{\alpha}$ is parallel.
(iv) $(d\psi \dot{\circ} \mathbf{X})(t) = d\psi(\dot{\mathbf{X}}(t))$ is a multiple of $\tilde{\mathbf{N}}(\psi(\alpha(t))) = d\psi(\mathbf{N}(\alpha(t)))$ if and only if $\dot{\mathbf{X}}(t)$ is a multiple of $\mathbf{N}(\alpha(t))$.
(v) This is contained in the proof of Theorem 2.
(vi) $\tilde{k}(d\psi(\mathbf{v})) = \mathscr{S}_{\psi(p)}(d\psi(\mathbf{v})) = \mathscr{S}_p(\mathbf{v}) = k(\mathbf{v})$. That the principal curvatures are the same follows from this or from (v), since L_p and $\tilde{L}_p = d\psi_p \circ L_p \circ d\psi_p^{-1}$ have the same eigenvalues.
(vii) Follows from (vi).

(viii) $(\psi(q) - \psi(p)) \cdot \tilde{N}(\psi(p)) = (\psi_1(q) - \psi_1(p)) \cdot \psi_1(N(p))$
$$= (q - p) \cdot N(p).$$

(ix) $\psi(p + (1/k_i(p))N(p)) = \psi_2(\psi_1(p) + (1/k_i(p))\psi_1(N(p)))$
$$= \psi_1(p) + (1/k_i(p))\psi_1(N(p)) + a$$
$$= \psi(p) + (1/\tilde{k}_i(\psi(p)))\tilde{N}(\psi(p)).$$

(x) For any local parametrization φ of S, $\psi \circ \varphi$ is a local parametrization $\tilde{S}$ and

$$\det(\mathbf{E}_i^{\psi \circ \varphi} \cdot \mathbf{E}_j^{\psi \circ \varphi}) = \det(d\psi(\mathbf{E}_i^{\varphi}) \cdot d\psi(\mathbf{E}_j^{\varphi})) = \det(\mathbf{E}_i^{\varphi} \cdot \mathbf{E}_j^{\varphi})$$

so the volume integrands are equal.

(xi) Follows from (vii).

(xii) $\psi \circ \exp = \widetilde{\exp} \circ d\psi$ where exp and $\widetilde{\exp}$ are the exponential maps of S and $\tilde{S}$ since, by (iii), ψ maps geodesics to geodesics. Hence ψ maps conjugate points along the geodesic $\exp(t\mathbf{v})$ in S to conjugate points along the geodesic $\widetilde{\exp}(td\psi(\mathbf{v}))$ in $\tilde{S}$. □

The converse of Theorem 2 is also true:

Theorem 3. *Let S and $\tilde{S}$ be connected oriented n-surfaces in $\mathbb{R}^{n+1}$. Suppose there exists a smooth map ψ from S onto $\tilde{S}$ such that*

(i) *$d\psi(\mathbf{v}) \cdot d\psi(\mathbf{w}) = \mathbf{v} \cdot \mathbf{w}$ for all $\mathbf{v}, \mathbf{w} \in S_p$, $p \in S$ and*
(ii) *the second fundamental forms $\mathscr{S}_p$ of S at p and $\tilde{\mathscr{S}}_{\psi(p)}$ of $\tilde{S}$ at $\psi(p)$ are related by $\mathscr{S}_p = \tilde{\mathscr{S}}_{\psi(p)} \circ d\psi$, for all $p \in S$.*

Then S and $\tilde{S}$ are congruent and, in fact, ψ is the restriction to S of a rigid motion of $\mathbb{R}^{n+1}$.

PROOF. Let $p_0 \in S$ and define a rigid motion $\bar{\psi}$ of $\mathbb{R}^{n+1}$ by $\bar{\psi} = \psi_2 \circ \psi_1$ where ψ_1 is the unique orthogonal transformation of $\mathbb{R}^{n+1}$ such that

(a) $\psi_1(N(p_0)) = \tilde{N}(\psi(p_0))$
(b) $(p_0, \psi_1(v)) = d\psi(p_0, v)$ for all $v \in \mathbb{R}^{n+1}$ such that $v \perp N(p_0)$ (i.e., such that $(p_0, v) \in S_{p_0}$),

and ψ_2 is the translation of $\mathbb{R}^{n+1}$ which sends $\psi_1(p_0)$ to $\psi(p_0)$. We shall show that $(\bar{\psi}^{-1} \circ \psi)(p) = p$ for all $p \in S$ thereby establishing that $\psi = \bar{\psi}|_S$.

Let $\varphi = \bar{\psi}^{-1} \circ \psi$. Then φ maps S onto the n-surface $\bar{S} = \bar{\psi}^{-1}(\tilde{S})$, and

(i) $d\varphi(\mathbf{v}) \cdot d\varphi(\mathbf{w}) = \mathbf{v} \cdot \mathbf{w}$ for all $\mathbf{v}, \mathbf{w} \in S_p$, $p \in S$,
(ii) $\mathscr{S}_p = \bar{\mathscr{S}}_{\varphi(p)} \circ d\varphi$ for all $p \in S$ ($\bar{\mathscr{S}}_{\varphi(p)}$ = second fundamental form of $\bar{S}$ at $\varphi(p)$),
(iii) $\varphi(p_0) = p_0$,
(iv) $d\varphi(\mathbf{v}) = \mathbf{v}$ for all $\mathbf{v} \in S_{p_0}$.

From (i) and (ii) we can conclude that the Weingarten maps L_p of S and $\bar{L}_{\varphi(p)}$ of $\bar{S}$ are related by $\bar{L}_{\varphi(p)} \circ d\varphi = d\varphi \circ L_p$, for all $p \in S$. Indeed,

$$\begin{aligned}\bar{L}_{\varphi(p)}(d\varphi(\mathbf{v})) \cdot d\varphi(\mathbf{w}) &= \tfrac{1}{2}(\bar{\mathscr{S}}_{\varphi(p)}(d\varphi(\mathbf{v}+\mathbf{w})) - \bar{\mathscr{S}}_{\varphi(p)}(d\varphi(\mathbf{v})) - \bar{\mathscr{S}}_{\varphi(p)}(d\varphi(\mathbf{w}))\\ &= \tfrac{1}{2}(\mathscr{S}_p(\mathbf{v}+\mathbf{w}) - \mathscr{S}_p(\mathbf{v}) - \mathscr{S}_p(\mathbf{w}))\\ &= L_p(\mathbf{v}) \cdot \mathbf{w} = d\varphi(L_p(\mathbf{v})) \cdot d\varphi(\mathbf{w})\end{aligned}$$

for all $\mathbf{v}, \mathbf{w} \in S_p$ and, since $d\varphi$ maps S_p onto $\bar{S}_{\varphi(p)}$ for all $p \in S$ ($d\varphi$ maps orthonormal bases for S_p to orthonormal bases for $\bar{S}_{\varphi(p)}$, by (i)), this implies that $\bar{L}_{\varphi(p)}(d\varphi(\mathbf{v})) = d\varphi(L_p(\mathbf{v}))$ for all $\mathbf{v} \in S_p$.

We shall show that if $\alpha: I \to S$ is any parametrized curve in S such that, for some $t_0 \in I$, $\varphi(\alpha(t_0)) = \alpha(t_0)$ and $d\varphi_{\alpha(t_0)} =$ identity (this will be the case, for example, if $\alpha(t_0) = p_0$) then $\varphi(\alpha(t)) = \alpha(t)$ and $d\varphi_{\alpha(t)} =$ identity for all $t \in I$. Let $\mathbf{X}_1, \ldots, \mathbf{X}_n$ be smooth vector fields parallel along α such that $\{\mathbf{X}_1(t), \ldots, \mathbf{X}_{n+1}(t)\}$ is an orthonormal basis for $S_{\alpha(t)}$ for each $t \in I$. Such vector fields can be obtained by choosing an orthonormal basis $\{\mathbf{x}_1, \ldots, \mathbf{x}_n\}$ for $S_{\alpha(t_0)}$ and defining $\mathbf{X}_i$ to be the unique vector field parallel along α with $\mathbf{X}_i(t_0) = \mathbf{x}_i$. Let $\mathbf{Y}_1, \ldots, \mathbf{Y}_n$ be the vector fields along $\varphi \circ \alpha$ defined by $\mathbf{Y}_i(t) = d\varphi(\mathbf{X}_i(t))$ for $t \in I$. Then $\{\mathbf{Y}_1(t), \ldots, \mathbf{Y}_n(t)\}$ is an orthonormal basis for $\bar{S}_{\varphi \circ \alpha(t)}$ for each $t \in I$ (by (i)), $\mathbf{Y}_i(t_0) = \mathbf{X}_i(t_0)$ for each i (since $d\varphi_{\alpha(t_0)} =$ identity), and

$$\begin{aligned}\varphi \dot{\circ} \alpha &= \sum_{i=1}^{n} ((\varphi \dot{\circ} \alpha) \cdot \mathbf{Y}_i)\mathbf{Y}_i = \sum_{i=1}^{n} (d\varphi(\dot{\alpha}) \cdot d\varphi(\mathbf{X}_i))\mathbf{Y}_i\\ &= \sum_{i=1}^{n} (\dot{\alpha} \cdot \mathbf{X}_i)\mathbf{Y}_i.\end{aligned}$$

If we could show that $Y_i = X_i$, where $Y_i: I \to \mathbb{R}^{n+1}$ is the vector part of $\mathbf{Y}_i$ and $X_i: I \to \mathbb{R}^{n+1}$ is the vector part of $\mathbf{X}_i$, we would have

$$\frac{d}{dt}(\varphi \circ \alpha) = \sum_{i=1}^{n} (\dot{\alpha} \cdot \mathbf{X}_i)Y_i = \sum_{i=1}^{n} (\dot{\alpha} \cdot \mathbf{X}_i)X_i = \frac{d\alpha}{dt}$$

so $\varphi \circ \alpha$ and α could differ at most by a constant. Since $(\varphi \circ \alpha)(t_0) = \alpha(t_0)$, we could then conclude that $(\varphi \circ \alpha)(t) = \alpha(t)$ for all $t \in I$ and, furthermore, that

$$d\varphi(\mathbf{X}_i(t)) = \mathbf{Y}_i(t) = (\varphi(\alpha(t)), Y_i(t)) = (\alpha(t), X_i(t)) = \mathbf{X}_i(t)$$

for each i so $d\varphi_{\alpha(t)} =$ identity, for $t \in I$, as required.

So we shall show that $X_i = Y_i$ for $i \in \{1, \ldots, n\}$. For this, set $\mathbf{X}_{n+1} = \mathbf{N} \circ \alpha$ and $\mathbf{Y}_{n+1} = \bar{N} \circ \varphi \circ \alpha$, so that $\{\mathbf{X}_1(t), \ldots, \mathbf{X}_{n+1}(t)\}$ is an orthonormal basis for $\mathbb{R}^{n+1}_{\alpha(t)}$ and $\{\mathbf{Y}_1(t), \ldots, \mathbf{Y}_{n+1}(t)\}$ is an orthonormal basis for $\mathbb{R}^{n+1}_{\varphi \circ \alpha(t)}$, for each $t \in I$. Then

$$\dot{\mathbf{X}}_i = \sum_{j=1}^{n+1} a_{ij}\mathbf{X}_j \quad \text{and} \quad \dot{\mathbf{Y}}_i = \sum_{j=1}^{n+1} b_{ij}\mathbf{Y}_j$$

where $a_{ij} = \dot{\mathbf{X}}_i \cdot \mathbf{X}_j$ and $b_{ij} = \dot{\mathbf{Y}}_i \cdot \mathbf{Y}_j$ are real valued functions on I. We shall show first that $a_{ij} = b_{ij}$ for all i and j. Since

$$0 = \frac{d}{dt}(\mathbf{X}_i \cdot \mathbf{X}_j) = \dot{\mathbf{X}}_i \cdot \mathbf{X}_j + \mathbf{X}_i \cdot \dot{\mathbf{X}}_j = a_{ij} + a_{ji}$$

we have that $a_{ji} = -a_{ij}$ for all i and j, and similarly $b_{ji} = -b_{ij}$ for all i and j, so it suffices to check that $a_{ij} = b_{ij}$ for $i < j$. But the vector fields $\mathbf{Y}_i = d\varphi \circ \mathbf{X}_i$, $i \in \{1, \ldots, n\}$, are parallel (in $\bar{S}$) along $\varphi \circ \alpha$. This follows from (i) and from the fact that the vector fields $\mathbf{X}_i$ are parallel along α; we will, however, delay the verification until the next chapter (Corollary 2 to Theorem 1, Chapter 23). Assuming this, we have that $\dot{\mathbf{Y}}_i$ is normal to $\bar{S}$, just as $\dot{\mathbf{X}}_i$ is normal to S, for $i \leq n$, and hence

$$b_{ij} = \dot{\mathbf{Y}}_i \cdot \mathbf{Y}_j = 0 = \dot{\mathbf{X}}_i \cdot \mathbf{X}_j = a_{ij}$$

for $1 \leq i, j \leq n$. Furthermore, for $i \leq n$,

$$\begin{aligned} b_{i,n+1} &= \dot{\mathbf{Y}}_i \cdot \mathbf{Y}_{n+1} = \dot{\mathbf{Y}}_i \cdot \bar{\mathbf{N}} \circ \varphi \circ \alpha = -\mathbf{Y}_i \cdot \dot{\bar{\mathbf{N}} \circ \varphi \circ \alpha} \\ &= \mathbf{Y}_i \cdot \bar{L}(\dot{\varphi \circ \alpha}) = \mathbf{Y}_i \cdot \bar{L}(d\varphi(\dot{\alpha})) = d\varphi(\mathbf{X}_i) \cdot d\varphi(L(\dot{\alpha})) \\ &= \mathbf{X}_i \cdot L(\dot{\alpha}) = -\mathbf{X}_i \cdot \dot{\mathbf{N} \circ \alpha} = \dot{\mathbf{X}}_i \cdot \mathbf{N} \circ \alpha = \dot{\mathbf{X}}_i \cdot \mathbf{X}_{n+1} = a_{i,n+1}, \end{aligned}$$

where $L(\dot{\alpha})$ is the vector field along α defined by $(L(\dot{\alpha}))(t) = L_{\alpha(t)}(\dot{\alpha}(t))$ and $\bar{L}(\dot{\varphi \circ \alpha})$ is defined similarly. We conclude, then, that $a_{ij} = b_{ij}$ whenever $1 \leq i < j \leq n + 1$ and hence $a_{ij} = b_{ij}$ for all i and j.

We can now complete the proof that $X_i = Y_i$ for all $i \in \{1, \ldots, n + 1\}$. For $Y_i = \sum_{j=1}^{n+1} c_{ij} X_j$, where $c_{ij} = Y_i \cdot X_j$: $I \to \mathbb{R}$, $(c_{ij}(t_0))$ is the identity matrix, and

$$\begin{aligned} \frac{dc_{ij}}{dt} &= \frac{dY_i}{dt} \cdot X_j + Y_i \cdot \frac{dX_j}{dt} \\ &= \left(\sum_{k=1}^{n+1} a_{ik} Y_k\right) \cdot X_j + Y_i \cdot \left(\sum_{k=1}^{n+1} a_{jk} X_k\right) \\ &= \sum_{k=1}^{n+1} (a_{ik} c_{kj} + a_{jk} c_{ik}). \end{aligned}$$

This system of first order differential equations, with initial conditions

$$c_{ij}(t_0) = \begin{cases} 1 & \text{if } i = j \\ 0 & \text{if } i \neq j \end{cases},$$

is satisfied by the functions

$$c_{ij}(t) = \begin{cases} 1 & \text{if } i = j \\ 0 & \text{if } i \neq j \end{cases},$$

since $a_{ij} + a_{ji} = 0$, so, by the uniqueness of solutions of first order differential equations, $(c_{ij}(t))$ is the identity matrix for all $t \in I$. In other words, $Y_i(t) = X_i(t)$ for all $i \in \{1, \ldots, n+1\}$ and all $t \in I$.

Finally, we must show that $\varphi(p) = p$ for all $p \in S$. Let

$$U = \{p \in S\colon \varphi(p) = p \quad \text{and} \quad d\varphi_p = \text{identity}\}.$$

Then U is an open set in S since each point $p \in U$ is contained in an open set of the form $\phi(V)$, where ϕ is a local parametrization of S and V is an open ball in $\mathbb{R}^n$, and p can be joined to any point of $\phi(V)$ by a parametrized curve in $\phi(V)$, so $\phi(V) \subset U$ by what has just been established. On the other hand, the complement $S - U$ of U in S is an open set in S since φ and $d\varphi$ are both continuous. Since S is connected, this can happen only if either U or $S - U$ is empty. (If $\alpha\colon [a, b] \to S$ were a continuous map with $\alpha(a) \in U$ and $\alpha(b) \in S - U$ then $\alpha(t_0)$ could belong neither to U nor to $S - U$, where t_0 is the least upper bound of the set $\{t \in [a, b] \in \alpha(t) \in U\}$). Since $p_0 \in S$, $S - U$ must be empty so $U = S$ and, in particular, $\varphi(p) = p$ for all $p \in S$. $\square$

Exercises

22.1. Verify that each of the maps described in Examples 1, 2 and 3 of this chapter are indeed rigid motions.

22.2. Show that if ψ_1 is an orthogonal transformation of $\mathbb{R}^{n+1}$ and ψ_2 is translation by a ($a \in \mathbb{R}^{n+1}$) then $\psi_1 \circ \psi_2 = \tilde{\psi}_2 \circ \psi_1$ where $\tilde{\psi}_2$ is translation by $\psi_1(a)$.

22.3. Show that each of the following statements is equivalent with the statement that the linear transformation $\psi\colon \mathbb{R}^{n+1} \to \mathbb{R}^{n+1}$ is an orthogonal transformation:

(a) $\psi(v) \cdot \psi(w) = v \cdot w$ for all $v, w \in \mathbb{R}^{n+1}$.
(b) ψ maps orthonormal bases to orthonormal bases; i.e., if $\{e_1, \ldots, e_{n+1}\}$ is an orthonormal basis for $\mathbb{R}^{n+1}$ then so is $\{\psi(e_1), \ldots, \psi(e_{n+1})\}$.
(c) the matrix for ψ relative to any orthonormal basis for $\mathbb{R}^{n+1}$ is an orthogonal matrix (i.e., its transpose is equal to its inverse).

22.4. A *rotation* of $\mathbb{R}^{n+1}$ is an orthogonal linear transformation with determinant $+1$.

(a) Show that a linear transformation $\psi\colon \mathbb{R}^2 \to \mathbb{R}^2$ is a rotation if and only if there exists a real number θ such that the matrix for ψ with respect to the standard basis for $\mathbb{R}^2$ is

$$\begin{pmatrix} \cos\theta & -\sin\theta \\ \sin\theta & \cos\theta \end{pmatrix}.$$

(b) Show that every rotation ψ of $\mathbb{R}^3$ leaves a direction fixed (i.e., ψ has a unit eigenvector with eigenvalue 1).
(c) Show that if ψ is a rotation of $\mathbb{R}^3$, e_1 is a unit vector with $\psi(e_1) = e_1$, and e_2 is a unit vector perpendicular to e_1, then the matrix for ψ with respect to

the orthonormal basis $\{e_1, e_2, e_1 \times e_2\}$ for $\mathbb{R}^3$ is of the form

$$\begin{pmatrix} 1 & 0 & 0 \\ 0 & \cos\theta & -\sin\theta \\ 0 & \sin\theta & \cos\theta \end{pmatrix}$$

for some $\theta \in \mathbb{R}$.

22.5. Show that the hyperbolas $x_1 x_2 = 1$ and $x_1^2 - x_2^2 = 2$ in $\mathbb{R}^2$ are congruent.

22.6. Let ψ be a rigid motion of $\mathbb{R}^{n+1}$ and let F denote the set of fixed points of ψ,

$$F = \{q \in \mathbb{R}^{n+1} \colon \psi(q) = q\}.$$

For $p \notin F$ let

$$H_p = \{x \in \mathbb{R}^{n+1} \colon \|x - \psi(p)\| = \|x - p\|\}.$$

(a) Show that H_p is an n-plane in $\mathbb{R}^{n+1}$.
(b) Show that $F \subset H_p$.
(c) Letting ψ_p be reflection through H_p, show that the set of fixed points of $\psi_p \circ \psi$ contains $\{p\} \cup F$.
(d) Show that if $0 \in F$ then $\sum_{i=1}^{k} c_i p_i \in F$ whenever $p_1, \ldots, p_k \in F$ and $c_1, \ldots, c_k \in \mathbb{R}$.
(e) Show that there exists a $k \leq n + 2$ and reflections $\psi_1, \ldots, \psi_k$ of $\mathbb{R}^{n+1}$ such that $\psi = \psi_1 \circ \cdots \circ \psi_k$.

22.7. A rigid motion ψ of $\mathbb{R}^{n+1}$ which maps an n-surface S onto itself is called a *symmetry* of S.

(a) Show that the set of rigid motions of $\mathbb{R}^{n+1}$ forms a group under composition, and that the symmetries of S form a subgroup.
(b) Show that the symmetry group of the unit n-sphere S^n is the group of all orthogonal transformations of $\mathbb{R}^{n+1}$.
(c) Describe the symmetry group of the cylinder $x_1^2 + x_2^2 = a^2$ in $\mathbb{R}^3$.
(d) Describe the symmetry group of the ellipsoid

$$\frac{x_1^2}{a^2} + \frac{x_2^2}{b^2} + \frac{x_3^2}{c^2} = 1$$

in $\mathbb{R}^3$, (i) when $b = c \neq a$ and (ii) when a, b, and c are distinct.

23 Isometries

As inhabitants of the earth, we are (or, at least, we were until the invention of air and space vehicles) forced to deduce the geometry of the earth from measurements made on the earth. We can measure distance along curves, and by taking a derivative with respect to time, we can measure velocity and speed. The geometry which can be derived from such measurements is called *intrinsic geometry*.

The primitive data needed to be able to compute distance along curves in an n-surface S is the dot product of tangent vectors. Indeed, given the dot product on each tangent space S_p, $p \in S$, the length $l(\alpha)$ of a parametrized curve $\alpha\colon [a, b] \to S$ can be computed from the formula

$$l(\alpha) = \int_a^b \|\dot{\alpha}(t)\|\, dt = \int_a^b (\dot{\alpha}(t) \cdot \dot{\alpha}(t))^{1/2}\, dt.$$

Conversely, if we can compute the length of arbitrary smooth curves in S then we can compute norms of vectors in S_p (for $\mathbf{v} \in S_p$, take $\alpha\colon [a, b] \to S$ with $\dot{\alpha}(t_0) = \mathbf{v}$ and let $s(t) =$ the length of α from a to t; then $\|\mathbf{v}\| = \|\dot{\alpha}(t_0)\| = (ds/dt)(t_0)$). From the norms we can then compute all dot products; e.g., via the identity

$$\mathbf{v} \cdot \mathbf{w} = \tfrac{1}{2}(\|\mathbf{v} + \mathbf{w}\|^2 - \|\mathbf{v}\|^2 - \|\mathbf{w}\|^2).$$

Thus the intrinsic geometry of an n-surface S, the geometry which can be derived from length measurements along curves in S, is the same as that part of the geometry of S which can be derived from a knowledge of the dot product on the tangent space at each point of S.

A smooth map ψ from one n-surface $S \subset \mathbb{R}^{n+k}$ to another $\tilde{S} \subset \mathbb{R}^{n+l}$ is called a *local isometry* if it preserves dot products of tangent vectors; that is, if

$$d\psi(\mathbf{v}) \cdot d\psi(\mathbf{w}) = \mathbf{v} \cdot \mathbf{w}$$

for all $\mathbf{v}, \mathbf{w} \in S_p$, $p \in S$. The differential $d\psi_p: S_p \to \tilde{S}_p$ at p of such a map is necessarily non-singular for each $p \in S$ so, by the inverse function theorem, ψ must map some open set about each point p in S one to one onto an open set about $\psi(p)$ in $\tilde{S}$. It follows from this that $\psi(S)$ is an open subset of $\tilde{S}$. But $\psi(S)$ need not be the whole of $\tilde{S}$, nor must $\psi: S \to \psi(S)$ be one to one. A local isometry which maps an n-surface S one to one onto an n-surface $\tilde{S}$ is called an *isometry* of S onto $\tilde{S}$.

Since an isometry is by definition a one to one map $\psi: S \to \tilde{S}$ which preserves dot products of tangent vectors, it preserves all the intrinsic geometry of the surface. Thus, for example, if $\alpha: [a, b] \to S$ is a parametrized curve in S then the corresponding curve $\psi \circ \alpha$ in $\tilde{S}$ has the same length:

$$l(\psi \circ \alpha) = \int_a^b \|(\psi \dot{\circ} \alpha)(t)\| \, dt = \int_a^b \|d\psi(\dot{\alpha}(t))\| \, dt$$

$$= \int_a^b \|\dot{\alpha}(t)\| \, dt = l(\alpha).$$

In fact, intrinsic geometry can be described as that part of the geometry of surfaces which is preserved by isometries. Two surfaces S and $\tilde{S}$ such that there exists an isometry $\psi: S \to \tilde{S}$ are said to be *isometric*; they necessarily have the same intrinsic geometry.

EXAMPLE 1. Let $\psi: \mathbb{R}^{n+k} \to \mathbb{R}^{n+k}$ be a rigid motion and let S be an n-surface in $\mathbb{R}^{n+k}$. Then $\psi|_S$ is an isometry of S onto $\psi(S)$.

EXAMPLE 2. Let $\psi: \mathbb{R}^2 \to \mathbb{R}^3$ be defined by $\psi(\theta, u) = (\cos\theta, \sin\theta, u)$. Thus ψ maps the plane around (and around) the cylinder $x_1^2 + x_2^2 = 1$ in $\mathbb{R}^3$ (see Figure 23.1). ψ is a local isometry since, for each $(\theta, u) \in \mathbb{R}^2$, $d\psi$ maps the orthonormal basis $\{(\theta, u, 1, 0), (\theta, u, 0, 1)\}$ for $\mathbb{R}^2_{(\theta, u)}$ to the orthonormal basis

$$\{\mathbf{E}_1(\theta, u), \mathbf{E}_2(\theta, u)\} = \{(\varphi(\theta, u), -\sin\theta, \cos\theta, 0), (\varphi(\theta, u), 0, 0, 1)\}$$

for Image $d\psi_{(\theta, u)}$ so $d\psi_{(\theta, u)}$ must preserve dot products. By restricting ψ to the open set $U = \{(\theta, u) \in \mathbb{R}^2: -\pi < \theta < \pi\}$ we obtain an isometry $\psi|_U$ from the infinite strip U in $\mathbb{R}^2$ onto the cylinder with a line removed.

EXAMPLE 3. The map φ from the plane to the torus $((x_1^2 + x_2^2)^{1/2} - a)^2 + x_3^2 = b^2$ $(a > b > 0)$ given by

$$\varphi(\phi, \theta) = ((a + b\cos\phi)\cos\theta, (a + b\cos\phi)\sin\theta, b\sin\phi)$$

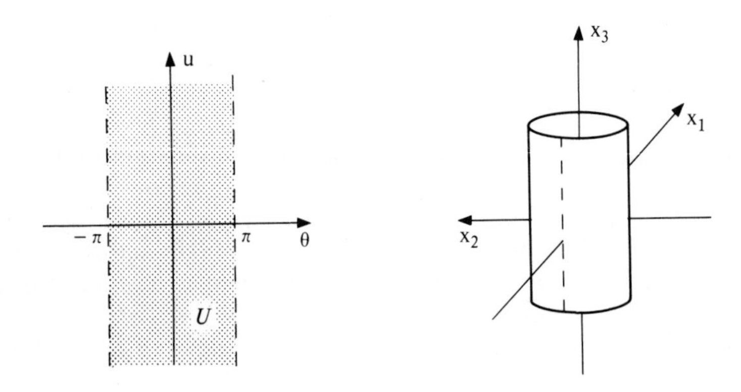

Figure 23.1 $\psi(\theta, u) = (\cos\theta, \sin\theta, u)$ is a local isometry from $\mathbb{R}^2$ onto the cylinder $x_1^2 + x_2^2 = 1$ in $\mathbb{R}^3$. The restriction of ψ to the infinite strip U is an isometry from U onto the cylinder with a line (dotted) removed.

is *not* a local isometry because, for example,

$$\begin{aligned}\|d\varphi(\phi, \theta, 0, 1)\| &= \|(\varphi(\phi, \theta), -(a + b\cos\phi)\sin\theta, (a + b\cos\phi)\cos\theta, 0)\| \\ &= a + b\cos\phi\end{aligned}$$

is not equal to $\|(\phi, \theta, 0, 1)\| = 1$ for all $(\phi, \theta) \in \mathbb{R}^2$. On the other hand, the map ψ which maps the plane onto the torus

$$\begin{cases} x_1^2 + x_2^2 = 1 \\ x_3^2 + x_4^2 = 1 \end{cases}$$

in $\mathbb{R}^4$ by $\psi(\phi, \theta) = (\cos\phi, \sin\phi, \cos\theta, \sin\theta)$ *is* a local isometry because, for each $(\phi, \theta) \in \mathbb{R}^2$, the vectors

$$d\psi(\phi, \theta, 1, 0) = (\psi(\phi, \theta), -\sin\phi, \cos\phi, 0, 0)$$

$$d\psi(\phi, \theta, 0, 1) = (\psi(\phi, \theta), 0, 0, -\sin\theta, \cos\theta)$$

do form an orthonormal basis for Image $d\psi_{(\phi, \theta)}$. By restricting ψ to the open set $U = \{(\phi, \theta) \in \mathbb{R}^2 : -\pi < \phi < \pi, -\pi < \theta < \pi\}$ we obtain an isometry $\psi|_U$ from the square U in $\mathbb{R}^2$ onto the torus in $\mathbb{R}^4$ with two circles removed.

Example 4. Let S be the punctured plane $\mathbb{R}^2 - \{(0, 0)\}$ and let $\tilde{S}$ be the cone $3x_1^2 + 3x_2^2 - x_3^2 = 0$, $x_3 > 0$, in $\mathbb{R}^3$. Define $\psi: S \to \tilde{S}$ by

$$\psi(r\cos\theta, r\sin\theta) = \left(\frac{r}{2}\cos 2\theta, \frac{r}{2}\sin 2\theta, \frac{\sqrt{3}}{2}r\right)$$

where (r, θ), $r > 0$, are polar coordinates on $\mathbb{R}^2 - \{(0, 0)\}$. Then ψ is a local isometry. For if $U = \{(r, \theta) \in \mathbb{R}^2 : r > 0\}$ then the maps $\varphi: U \to S$ and

$\tilde{\varphi}: U \to \tilde{S}$ defined by

$$\varphi(r, \theta) = (r \cos \theta, r \sin \theta)$$
$$\tilde{\varphi}(r, \theta) = \left(\frac{r}{2} \cos 2\theta, \frac{r}{2} \sin 2\theta, \frac{\sqrt{3}}{2} r\right)$$

are parametrized 2-surfaces mapping U onto S and $\tilde{S}$ respectively, and $\tilde{\varphi} = \psi \circ \varphi$ (see Figure 23.2). The coordinate vector fields $\mathbf{E}_i$ along φ and $\tilde{\mathbf{E}}_i$ along $\tilde{\varphi}$ are given by

$$\mathbf{E}_1(r, \theta) = \left(\varphi(r, \theta), \frac{\partial \varphi}{\partial r}(r, \theta)\right) = (\varphi(r, \theta), \cos \theta, \sin \theta)$$

$$\mathbf{E}_2(r, \theta) = \left(\varphi(r, \theta), \frac{\partial \varphi}{\partial \theta}(r, \theta)\right) = (\varphi(r, \theta), -r \sin \theta, r \cos \theta)$$

$$\tilde{\mathbf{E}}_1(r, \theta) = \left(\tilde{\varphi}(r, \theta), \frac{\partial \tilde{\varphi}}{\partial r}(r, \theta)\right) = \left(\tilde{\varphi}(r, \theta), \tfrac{1}{2} \cos 2\theta, \tfrac{1}{2} \sin 2\theta, \frac{\sqrt{3}}{2}\right)$$

$$\tilde{\mathbf{E}}_2(r, \theta) = \left(\tilde{\varphi}(r, \theta), \frac{\partial \tilde{\varphi}}{\partial \theta}(r, \theta)\right) = (\tilde{\varphi}(r, \theta), -r \sin 2\theta, r \cos 2\theta, 0).$$

Moreover, for each $p \in U$, $d\psi(\mathbf{E}_i(p)) = \tilde{\mathbf{E}}_i(p)$ for $i \in \{1, 2\}$, since

$$d\psi(\mathbf{E}_i(p)) = d\psi(d\varphi(\mathbf{e}_i)) = d(\psi \circ \varphi)(\mathbf{e}_i) = d\tilde{\varphi}(\mathbf{e}_i) = \tilde{\mathbf{E}}_i(p)$$

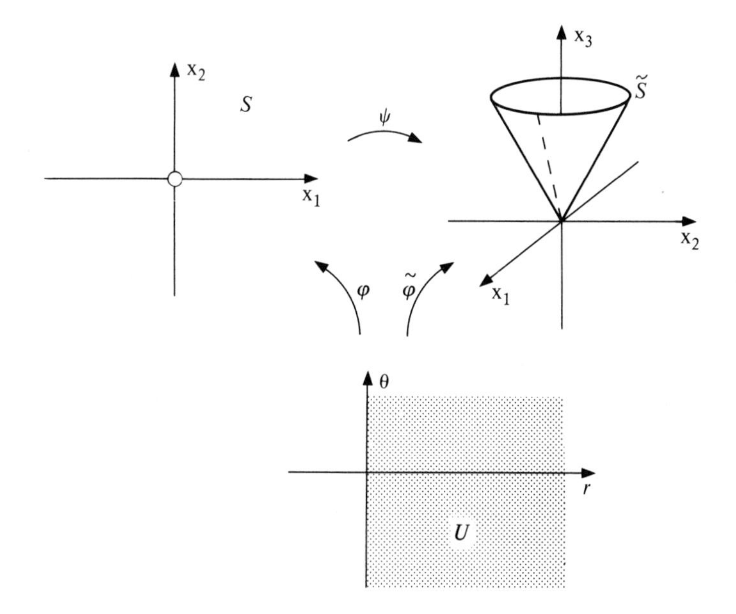

Figure 23.2 ψ is a local isometry from the punctured plane S onto the cone $\tilde{S}$. The restriction of ψ to the upper half plane $x_2 > 0$ is an isometry onto the cone with a line removed.

where $\mathbf{e}_1 = (p, 1, 0)$ and $\mathbf{e}_2 = (p, 0, 1)$. Finally, dot products of tangent vectors are preserved by ψ because dot products of the basis vectors are preserved:

$$\begin{aligned} d\psi(\mathbf{E}_i(p)) \cdot d\psi(\mathbf{E}_j(p)) &= \tilde{\mathbf{E}}_i(p) \cdot \tilde{\mathbf{E}}_j(p) \\ &= \begin{cases} 1 & \text{if } i = j = 1 \\ 0 & \text{if } (i, j) = (1, 2) \text{ or } (i, j) = (2, 1) \\ r^2 & \text{if } i = j = 2 \end{cases} \\ &= \mathbf{E}_i(p) \cdot \mathbf{E}_j(p) \end{aligned}$$

for all $p = (r, \theta) \in U$.

Remark 1. Example 2 is an example of a special class of isometries called bendings. Roughly, the cylinder with a line deleted is obtained by bending the strip $\{(x_1, x_2, x_3) \in \mathbb{R}^3 : -\pi < x_1 < \pi, x_3 = 0\}$, without stretching, tearing, or glueing, into the shape of a cylinder. This process clearly leaves length measurements along curves in the surface unchanged; i.e., two surfaces which can be obtained from one another by bending are isometric. A precise definition of bending is as follows. An n-surface $\tilde{S}$ in $\mathbb{R}^{n+k}$ *is obtained from an n-surface S in $\mathbb{R}^{n+k}$ by bending* if there exists a smooth map $\psi: [a, b] \times S \to \mathbb{R}^{n+k}$ such that

(i) for each $t \in [a, b]$, the map $\psi_t: S \to \mathbb{R}^{n+k}$ defined by $\psi_t(p) = \psi(t, p)$ is an isometry,
(ii) $\psi_a(p) = p$ for all $p \in S$; i.e., $\psi_a =$ identity map on S, and
(iii) ψ_b is an isometry of S onto $\tilde{S}$.

The 2-torus in $\mathbb{R}^4$ with two circles deleted and the cone in $\mathbb{R}^3$ with a line deleted (Examples 3 and 4) are also obtained by bending portions of 2-planes.

Remark 2. Example 4 illustrates a useful technique for checking that a map $\psi: S \to \tilde{S}$ of n-surfaces is a local isometry. For $p \in S$, let $\varphi: U \to S$ be a local parametrization of an open set of S about p. If ψ is a local isometry, then $\tilde{\varphi} = \psi \circ \varphi$ will be a parametrized n-surface, with Image $\tilde{\varphi} \subset \tilde{S}$, and the coordinate vector fields $\mathbf{E}_i$ along φ and $\tilde{\mathbf{E}}_i$ along $\tilde{\varphi}$ will be such that $\mathbf{E}_i \cdot \mathbf{E}_j = \tilde{\mathbf{E}}_i \cdot \tilde{\mathbf{E}}_j$. Conversely, if for each $p \in S$ there is such a $\varphi: U \to S$, with $p \in$ Image φ, then ψ is a local isometry.

Those features of the geometry of an n-surface S which are part of the intrinsic geometry of S (i.e., which can be determined from measurements on the surface) are the ones that are preserved, or *invariant*, under isometries. We have seen that the length of curves is an isometry invariant. It follows then that geodesics are isometry invariants, since a parametrized curve α is a geodesic if and only if it has constant speed and is such that the length integral is stationary at α with respect to fixed endpoint variations. Volume is also an isometry invariant since the volume integral $V(\varphi) = \int_U \det(\mathbf{E}_i \cdot \mathbf{E}_j)^{1/2}$ of a parametrized n-surface φ depends only on the dot products of the coordinate vector fields. On the other hand, the spherical image, the principal curvatures, the mean curvature, the focal locus, and the property of being a minimal surface in $\mathbb{R}^{n+1}$ are not isometry invariants since, for example, none of these features are preserved under the isometry $\psi(x_1, x_2, 0) = (\cos x_1, \sin x_1, x_2)$ mapping the strip $x_3 = 0$, $-\pi < x_1 < \pi$, onto the cylinder $x_1^2 + x_2^2 = 1$ with a line removed (Figure 23.1). We shall see that, contrary to intuition, the Gauss-Kronecker curvature K of an n-surface in $\mathbb{R}^{n+1}$ *is* an isometry invariant, whenever n is even. Thus, even though the principal curvatures are not invariant under isometries, their product is, in even dimensions. This theorem so pleased Gauss, who discovered it for $n = 2$, that he called it a "theorema egregium" (a "most excellent theorem"). We shall see also that parallel transport is invariant under isometries. The key to all of these facts is the observation that covariant differentiation is intrinsic.

Let us recall the concept of *covariant differentiation* and extend it to n-surfaces in $\mathbb{R}^{n+k}$. Given an n-surface S in $\mathbb{R}^{n+k}$ and a smooth vector field $\mathbf{X}$ tangent to S along a parametrized curve $\alpha: I \to S$, the covariant derivative of $\mathbf{X}$ along α is the vector field $\mathbf{X}'$, tangent to S along α, obtained by projecting $\dot{\mathbf{X}}(t)$ orthogonally onto the tangent space $S_{\alpha(t)}$, for each $t \in I$. Thus $\mathbf{X}'(t)$ is the tangential component of $\dot{\mathbf{X}}(t)$. For $\mathbf{X}$ a smooth tangent vector field on the n-surface $S \subset \mathbb{R}^{n+k}$, and $\mathbf{v} \in S_p$, $p \in S$, the covariant derivative $D_{\mathbf{v}}\mathbf{X}$ of $\mathbf{X}$ with respect to $\mathbf{v}$ is the tangential component of the derivative $\nabla_{\mathbf{v}}\mathbf{X}$. Similarly, for $\mathbf{X}$ a smooth tangent vector field along a parametrized n-surface $\varphi: U \to \mathbb{R}^{n+k}$, the covariant derivative $D_{\mathbf{v}}\mathbf{X}$ of $\mathbf{X}$ with respect to $\mathbf{v} \in \mathbb{R}^n_p$, $p \in U$, is the tangential component of $\nabla_{\mathbf{v}}\mathbf{X}$; that is, $D_{\mathbf{v}}\mathbf{X}$ is the orthogonal projection of $\nabla_{\mathbf{v}}\mathbf{X}$ into the tangent space Image $d\varphi_p$. These covariant differentiation operations are related to each other as follows.

If $\mathbf{X}$ is a smooth tangent vector field on the n-surface $S \subset \mathbb{R}^{n+k}$ and α is a parametrized curve in S, or if $\mathbf{X}$ is a smooth tangent vector field along the parametrized n-surface $\varphi: U \to \mathbb{R}^{n+k}$ and α is a parametrized curve in U, then

$$D_{\dot{\alpha}(t)}\mathbf{X} = (\mathbf{X} \circ \alpha)'(t)$$

for all t in the domain of α.

If $\mathbf{X}$ is a smooth vector field on the n-surface $S \subset \mathbb{R}^{n+k}$ and $\varphi: U \to S$ is a parametrized n-surface in $\mathbb{R}^{n+k}$ whose image is contained in S, then

$$D_{\dot{\alpha}(t)}\mathbf{X} = (\mathbf{X} \circ \alpha)'(t)$$

for all $\mathbf{v} \in \mathbb{R}^{n+1}_p$, $p \in U$.

Given two smooth tangent vector fields $\mathbf{X}$ and $\mathbf{Y}$ on an n-surface $S \subset \mathbb{R}^{n+k}$, the covariant derivative of $\mathbf{Y}$ with respect to $\mathbf{X}$ is the tangent vector field $D_{\mathbf{X}}\mathbf{Y}$ on S defined by $(D_{\mathbf{X}}\mathbf{Y})(p) = D_{\mathbf{X}(p)}\mathbf{Y}$ for $p \in S$. Similarly, given the smooth tangent vector fields $\mathbf{X}$ and $\mathbf{Y}$ along a parametrized n-

surface $\varphi: U \to \mathbb{R}^{n+1}$, the covariant derivative $D_{\mathbf{X}}\mathbf{Y}$ is the tangent vector field along φ defined by $(D_{\mathbf{X}}\mathbf{Y})(p) = D_{d\varphi_p{}^{-1}(\mathbf{X}(p))}\mathbf{Y}$, for $p \in U$.

The covariant derivative of one smooth tangent vector field with respect to another is a smooth tangent vector field. Furthermore, covariant differentiation has the following familiar properties:

(i) $D_{(\mathbf{X}+\mathbf{Y})}\mathbf{Z} = D_{\mathbf{X}}\mathbf{Z} + D_{\mathbf{Y}}\mathbf{Z}$
(ii) $D_{f\mathbf{X}}\mathbf{Y} = fD_{\mathbf{X}}\mathbf{Y}$
(iii) $D_{\mathbf{X}}(\mathbf{Y} + \mathbf{Z}) = D_{\mathbf{X}}\mathbf{Y} + D_{\mathbf{X}}\mathbf{Z}$
(iv) $D_{\mathbf{X}}(f\mathbf{Y}) = (\nabla_{\mathbf{X}} f)\mathbf{Y} + fD_{\mathbf{X}}\mathbf{Y}$
(v) $\nabla_{\mathbf{X}}(\mathbf{Y} \cdot \mathbf{Z}) = (D_{\mathbf{X}}\mathbf{Y}) \cdot \mathbf{Z} + \mathbf{Y} \cdot (D_{\mathbf{X}}\mathbf{Z})$

for all smooth tangent vector fields $\mathbf{X}$, $\mathbf{Y}$, and $\mathbf{Z}$ on S (or along φ) and all smooth functions f on S (or along φ). In (v), the derivative of $\nabla_{\mathbf{X}}h$ of the smooth function on $h = \mathbf{Y} \cdot \mathbf{Z}$ is defined to be the smooth function on S given by $(\nabla_{\mathbf{X}}h)(p) = \nabla_{\mathbf{X}(p)}h$ (or along φ given by $(\nabla_{\mathbf{X}}h)(p) = \nabla_{d\varphi_p{}^{-1}(\mathbf{X}(p))}h$). Verification of these properties of covariant differentiation is left as an exercise.

Theorem 1. *Let $\varphi: U \to \mathbb{R}^{n+k}$ be a parametrized n-surface and let $\mathbf{E}_i$, $i \in \{1, \ldots, n\}$, be the coordinate vector fields along φ. Then, for $p \in U$ and i, $j, k \in \{1, \ldots, n\}$.*

$$(D_{\mathbf{E}_i}\mathbf{E}_j) \cdot \mathbf{E}_k = \frac{1}{2}\left(\frac{\partial g_{ik}}{\partial x_j} + \frac{\partial g_{jk}}{\partial x_i} - \frac{\partial g_{ij}}{\partial x_k}\right),$$

where the g_{ij}: $U \to \mathbb{R}$ are defined by $g_{ij} = \mathbf{E}_i \cdot \mathbf{E}_j$.

PROOF. Note first that $D_{\mathbf{E}_i}\mathbf{E}_j = D_{\mathbf{E}_j}\mathbf{E}_i$ for all i and j. Indeed, for each $p \in U$,

$$\mathbf{E}_j(p) = \left(\varphi(p), \frac{\partial \varphi}{\partial x_j}(p)\right)$$

and

$$(\nabla_{\mathbf{E}_i}\mathbf{E}_j)(p) = \left(\varphi(p), \frac{\partial^2 \varphi}{\partial x_i\, \partial x_j}(p)\right).$$

By the symmetry of the second partials, we have $(\nabla_{\mathbf{E}_i}\mathbf{E}_j)(p) = (\nabla_{\mathbf{E}_j}\mathbf{E}_i)(p)$. Projecting orthogonally onto Image $d\varphi_p$ then establishes that $D_{\mathbf{E}_i}\mathbf{E}_j = D_{\mathbf{E}_j}\mathbf{E}_i$.

Using this symmetry, we compute the partial derivatives of the g_{ik}:

$$\frac{\partial g_{ik}}{\partial x_j} = \nabla_{\mathbf{E}_j}(\mathbf{E}_i \cdot \mathbf{E}_k) = (D_{\mathbf{E}_j}\mathbf{E}_i) \cdot \mathbf{E}_k + \mathbf{E}_i \cdot (D_{\mathbf{E}_j}\mathbf{E}_k)$$

$$\frac{\partial g_{jk}}{\partial x_i} = \nabla_{\mathbf{E}_i}(\mathbf{E}_j \cdot \mathbf{E}_k) = (D_{\mathbf{E}_i}\mathbf{E}_j) \cdot \mathbf{E}_k + \mathbf{E}_j \cdot (D_{\mathbf{E}_i}\mathbf{E}_k)$$

$$\frac{\partial g_{ij}}{\partial x_k} = (D_{\mathbf{E}_k}\mathbf{E}_i) \cdot \mathbf{E}_j + \mathbf{E}_i \cdot (D_{\mathbf{E}_k}\mathbf{E}_j) = (D_{\mathbf{E}_i}\mathbf{E}_k) \cdot \mathbf{E}_j + \mathbf{E}_i \cdot (D_{\mathbf{E}_j}\mathbf{E}_k).$$

Thus

$$\frac{\partial g_{ik}}{\partial x_j} + \frac{\partial g_{jk}}{\partial x_i} - \frac{\partial g_{ij}}{\partial x_k} = (D_{\mathbf{E}_j}\mathbf{E}_i) \cdot \mathbf{E}_k + (D_{\mathbf{E}_i}\mathbf{E}_j) \cdot \mathbf{E}_k$$
$$= 2(D_{\mathbf{E}_i}\mathbf{E}_j) \cdot \mathbf{E}_k. \qquad \square$$

Corollary 1. *Covariant differentiation is intrinsic.*

Proof. It suffices to check this along a parametrized n-surface $\varphi: U \to \mathbb{R}^{n+k}$. Given a smooth vector field $\mathbf{X}$ along φ, we may express $\mathbf{X}$ as a linear combination of the coordinate vector fields $\mathbf{E}_i$ of φ,

$$\mathbf{X} = \sum_{i=1}^{n} f_i \mathbf{E}_i$$

where $f_i: U \to \mathbb{R}$ for $i \in \{1, \dots, n\}$. Thus, for $\mathbf{v} \in \mathbb{R}^n_p$, $p \in U$,

$$D_\mathbf{v}\mathbf{X} = D_\mathbf{v}\left(\sum_{i=1}^{n} f_i \mathbf{E}_i\right)$$
$$= \sum_{i=1}^{n} ((\nabla_\mathbf{v} f_i)\mathbf{E}_i(p) + f_i(p) D_\mathbf{v}\mathbf{E}_i)$$
$$= \sum_{i=1}^{n}\left((\nabla_\mathbf{v} f_i)\mathbf{E}_i(p) + f_i(p)\sum_{j=1}^{n} v_j D_{\mathbf{E}_j(p)}\mathbf{E}_i\right)$$

where $\mathbf{v} = (p, v_1, \dots, v_n)$. Since all the quantities in this last expression can be computed from intrinsic information along φ ($D_{\mathbf{E}_j(p)}\mathbf{E}_i$ can be determined from the set of dot products $\{(D_{\mathbf{E}_j(p)}\mathbf{E}_i) \cdot \mathbf{E}_k(p)\}$ which can be calculated from the formula of Theorem 1), $D_\mathbf{v}\mathbf{X}$ is intrinsic. $\square$

Corollary 2. *Parallel transport is intrinsic.*

Proof. Immediate from Corollary 1 since $\mathbf{X}$ is parallel along α if and only if $\mathbf{X}' = \mathbf{0}$. $\square$

Theorem 2. *The Gauss-Kronecker curvature of an oriented n-surface S in $\mathbb{R}^{n+1}$ is intrinsic, for n even.*

Proof. It suffices to work with a parametrized n-surface $\varphi: U \to S$. For $\mathbf{E}_j$ the coordinate vector fields along φ and $\mathbf{Z}$ any smooth tangent vector field along φ, we have

$$(\nabla_{\mathbf{E}_i}\nabla_{\mathbf{E}_j}\mathbf{Z})(p) = \left(\varphi(p), \frac{\partial^2 Z}{\partial x_i\, \partial x_j}(p)\right) = (\nabla_{\mathbf{E}_j}\nabla_{\mathbf{E}_i}\mathbf{Z})(p),$$

so

$$\nabla_{\mathbf{E}_i}\nabla_{\mathbf{E}_j}\mathbf{Z} - \nabla_{\mathbf{E}_j}\nabla_{\mathbf{E}_i}\mathbf{Z} = 0$$

for all i and j. We shall compute the tangential component of the left hand side of this equation; the theorem will follow from the fact that this tangential component must be zero. Since

$$\begin{aligned}\nabla_{\mathbf{E}_i}\nabla_{\mathbf{E}_j}\mathbf{Z} &= \nabla_{\mathbf{E}_i}(D_{\mathbf{E}_j}\mathbf{Z} + ((\nabla_{\mathbf{E}_j}\mathbf{Z})\cdot\mathbf{N})\mathbf{N}) \\ &= D_{\mathbf{E}_i}D_{\mathbf{E}_j}\mathbf{Z} + ((\nabla_{\mathbf{E}_j}\mathbf{Z})\cdot\mathbf{N})\nabla_{\mathbf{E}_i}\mathbf{N} + (\text{a multiple of } \mathbf{N}) \\ &= D_{\mathbf{E}_i}D_{\mathbf{E}_j}\mathbf{Z} - (\mathbf{Z}\cdot\nabla_{\mathbf{E}_j}\mathbf{N})\nabla_{\mathbf{E}_i}\mathbf{N} + (\text{a multiple of } \mathbf{N}),\end{aligned}$$

we find that, for $p \in U$,

$$\begin{aligned}(\nabla_{\mathbf{E}_i}\nabla_{\mathbf{E}_j}\mathbf{Z})(p) = (D_{\mathbf{E}_i}D_{\mathbf{E}_j}\mathbf{Z})(p) - (L_p(\mathbf{E}_j(p))\cdot\mathbf{Z}(p))L_p(\mathbf{E}_i(p)) \\ + (\text{a multiple of } \mathbf{N}(p)).\end{aligned}$$

Interchanging i and j, subtracting, and using the fact that the tangential component of the result must be zero, we obtain the equation

$$\begin{aligned}(D_{\mathbf{E}_i}D_{\mathbf{E}_j}\mathbf{Z} - D_{\mathbf{E}_j}D_{\mathbf{E}_i}\mathbf{Z})(p) = (L_p(\mathbf{E}_j(p))\cdot\mathbf{Z}(p))L_p(\mathbf{E}_i(p)) \\ - (L_p(\mathbf{E}_i(p))\cdot\mathbf{Z}(p))L_p(\mathbf{E}_j(p)).\end{aligned}$$

Since the left hand side of this last equation is intrinsic, so is the right, for all i and j. Using the linearity of L_p, we see that given any three vectors $\mathbf{x}$, $\mathbf{y}$, $\mathbf{z} \in S_p$, $p \in S$, the vector $R(\mathbf{x}, \mathbf{y}, \mathbf{z}) \in S_p$ defined by

$$R(\mathbf{x}, \mathbf{y}, \mathbf{z}) = [L_p(\mathbf{y})\cdot\mathbf{z}]L_p(\mathbf{x}) - [L_p(\mathbf{x})\cdot\mathbf{z}]L_p(\mathbf{y})$$

is intrinsic. This map R which assigns to each triple $(\mathbf{x}, \mathbf{y}, \mathbf{z})$ of vectors in S_p, $p \in S$, the vector $R(\mathbf{x}, \mathbf{y}, \mathbf{z})$ in S_p is called the *Riemann tensor* of S; it belongs to the intrinsic geometry of S.

Now, if $n = 2$ and $\{\mathbf{e}_1, \mathbf{e}_2\}$ is an orthonormal basis for S_p, $p \in S$, then the Gaussian curvature K at p is given by

$$\begin{aligned}K(p) = \det L_p &= [L_p(\mathbf{e}_1)\cdot\mathbf{e}_1][L_p(\mathbf{e}_2)\cdot\mathbf{e}_2] - [L_p(\mathbf{e}_2)\cdot\mathbf{e}_1][L_p(\mathbf{e}_1)\cdot\mathbf{e}_2] \\ &= R(\mathbf{e}_2, \mathbf{e}_1, \mathbf{e}_1)\cdot\mathbf{e}_2\end{aligned}$$

so K is intrinsic, as claimed. If $n > 2$, but n even, and $\{\mathbf{e}_1, \ldots, \mathbf{e}_n\}$ is an orthonormal basis for S_p, expansion of the determinant

$$K(p) = \det L_p = \det[L_p(\mathbf{e}_j)\cdot\mathbf{e}_i]$$

in terms of its 2×2 minors yields

$$\begin{aligned}K(p) = ((-1)^{n/2}/2^{n/2}n!)\sum_{\sigma,\tau}\varepsilon(\sigma)\varepsilon(\tau)[R(\mathbf{e}_{\sigma(1)}, \mathbf{e}_{\sigma(2)}, \mathbf{e}_{\tau(1)})\cdot\mathbf{e}_{\tau(2)}] \\ \cdots[R(\mathbf{e}_{\sigma(n-1)}, \mathbf{e}_{\sigma(n)}, \mathbf{e}_{\tau(n-1)})\cdot\mathbf{e}_{\tau(n)}]\end{aligned}$$

where the sum is over all permutations σ and τ of $\{1, \ldots, n\}$ and $\varepsilon(\sigma)$ denotes the sign of the permutation σ. Hence K is intrinsic, for n even. □

Exercises

23.1. Show that if $\psi: S \to \tilde{S}$ is an isometry, then so is $\psi^{-1}: \tilde{S} \to S$.

23.2. Which of the following maps are local isometries?

(a) The map ψ defined by $\psi(p) = 2p$, mapping the n-sphere $x_1^2 + \cdots + x_{n+1}^2 = 1$ onto the n-sphere $x_1^2 + \cdots + x_{n+1}^2 = 4$.
(b) The map ψ defined by $\psi(p) = -p$, mapping the n-sphere $x_1^2 + \cdots + x_{n+1}^2 = 1$ onto itself.
(c) The map defined by

$$\psi(\cos\theta, \sin\theta, u) = ((a + b\cos u)\cos\theta, (a + b\cos u)\sin\theta, b\sin u),$$

mapping the cylinder $x_1^2 + x_2^2 = 1$ in $\mathbb{R}^3$ onto the torus $((x_1^2 + x_2^2)^{1/2} - a)^2 + x_3^2 = b^2$ in $\mathbb{R}^3$ $(a > b > 0)$.
(d) The map ψ defined by

$$\psi(\cos\theta, \sin\theta, u) = (\cos\theta, \sin\theta, \cos u, \sin u),$$

mapping the cylinder $x_1^2 + x_2^2 = 1$ in $\mathbb{R}^3$ onto the torus

$$\begin{cases} x_1^2 + x_2^2 = 1 \\ x_3^2 + x_4^2 = 1 \end{cases}$$

in $\mathbb{R}^4$.

23.3. Show that the cylinders $x_1^2 + x_2^2 = 4$ and $x_1^2 + x_2^2 = 1$ in $\mathbb{R}^3$ are not isometric, but that the map ψ defined by $\psi(2\cos\theta, 2\sin\theta, t) = (\cos 2\theta, \sin 2\theta, t)$ is a local isometry from the first cylinder onto the second.

23.4. Show that if $V = \{(x_1, x_2) \in \mathbb{R}^2 : x_2 > 0\}$, then the restriction $\psi|_V$ of the map ψ in Example 4 to V is an isometry of the upper half plane onto the cone $3x_1^2 + 3x_2^2 - x_3^2 = 0$, $x_3 > 0$ in $\mathbb{R}^3$ with a line removed.

23.5. Sketch the images of the parametrized 2-surfaces ψ and $\tilde{\psi}$ in $\mathbb{R}^3$ defined by $\psi(\theta, \phi) = (\sinh\theta\cos\phi, \sinh\theta\sin\phi, \phi)$ (helicoid) and $\tilde{\psi}(\theta, \phi) = (\cosh\theta\cos\phi, \cosh\theta\sin\phi, \theta)$ (catenoid). Show that the map which sends $\psi(\phi, \theta)$ to $\tilde{\psi}(\phi, \theta)$ is a local isometry from the first onto the second.

23.6. (a) Show that given any connected plane curve C, there exists a local isometry $\psi: I \to C$, for some open interval $I \subset \mathbb{R}$.
(b) Show that two compact connected plane curves are isometric if and only if they have the same length.

23.7. Let $\mathbf{X} = \sum_{i=1}^n f_i \mathbf{E}_i$ be a tangent vector field along the parametrized n-surface $\varphi: U \to \mathbb{R}^{n+k}$, and let $\alpha: I \to U$. Show that

$$(X \circ \alpha)' = \sum_{k=1}^n \left[\frac{d}{dt}(f_k \circ \alpha) + \sum_{i,j=1}^n (\Gamma_{ij}^k \circ \alpha)(f_i \circ \alpha)\frac{dx_j}{dt}\right](\mathbf{E}_k \circ \alpha)$$

where $\alpha(t) = (x_1(t), \ldots, x_n(t))$ and the $\Gamma_{ij}^k: U \to \mathbb{R}$ are such that $D_{\mathbf{E}_j}\mathbf{E}_i = \sum_{k=1}^n \Gamma_{ij}^k \mathbf{E}_k$. (The Γ_{ij}^k are called *Christoffel symbols* along φ).

23.8. Show that if $h: S \to \mathbb{R}$ is a smooth function on the n-surface $S \subset \mathbb{R}^{n+1}$ then the vector field grad h and the Hessian $\mathscr{H}_p$ at a critical point p of h are both part of the intrinsic geometry of S.

23.9. Show that the Gauss-Kronecker curvature of an oriented 1-surface in $\mathbb{R}^2$ is not intrinsic.

23.10. Let S be an oriented n-surface in $\mathbb{R}^{n+1}$. For $\mathbf{X}$ and $\mathbf{Y}$ smooth tangent vector fields on S, let $[\mathbf{X}, \mathbf{Y}]$ denote the Lie bracket of $\mathbf{X}$ and $\mathbf{Y}$ (see Exercise 9.12) defined by

$$[\mathbf{X}, \mathbf{Y}] = \nabla_{\mathbf{X}}\mathbf{Y} - \nabla_{\mathbf{Y}}\mathbf{X}.$$

Show that $[\mathbf{X}, \mathbf{Y}]$ is also given by the formula

$$[\mathbf{X}, \mathbf{Y}] = D_{\mathbf{X}}\mathbf{Y} - D_{\mathbf{Y}}\mathbf{X}$$

and hence the Lie bracket is part of the intrinsic geometry of S.

23.11. Show that the Riemann tensor R of an oriented n-surface $S \subset \mathbb{R}^{n+1}$ has the following properties:

(a) $R(\mathbf{x}, \mathbf{y}, \mathbf{z}) \cdot \mathbf{w} = R(\mathbf{z}, \mathbf{w}, \mathbf{x}) \cdot \mathbf{y}$,
(b) $R(\mathbf{x}, \mathbf{y}, \mathbf{z}) \cdot \mathbf{w} = -R(\mathbf{y}, \mathbf{x}, \mathbf{z}) \cdot \mathbf{w} = -R(\mathbf{x}, \mathbf{y}, \mathbf{w}) \cdot \mathbf{z}$, and
(c) $R(\mathbf{x}, \mathbf{y}, \mathbf{z}) + R(\mathbf{y}, \mathbf{z}, \mathbf{x}) + R(\mathbf{z}, \mathbf{x}, \mathbf{y}) = \mathbf{0}$

for all $\mathbf{x}, \mathbf{y}, \mathbf{z}, \mathbf{w} \in S_p$, $p \in S$.

23.12. Let S be an oriented n-surface in $\mathbb{R}^{n+1}$ and let $\mathbf{x}, \mathbf{y}, \mathbf{z} \in S_p$, $p \in S$. Show that the value on $\mathbf{x}, \mathbf{y}, \mathbf{z}$ of the Riemann tensor R of S at p is given by the intrinsic formula

$$R(\mathbf{x}, \mathbf{y}, \mathbf{z}) = D_{\mathbf{X}}D_{\mathbf{Y}}\mathbf{Z} - D_{\mathbf{Y}}D_{\mathbf{X}}\mathbf{Z} - D_{[\mathbf{X}, \mathbf{Y}]}\mathbf{Z}$$

where $\mathbf{X}$, $\mathbf{Y}$, and $\mathbf{Z}$ are any smooth tangent vector fields on S such that $\mathbf{X}(p) = \mathbf{x}$, $\mathbf{Y}(p) = \mathbf{y}$, and $\mathbf{Z}(p) = \mathbf{z}$, and $[\mathbf{X}, \mathbf{Y}]$ is the Lie bracket of $\mathbf{X}$ and $\mathbf{Y}$ (Exercise 23.10). [*Hint*: Choose a local parametrization φ of S with $p \in U =$ Image φ, express the restrictions to U of $\mathbf{X}$, $\mathbf{Y}$, and $\mathbf{Z}$ as linear combinations (with smooth coefficients) of the coordinate vector fields $\mathbf{E}_i$ of φ, and compute.]

23.13. Let S be an oriented n-surface in $\mathbb{R}^{n+1}$ ($n > 1$), let $p \in S$, and let P be a 2-dimensional subspace of S_p.

(a) Show that the real number $\sigma(P)$ defined by

$$\sigma(P) = R(\mathbf{e}_1, \mathbf{e}_2, \mathbf{e}_2) \cdot \mathbf{e}_1,$$

where $\{\mathbf{e}_1, \mathbf{e}_2\}$ is an orthonormal basis for P, is independent of the choice of orthonormal basis. [*Hint*: Use Exercise 23.11 to show that if $\{\tilde{\mathbf{e}}_1, \tilde{\mathbf{e}}_2\}$ is another basis for P then $R(\tilde{\mathbf{e}}_1, \tilde{\mathbf{e}}_2, \tilde{\mathbf{e}}_2) \cdot \tilde{\mathbf{e}}_1 = (\det a_{ij})^2 R(\mathbf{e}_1, \mathbf{e}_2, \mathbf{e}_2) \cdot \mathbf{e}_1$ where (a_{ij}) is the change of basis matrix.]
(b) Show that if $n = 2$ then $\sigma(S_p)$ is equal to the Gaussian curvature of S at p.

The number $\sigma(P)$ is called the *Riemannian curvature*, or *sectional curvature*, of S on P.

Riemannian Metrics 24

The intrinsic geometric features of an n-surface S depend only on dot products of vectors tangent to S and derivatives along parametrized curves in S of functions obtained as dot products of vector fields tangent to S along these curves. In other words, given the dot product on each tangent space S_p, $p \in S$, the intrinsic geometry of S can be studied without reference to the way in which S sits in $\mathbb{R}^{n+1}$. If we are told what the dot product is on each S_p then we can compute, for example, the lengths of curves in S, the volume of S, the geodesics in S, parallel transport along curves in S, and the Gauss-Kronecker curvature of S (if n is even) without any knowledge of how S curves around in $\mathbb{R}^{n+1}$. In fact, if we are given a dot product on each tangent space S_p different from the one which comes from $\mathbb{R}_p^{n+1}$, we can still do these intrinsic computations but of course the results of our computations will depend on the dot products used, and the geometry we find will in general be quite different from the geometry we are familiar with. The geometry obtained from such dot products is called *Riemannian geometry*; the collection of dot products on the tangent spaces S_p from which the geometry is derived is called a Riemannian metric.

A *Riemannian metric* on an n-surface S is a function g which assigns to each pair $\{\mathbf{v}, \mathbf{w}\}$ of vectors in S_p, $p \in S$, a real number $g(\mathbf{v}, \mathbf{w})$ such that for each $\mathbf{v}$, $\mathbf{w}$, and $\mathbf{x} \in S_p$, $p \in S$, and $\lambda \in \mathbb{R}$,

(i) $g(\mathbf{v}, \mathbf{w}) = g(\mathbf{w}, \mathbf{v})$
(ii) $g(\mathbf{v} + \mathbf{w}, \mathbf{x}) = g(\mathbf{v}, \mathbf{x}) + g(\mathbf{w}, \mathbf{x})$, $g(\mathbf{v}, \mathbf{w} + \mathbf{x}) = g(\mathbf{v}, \mathbf{w}) + g(\mathbf{v}, \mathbf{x})$
(iii) $g(\lambda\mathbf{v}, \mathbf{w}) = \lambda g(\mathbf{v}, \mathbf{w})$, $g(\mathbf{v}, \lambda\mathbf{w}) = \lambda g(\mathbf{v}, \mathbf{w})$
(iv) $g(\mathbf{v}, \mathbf{v}) \geq 0$, $g(\mathbf{v}, \mathbf{v}) = 0$ if and only if $\mathbf{v} = \mathbf{0}$

and such that for each pair $\{\mathbf{X}, \mathbf{Y}\}$ of smooth tangent vector fields defined on an open subset U of S the function $g(\mathbf{X}, \mathbf{Y})\colon U \to \mathbb{R}$ defined by $[g(\mathbf{X}, \mathbf{Y})](p) = g(\mathbf{X}(p), \mathbf{Y}(p))$ is smooth.

Properties (i)–(iv) are familiar properties of the dot product and, in fact, given g we can define a dot product on each S_p by $\mathbf{v} \cdot \mathbf{w} = g(\mathbf{v}, \mathbf{w})$. The smoothness property of g assures that we can do the differential calculus computations required to investigate the geometry of (S, g).

EXAMPLE 1. Let S be an n-surface in $\mathbb{R}^{n+k}$. For $p \in S$ and $\mathbf{v}, \mathbf{w} \in S_p$, define $g(\mathbf{v}, \mathbf{w})$ by $g(\mathbf{v}, \mathbf{w}) = \mathbf{v} \cdot \mathbf{w}$ (usual dot product of vectors in $\mathbb{R}^{n+k}_p$). Then g is a Riemannian metric on S. This g is called the *usual metric* on S.

EXAMPLE 2. Let $\psi\colon S^n - \{q\} \to \mathbb{R}^n$ denote stereographic projection from the north pole $\{q\}$ of the unit n-sphere $S^n \subset \mathbb{R}^{n+1}$ onto the equatorial hyperplane $\mathbb{R}^n \subset \mathbb{R}^{n+1}$. Define a Riemannian metric on $S^n - \{q\}$ by

$$g(\mathbf{v}, \mathbf{w}) = d\psi(\mathbf{v}) \cdot d\psi(\mathbf{w}) \qquad (\mathbf{v}, \mathbf{w} \in S^n_p,\ p \in S^n - \{q\})$$

where the dot product on the right hand side is the usual dot product in $\mathbb{R}^n_{\psi(p)}$. Thus the metric g is defined precisely so that ψ is an isometry ($d\psi$ preserves dot products) from $S^n - \{q\}$ with the metric g to $\mathbb{R}^n$ with its usual metric. From the fact that ψ is an isometry, and hence preserves all intrinsic geometric features of the surface, we can deduce the following facts about the geometry of $(S^n - \{q\}, g)$:

(i) The geodesics of $(S^n - \{q\}, g)$ will be the images under the isometry $\varphi = \psi^{-1}$ of the geodesics of $\mathbb{R}^n$ (see Figure 24.1). Hence the family of

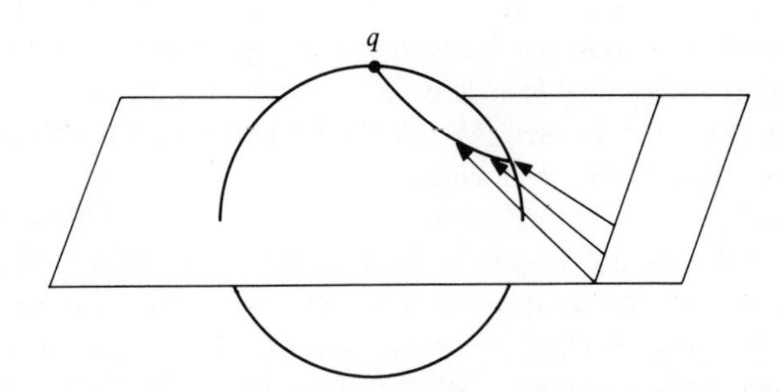

Figure 24.1 A typical geodesic on the 2-sphere S^2 (north pole deleted) with its stereographic metric.

maximal geodesics in $(S^n - \{q\}, g)$ will be the family of (appropriately parametrized) circles in S^n passing through q, with the point q removed.

(ii) The length of each parametrized curve $\alpha\colon (a, b) \to S^n - \{q\}$ with $\lim_{t\to b} \alpha(t) = q$ will be infinite since $l(\psi \circ \alpha) = \infty$ for all such α.

(iii) For n even, the Gauss-Kronecker curvature K of $(S^n - \{q\}, g)$ will be identically zero.

EXAMPLE 3. Let $\varphi\colon \mathbb{R}^n \to \mathbf{S}^n$ denote the inverse of stereographic projection from the north pole q of the unit n-sphere $\mathbf{S}^n$ to the equatorial hyperplane. Define a Riemannian metric g on $\mathbb{R}^n$ by

$$g(\mathbf{v}, \mathbf{w}) = d\varphi(\mathbf{v}) \cdot d\varphi(\mathbf{w}) \qquad (\mathbf{v}, \mathbf{w} \in \mathbb{R}^n_p,\ p \in \mathbb{R}^n)$$

where the dot product on the right hand side is the usual dot product in $\mathbf{S}^n_{\varphi(p)} \subset \mathbb{R}^{n+1}_{\varphi(p)}$. Thus φ is an isometry from $(\mathbb{R}^n, g)$ to $\mathbf{S}^n - \{q\}$ with its usual metric. From the fact that φ is an isometry we can deduce that:

(i) The geodesics of $(\mathbb{R}^n, g)$ will be the images under the isometry $\psi = \varphi^{-1}$ of the geodesics of $\mathbf{S}^n$ (see Figure 24.2). The maximal geodesics radiating

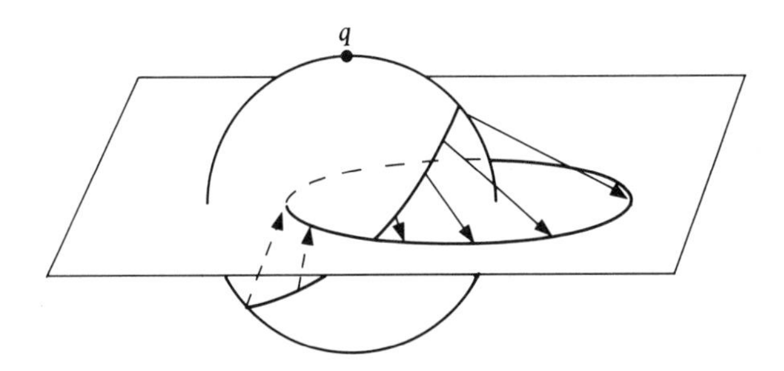

Figure 24.2 A typical geodesic on the plane $\mathbb{R}^2$ with its stereographic metric.

from the origin will be straight lines in $\mathbb{R}^n$, suitably parametrized; each of these geodesics will have finite length (2π) relative to the metric g. All other maximal geodesics will be circles in $\mathbb{R}^n$, suitably parametrized (see Figure 24.3 and Exercise 24.1); each of these geodesics will be periodic with period 2π.

(ii) For n even, the Gauss-Kronecker curvature K of $(\mathbb{R}^n, g)$ will be constant, equal to 1.

The most interesting Riemannian metrics will not be related by isometries to "usual metrics". We shall consider now one of the most important of these, the hyperbolic metric on the unit disc in $\mathbb{R}^2$. This metric is suggested by the stereographic metrics on $\mathbb{R}^n$.

Let S be the n-sphere $x_1^2 + \cdots + x_{n+1}^2 = r^2$ of radius $r > 0$ in $\mathbb{R}^{n+1}$. Just as with the unit n-sphere $\mathbf{S}^n$, we can use the usual metric on the n-sphere S together with stereographic projection to define a Riemannian metric on $\mathbb{R}^n$. Let us derive an explicit formula for this metric. For $p \in \mathbb{R}^n$, the line through $(p, 0)$ and the north pole $q = (0, \ldots, 0, r)$ in S is $\alpha(t) = (tp, (1 - t)r)$. This line cuts S when $\|\alpha(t)\|^2 = r^2$; that is, when $t = 2r^2/(\|p\|^2 + r^2)$, so the map $\varphi\colon \mathbb{R}^2 \to S$ inverse to stereographic projection onto the hyperplane $x_{n+1} = 0$ is given by

$$\varphi(p) = (2r^2 p, r(\|p\|^2 - r^2))/(\|p\|^2 + r^2).$$

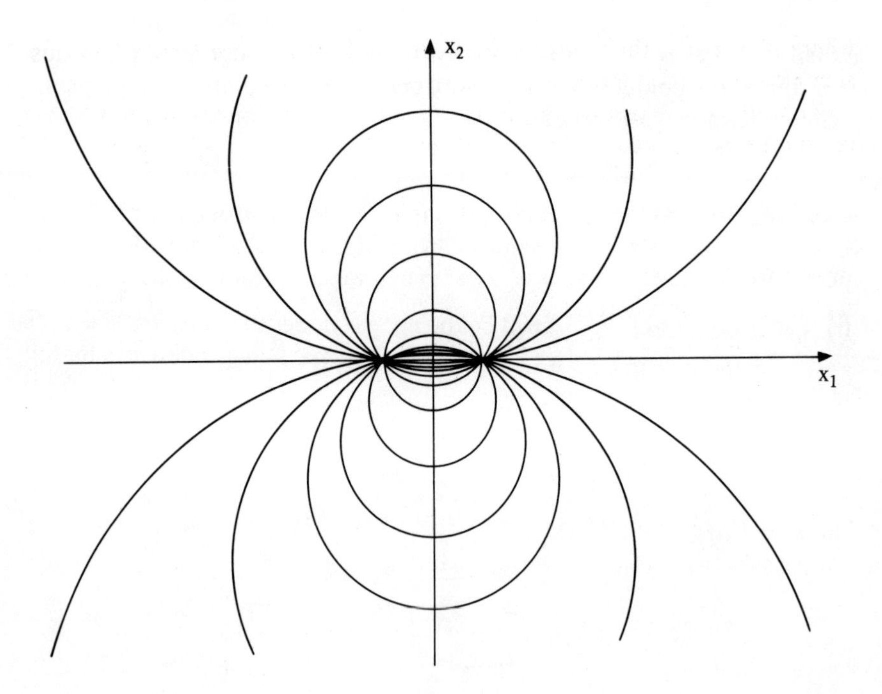

Figure 24.3 Geodesics passing through (1, 0) in the plane $\mathbb{R}^2$ with its stereographic metric.

For $\mathbf{v} = (p, v) \in \mathbb{R}^n_p$, $p \in \mathbb{R}^n$,

$$d\varphi(\mathbf{v}) = \left(\varphi(p), \left.\frac{d}{dt}\right|_0 \varphi(p + tv)\right)$$
$$= (2r^2/(\|p\|^2 + r^2)^2)(\varphi(p), (\|p\|^2 + r^2)v - 2(p \cdot v)p, 2r(p \cdot v))$$

so, for $\mathbf{v}, \mathbf{w} \in \mathbb{R}^n_p$ we find

$$d\varphi(\mathbf{v}) \cdot d\varphi(\mathbf{w}) = \frac{4}{(1 + (\|p\|^2/r^2))^2} \mathbf{v} \cdot \mathbf{w}.$$

Thus *the Riemannian metric g on $\mathbb{R}^n$ obtained from the usual metric on the sphere S of radius r via stereographic projection into the equatorial hyperplane is given by*

$$g(\mathbf{v}, \mathbf{w}) = \frac{4}{(1 + (\|p\|^2/r^2))^2} \mathbf{v} \cdot \mathbf{w} \qquad (\mathbf{v}, \mathbf{w} \in \mathbb{R}^n_p,\ p \in \mathbb{R}^n)$$

where the dot product on the right hand side is the usual dot product on $\mathbb{R}^n_p$. When $n = 2$, this formula can be rewritten as

$$g(\mathbf{v}, \mathbf{w}) = \frac{4}{(1 + K\|p\|^2)^2} \mathbf{v} \cdot \mathbf{w} \qquad (\mathbf{v}, \mathbf{w} \in \mathbb{R}^2_p,\ p \in \mathbb{R}^2)$$

where $K = 1/r^2$ is the Gaussian curvature of $(\mathbb{R}^2, g)$ (since $1/r^2$ is the Gaussian curvature of S and φ is an isometry).

The above discussion shows that a Riemannian metric g on $\mathbb{R}^2$ with constant Gaussian curvature $K > 0$ can be obtained by defining

$$g(\mathbf{v}, \mathbf{w}) = \frac{4}{(1 + K\|p\|^2)^2}\mathbf{v} \cdot \mathbf{w} \qquad (\mathbf{v}, \mathbf{w} \in \mathbb{R}^2_p, \, p \in \mathbb{R}^2).$$

If we take $K = 0$ in this formula we obtain a constant multiple of the usual metric on $\mathbb{R}^2$ and it is easy to check that $(\mathbb{R}^2, g)$ for this g has Gaussian curvature identically zero. One would hope that if we took K to be a constant < 0 in this formula then we would find that $(\mathbb{R}^2, g)$ had constant negative Gaussian curvature K. This is indeed the case, except that this formula defines a Riemannian metric not on $\mathbb{R}^2$ but only on the disc $\{(x_1, x_2) \in \mathbb{R}^2 \colon x_1^2 + x_2^2 < 1/|K|\}$.

Theorem 1. *Given $K \in \mathbb{R}$, $K < 0$, let $U = \{(x_1, x_2) \in \mathbb{R}^2 \colon x_1^2 + x_2^2 < 1/|K|\}$, and let g be the Riemannian metric on U defined by*

$$g(\mathbf{v}, \mathbf{w}) = \frac{4}{(1 + K\|p\|^2)^2}\mathbf{v} \cdot \mathbf{w} \qquad (\mathbf{v}, \mathbf{w} \in \mathbb{R}^2_p, \, p \in U)$$

where the dot product on the right hand side is the usual dot product in $\mathbb{R}^2_p$. Then (U, g) has constant Gaussian curvature $K < 0$.

PROOF. Let $h \colon U \to \mathbb{R}$ be defined by $h(p) = \frac{1}{2}(1 + K\|p\|^2)$ so that $g(\mathbf{v}, \mathbf{w}) = (1/(h(p))^2)\mathbf{v} \cdot \mathbf{w}$ for all $\mathbf{v}, \mathbf{w} \in \mathbb{R}^2_p$, $p \in U$. Using the intrinsic formulas of Chapter 23, we shall derive a formula for the Gaussian curvature of (U, g) in terms of the function h and its derivatives.

Note that the identity map from U into itself is a global parametrization of U with coordinate vector fields given by $\mathbf{E}_1(p) = (p, 1, 0)$ and $\mathbf{E}_2(p) = (p, 0, 1)$ for $p \in U$. The metric coefficients $g_{ij} \colon U \to \mathbb{R}$ of g are given by

$$g_{11} = g(\mathbf{E}_1, \mathbf{E}_1) = 1/h^2 \qquad g_{12} = g(\mathbf{E}_1, \mathbf{E}_2) = 0$$
$$g_{21} = g(\mathbf{E}_2, \mathbf{E}_1) = 0 \qquad g_{22} = g(\mathbf{E}_2, \mathbf{E}_2) = 1/h^2$$

so, using the formula of Theorem 1, Chapter 23, we find

$$g((D_{\mathbf{E}_1}\mathbf{E}_1), \mathbf{E}_1) = -\frac{1}{h^3}\frac{\partial h}{\partial x_1} \qquad g((D_{\mathbf{E}_1}\mathbf{E}_1), \mathbf{E}_2) = \frac{1}{h^3}\frac{\partial h}{\partial x_2}$$

$$g((D_{\mathbf{E}_1}\mathbf{E}_2), \mathbf{E}_1) = -\frac{1}{h^3}\frac{\partial h}{\partial x_2} \qquad g((D_{\mathbf{E}_1}\mathbf{E}_2), \mathbf{E}_2) = -\frac{1}{h^3}\frac{\partial h}{\partial x_1}$$

$$g(D_{\mathbf{E}_2}\mathbf{E}_1), \mathbf{E}_1) = -\frac{1}{h^3}\frac{\partial h}{\partial x_2} \qquad g((D_{\mathbf{E}_2}\mathbf{E}_1), \mathbf{E}_2) = -\frac{1}{h^3}\frac{\partial h}{\partial x_1}$$

$$g((D_{\mathbf{E}_2}\mathbf{E}_2), \mathbf{E}_1) = \frac{1}{h^3}\frac{\partial h}{\partial x_1} \qquad g((D_{\mathbf{E}_2}\mathbf{E}_2), \mathbf{E}_2) = -\frac{1}{h^3}\frac{\partial h}{\partial x_2}.$$

Since $\{\mathbf{E}_1, \mathbf{E}_2\}$ is orthogonal (but not orthonormal!) with respect to the metric g ($g_{12} = g(\mathbf{E}_1, \mathbf{E}_2) = 0$), each vector field $\mathbf{X}$ on U can be expressed as $\mathbf{X} = f_1 \mathbf{E}_1 + f_2 \mathbf{E}_2$ where $f_1 = (1/g_{11})g(\mathbf{X}, \mathbf{E}_1)$ and $f_2 = (1/g_{22})g(\mathbf{X}, \mathbf{E}_2)$. In particular,

$$D_{\mathbf{E}_1}\mathbf{E}_1 = -\frac{1}{h}\frac{\partial h}{\partial x_1}\mathbf{E}_1 + \frac{1}{h}\frac{\partial h}{\partial x_2}\mathbf{E}_2$$

$$D_{\mathbf{E}_1}\mathbf{E}_2 = D_{\mathbf{E}_2}\mathbf{E}_1 = -\frac{1}{h}\frac{\partial h}{\partial x_2}\mathbf{E}_1 - \frac{1}{h}\frac{\partial h}{\partial x_1}\mathbf{E}_2$$

$$D_{\mathbf{E}_2}\mathbf{E}_2 = \frac{1}{h}\frac{\partial h}{\partial x_1}\mathbf{E}_1 - \frac{1}{h}\frac{\partial h}{\partial x_2}\mathbf{E}_2.$$

Hence the Gaussian curvature of (U, g), according to the intrinsic formula for Gaussian curvature derived in Chapter 23, is equal to

$$\begin{aligned}
&g(R(\mathbf{E}_2/\|\mathbf{E}_2\|, \mathbf{E}_1/\|\mathbf{E}_1\|, \mathbf{E}_1/\|\mathbf{E}_1\|), \mathbf{E}_2/\|\mathbf{E}_2\|)\\
&= (1/\|\mathbf{E}_1\|^2\|\mathbf{E}_2\|^2)g(R(\mathbf{E}_2, \mathbf{E}_1, \mathbf{E}_1), \mathbf{E}_2)\\
&= (1/g_{11}g_{22})g(D_{\mathbf{E}_2}D_{\mathbf{E}_1}\mathbf{E}_1 - D_{\mathbf{E}_1}D_{\mathbf{E}_2}\mathbf{E}_1, \mathbf{E}_2)\\
&= (1/g_{11}g_{22})g\left(D_{\mathbf{E}_2}\left(-\frac{1}{h}\frac{\partial h}{\partial x_1}\mathbf{E}_1 + \frac{1}{h}\frac{\partial h}{\partial x_2}\mathbf{E}_2\right)\right.\\
&\qquad \left. - D_{\mathbf{E}_1}\left(-\frac{1}{h}\frac{\partial h}{\partial x_2}\mathbf{E}_1 - \frac{1}{h}\frac{\partial h}{\partial x_1}\mathbf{E}_2\right), \mathbf{E}_2\right)\\
&= h\left(\frac{\partial^2 h}{\partial x_1^2} + \frac{\partial^2 h}{\partial x_2^2}\right) - \left(\left(\frac{\partial h}{\partial x_1}\right)^2 + \left(\frac{\partial h}{\partial x_2}\right)^2\right).
\end{aligned}$$

Finally, since $h(x_1, x_2) = \frac{1}{2}(1 + K(x_1^2 + x_2^2))$, we find that the Gaussian curvature of (U, g) is precisely K. □

When $K = -1$, Theorem 1 describes a Riemannian metric g with constant Gaussian curvature -1 on the unit disc $x_1^2 + x_2^2 < 1$ in $\mathbb{R}^2$ by $g(\mathbf{v}, \mathbf{w}) = 4\mathbf{v} \cdot \mathbf{w}/(1 - \|p\|^2)^2$. This metric is called the *hyperbolic metric*. In order to gain insight into the geometry of this metric, it is convenient, if not absolutely necessary, to use some of the ideas from the elementary theory of functions of a complex variable. Rather than do that here (see, however, Exercises 24.5 and 24.6) we shall study a related metric on the upper half plane.

For $p = (x, y) \in \mathbb{R}^2$ with $y > 0$ and $\mathbf{v}, \mathbf{w} \in \mathbb{R}_p^2$, define $g(\mathbf{v}, \mathbf{w}) = \mathbf{v} \cdot \mathbf{w}/y^2$ where the dot product on the right hand side is the usual dot product on $\mathbb{R}_p^2$. Then g is a Riemannian metric on the upper half plane $U = \{(x, y) \in \mathbb{R}^2 \colon y > 0\}$. This metric g called the *Poincaré metric* on U. Note that (U, g) has constant Gaussian curvature -1 since, taking $h(x_1, x_2) = x_2$, we have $g(\mathbf{v}, \mathbf{w}) = (1/h(p)^2)\mathbf{v} \cdot \mathbf{w}$ for $\mathbf{v}, \mathbf{w} \in \mathbb{R}_p^2$, $p \in U$, and hence, just as in the proof of Theorem 1, the Gaussian curvature is given by

$$K = h\left(\frac{\partial^2 h}{\partial x_1^2} + \frac{\partial^2 h}{\partial x_2^2}\right) - \left(\left(\frac{\partial h}{\partial x_1}\right)^2 + \left(\frac{\partial h}{\partial x_2}\right)^2\right) = -1.$$

Each of the following maps $\psi: U \to U$ is an isometry from the upper half plane U with its Poincaré metric onto itself:

(i) $\psi(x, y) = (x + \lambda, y)$ where λ is any real number,
(ii) $\psi(x, y) = (\lambda x, \lambda y)$ where λ is any positive real number,
(iii) $\psi(x, y) = (-x, y)$, and
(iv) $\psi(x, y) = (x/(x^2 + y^2), y/(x^2 + y^2))$.

Indeed, let $\mathbf{v}, \mathbf{w} \in \mathbb{R}_p^2$, $p \in U$. For ψ defined as in (i) or (iii) we have

$$g(d\psi(\mathbf{v}), d\psi(\mathbf{w})) = d\psi(\mathbf{v}) \cdot d\psi(\mathbf{w})/y^2 = \mathbf{v} \cdot \mathbf{w}/y^2 = g(\mathbf{v}, \mathbf{w}),$$

and for ψ defined as in (ii) we have

$$\begin{aligned} g(d\psi(\mathbf{v}), d\psi(\mathbf{w})) &= d\psi(\mathbf{v}) \cdot d\psi(\mathbf{w})/(\lambda y)^2 \\ &= \lambda\mathbf{v} \cdot \lambda\mathbf{w}/\lambda^2 y^2 = \mathbf{v} \cdot \mathbf{w}/y^2 = g(\mathbf{v}, \mathbf{w}), \end{aligned}$$

as required for an isometry. For ψ defined as in (iv) we have, where $\mathbf{E}_1(x, y) = (x, y, 1, 0)$ and $\mathbf{E}_2(x, y) = (x, y, 0, 1)$,

$$d\psi(\mathbf{E}_1(x, y)) = \left(\psi(x, y), \frac{y^2 - x^2}{(x^2 + y^2)^2}, \frac{-2xy}{(x^2 + y^2)^2}\right)$$

$$d\psi(\mathbf{E}_2(x, y)) = \left(\psi(x, y), \frac{-2xy}{(x^2 + y^2)^2}, \frac{x^2 - y^2}{(x^2 + y^2)^2}\right)$$

so

$$\begin{aligned} g(d\psi(\mathbf{E}_i(x, y)), d\psi(\mathbf{E}_i(x, y))) &= \frac{1}{(x^2 + y^2)^2} \Big/ \left(\frac{y}{x^2 + y^2}\right)^2 = \frac{1}{y^2} \\ &= g(\mathbf{E}_i(x, y), \mathbf{E}_i(x, y)) \end{aligned}$$

for $i = 1$ or 2, and

$$g(d\psi(\mathbf{E}_1(x, y)), d\psi(\mathbf{E}_2(x, y))) = 0 = g(\mathbf{E}_1(x, y), \mathbf{E}_2(x, y));$$

thus $d\psi$ preserves dot products of basis vectors, and hence of all pairs of tangent vectors, as required for an isometry.

One consequence of the fact that the maps (i) and (ii) are isometries is that, just as the geometry of S^n looks the same from every point of S^n as it does from the north pole (0, 0, 1), the geometry of U with the Poincaré metric looks the same from every point of U as it does from the point $(0, 1) \in U$. This is because given any point $p \in U$ there is an isometry $\psi_2 \circ \psi_1$ of U, where ψ_1 is an isometry of type (i) and ψ_2 is one of type (ii), which sends (0, 1) onto p (see Figure 24.4). In particular, all points of U must

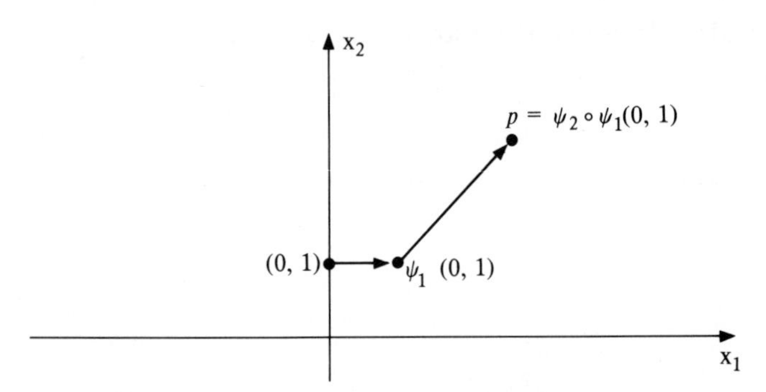

Figure 24.4 For each point p in the upper half plane U there is an isometry from U with its Poincaré metric onto itself which sends (0, 1) to p.

be the same (intrinsic) distance from the edge of U (the x_1-axis). This perhaps surprising phenomenon is due to the fact that, in the geometry of the metric g, these distances are infinite. Thus for example, if we compute the length, relative to the Poincaré metric, of the curve α: $[0, 1) \to U$ defined by $\alpha(t) = (0, 1 - t)$ we find

$$l(\alpha) = \int_0^1 \|\dot{\alpha}(t)\| \, dt = \int_0^1 (g(\dot{\alpha}(t), \dot{\alpha}(t)))^{1/2} \, dt = \int_0^1 \frac{1}{1 - t} \, dt = \infty.$$

We can use the fact that the maps (i)–(iv) above are isometries also to identify the geodesics in U relative to the Poincaré metric.

Theorem. *Let U be an open set in $\mathbb{R}^2$, let g be a Riemannian metric on U, and let ψ: $U \to U$ be an isometry of (U, g). Suppose the fixed point set $F = \{p \in U: \psi(p) = p\}$ of ψ is a connected plane curve. Then F is (the image of) a geodesic of (U, g).*

PROOF. Let α: $I \to F$ be a unit speed $(g(\dot{\alpha}, \dot{\alpha}) = 1)$ global parametrization of F, constructed as in Chapter 11. We shall show that α is a geodesic, thereby establishing the theorem.

It will suffice to show that for each $t_0 \in I$ the restriction of α to some interval about t_0 is a geodesic. So let $t_0 \in I$, let $\mathbf{v} = \dot{\alpha}(t_0)$, and let $\alpha_{\mathbf{v}}$ denote the maximal geodesic of (U, g) with initial velocity $\mathbf{v}$. We shall show that $\alpha(t) = \alpha_{\mathbf{v}}(t - t_0)$ for all $t \in I$ such that $t - t_0$ is in the domain of $\alpha_{\mathbf{v}}$.

Note first that Image $\alpha_{\mathbf{v}} \subset F$. This is because $\psi \circ \alpha_{\mathbf{v}}$ is a geodesic (since ψ is an isometry) with initial velocity $\mathbf{v}$ (since $\psi|_F$ is the identity map and hence $d\psi|_{F_p}$ is the identity map so $\psi \dot{\circ} \alpha(0) = d\psi(\dot{\alpha}(0)) = d\psi(\mathbf{v}) = \mathbf{v}$) so $\psi \circ \alpha_{\mathbf{v}}$ and $\alpha_{\mathbf{v}}$ are geodesics with the same initial velocity. By uniqueness of geodesics, $\psi \circ \alpha_{\mathbf{v}} = \alpha_{\mathbf{v}}$ and so Image $\alpha_{\mathbf{v}} \subset F$.

Now α and $\alpha_{\mathbf{v}}$, being unit speed curves in F with $\dot{\alpha}(t_0) = \dot{\alpha}_{\mathbf{v}}(0)$, are integral curves of the same unit tangent vector field on F. By uniqueness of integral curves, $\alpha(t) = \alpha_{\mathbf{v}}(t - t_0)$ for all t in the interval

$$\{t \in I: t - t_0 \in \text{domain } \alpha_{\mathbf{v}}\}$$

about t_0. It follows that the restriction of α to this interval is a geodesic, and, since $t_0 \in I$ was arbitrary, that α is a geodesic. $\square$

Remark. We have implicitly used, in this proof, the fact that the existence and uniqueness theorem for geodesics is valid for surfaces with arbitrary Riemannian metrics. Although our proof of this theorem in Chapter 7 is valid only for the usual metric on an n-surface, it can be modified to work for any Riemannian metric. The important observation here is that the intrinsic geodesic equation $\dot{\alpha}' = \mathbf{0}$ is still a second order ordinary differential equation.

Corollary. *The geodesics in the upper half plane U relative to the Poincaré metric are* (i) *vertical lines, and* (ii) *semi-circles centered on the x_1-axis, suitably parametrized* (*see Figure 24.5*).

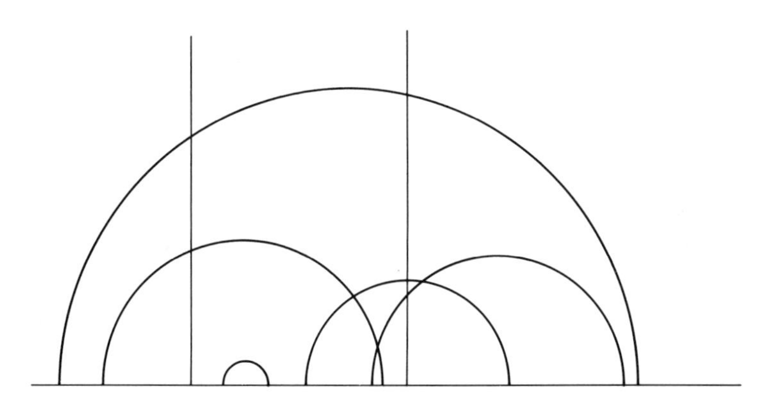

Figure 24.5 Typical geodesics in the upper half plane with its Poincaré metric.

PROOF. Applying the theorem to the isometry $\psi(x_1, x_2) = (-x_1, x_2)$ we see that the line $x_1 = 0$, $x_2 > 0$, suitably parametrized, is a geodesic. Similarly, from the isometry $\psi(x_1, x_2) = (x_1/(x_1^2 + x_2^2), x_2/(x_1^2 + x_2^2))$ we see that the semicircle $x_1^2 + x_2^2 = 1$, $x_2 > 0$, suitably parametrized, is a geodesic. Since every vertical line in U is the image of the line $x_1 = 0$, $x_2 > 0$ under an isometry $\psi(x_1, x_2) = (x_1 + \lambda, x_2)$, it follows that every vertical line in U is a geodesic. Similarly, since every semi-circle in U centered at the origin is the image of the semi-circle $x_1^2 + x_2^2 = 1$, $x_2 > 0$ under an isometry $\psi(x_1, x_2) = (\lambda x_1, \lambda x_2)$, $\lambda > 0$, these semicircles are geodesics. Finally, every semi-circle in U centered on the x_1-axis is the image of a semi-circle in U centered at the origin under an isometry $\psi(x_1, x_2) = (x_1 + \lambda, x_2)$ so they too are geodesics.

Note that every geodesic in U relative to the Poincaré metric must belong to this family since given any point p of U and any tangent direction $\mathbf{v}$ at p there is a geodesic in this family passing through p in the direction $\mathbf{v}$. $\square$

Remark 1. The upper half plane U with its Poincaré metric is geodesically complete; that is, every maximal geodesic in U has domain the whole real line. See Exercise 24.4.

Remark 2. The upper half plane U with its Poincaré metric g provides an example of a geometry in which Euclid's parallel postulate fails. The *parallel postulate* asserts that given any straight line l and any point p not on l there is a unique straight line through p which does not meet l. If we define the straight lines of (U, g) to be the images of maximal geodesics, then all of Euclid's axioms for geometry except the parallel postulate are valid for (U, g). But, given any straight line l in (U, g) and point p not on l, there are in fact infinitely many straight lines of (U, g) through p which do not meet l (see Figure 24.6).

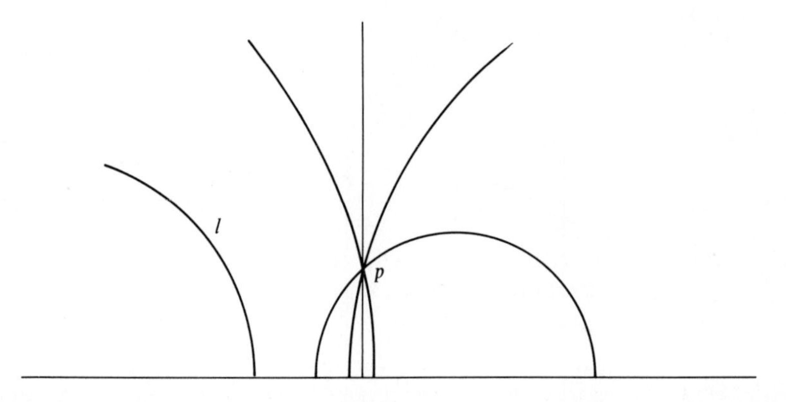

Figure 24.6 Euclid's parallel postulate fails in the upper half plane with its Poincaré metric.

Exercises

24.1. Let $\psi\colon \mathbf{S}^2 - \{q\} \to \mathbb{R}$ denote stereographic projection from the north pole q of the unit 2-sphere $\mathbf{S}^2$ onto the equatorial plane.

(a) Show that $\psi(x, y, z) = (x/(1 - z), y/(1 - z))$ for all $(x, y, z) \in \mathbf{S}^2 - \{q\}$.

(b) Let $e_1 = (1, 0, 0)$, $e_2 = (0, \cos\phi, \sin\phi)$ where $-\pi/2 < \phi < \pi/2$, and let $\alpha\colon \mathbb{R} \to \mathbf{S}^2$ be the geodesic in $\mathbf{S}^2$ given by $\alpha(t) = (\cos t)e_1 + (\sin t)e_2$. Show that the image of the parametrized curve $\psi \circ \alpha\colon \mathbb{R} \to \mathbb{R}^2$ is the circle

$$x_1^2 + (x_2 - \tan\phi)^2 = \sec^2\phi.$$

(c) Use the symmetries of $\mathbf{S}^2$ to show that each great circle in $\mathbf{S}^2$ not passing through q is mapped by ψ onto a circle in $\mathbb{R}^2$ obtained by rotating about the origin one of the circles described in (b). Conclude that the geodesics of $\mathbb{R}^2$ relative to the stereographic metric of Example 3 are as in Figure 24.3.

24.2. Find the length of the circle $\alpha_r(t) = (r \cos t, r \sin t), 0 \le t \le 2\pi$, relative to the stereographic metric

$$g(\mathbf{v}, \mathbf{w}) = \frac{4}{(1 + \|p\|^2)^2} \mathbf{v} \cdot \mathbf{w} \qquad (\mathbf{v}, \mathbf{w} \in \mathbb{R}_p^2, p \in \mathbb{R}^2)$$

on $\mathbb{R}^2$ and show that $\lim_{r\to\infty} l(\alpha_r) = 0$.

24.3. Let S be the 2-sphere $x_1^2 + x_2^2 + x_3^2 = r^2$ of radius $r > 0$ in $\mathbb{R}^3$ and let $\varphi: \mathbb{R}^2 \to S - \{q\}$ be the inverse of stereographic projection from the north pole $q = (0, 0, r)$ of S onto the tangent plane $x_3 = -r$ at the south pole $(0, 0, -r)$ of S. (Thus, for $p \in \mathbb{R}^2$, $\varphi(p)$ is the point of S different from q which lies on the line through q and $(p, -r)$ in $\mathbb{R}^3$.) Show that the Riemannian metric on $\mathbb{R}^2$ defined by

$$g(\mathbf{v}, \mathbf{w}) = d\varphi(\mathbf{v}) \cdot d\varphi(\mathbf{w}) \qquad (\mathbf{v}, \mathbf{w} \in \mathbb{R}_p^2, p \in \mathbb{R}^2),$$

where the dot product on the right hand side is the usual dot product in $S_{\varphi(p)} \subset \mathbb{R}^3_{\varphi(p)}$, can be described explicitly by the formula

$$g(\mathbf{v}, \mathbf{w}) = \frac{1}{\left(1 + \frac{K}{4}\|p\|^2\right)^2} \mathbf{v} \cdot \mathbf{w} \qquad (\mathbf{v}, \mathbf{w} \in \mathbb{R}_p^2, p \in \mathbb{R}^2)$$

where K is the Gaussian curvature of S.

24.4. Let U be the upper half plane and let $\alpha: [0, 1) \to U$ and $\beta: [0, \pi/2) \to U$ be defined by $\alpha(t) = (0, 1 - t)$ and $\beta(t) = (\sin t, \cos t)$. Show that both α and β have infinite length relative to the Poincaré metric on U and use this fact to show that the upper half plane with its Poincaré metric is geodesically complete.

24.5. Show that each of the following maps is an isometry from the unit disc $x_1^2 + x_2^2 < 1$ with its hyperbolic metric onto itself:

(i) $\psi(x, y) = (x \cos \theta - y \sin \theta, x \sin \theta + y \cos \theta) \qquad (\theta \in \mathbb{R})$
(ii) $\psi(x, y) = (x, -y)$.

Using the isometries (i) and (ii), show that radial straight lines in the unit disc, suitably parametrized, are geodesics relative to the hyperbolic metric.

24.6. (a) Viewing $\mathbb{R}^2$ as the set $\mathbb{C}$ of complex numbers by identifying (a, b) with $a + bi$, show that, for each $\lambda \in \mathbb{R}$ with $-1 < \lambda < 1$, the function

$$\psi(z) = \frac{z + \lambda}{1 + \lambda z}$$

is an isometry from the unit disc U with its hyperbolic metric onto itself.

(b) Show that the image under ψ of the radial line $x_1 = 0$ in U is the intersection with U of a circle centered on the x_1-axis, passing through $(\lambda, 0)$, and meeting the unit circle $x_1^2 + x_2^2 = 1$ orthogonally.

(c) Combining (b) with the result of Exercise 24.5, show that the geodesics in the unit disc U relative to its hyperbolic metric are the intersections with U of (i) radial straight lines and (ii) circles meeting the unit circle $x_1^2 + x_2^2 = 1$ orthogonally (see Figure 24.7).

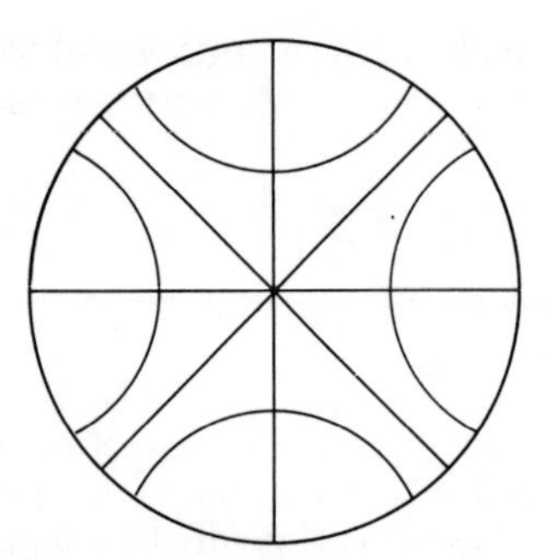

Figure 24.7 Typical geodesics in the unit disc with its hyperbolic metric.

24.7. Viewing $\mathbb{R}^2$ as the set $\mathbb{C}$ of complex numbers by identifying (a, b) with $a + bi$, show that the function

$$\psi(z) = \frac{z + i}{iz + 1}$$

is an isometry from the unit disc with its hyperbolic metric onto the upper half plane with its Poincaré metric.

24.8. Let g be a Riemannian metric on an open set U of $\mathbb{R}^n$. Suppose $\psi: U \to U$ is an isometry of (U, g) whose fixed point set $F = \{p \in U: \psi(p) = p\}$ is an $(n-1)$-surface in $\mathbb{R}^n$. Show that F is *totally geodesic* in U; that is, show that if $\alpha: I \to U$ is any geodesic of (U, g) with $\alpha(t_0) \in F$ and $\dot{\alpha}(t_0) \in F_{\alpha(t_0)}$ for some $t_0 \in I$ then $\alpha(t) \in F$ for all $t \in I$.

24.9. The *Poincaré metric* on the half space $U = \{(x_1, \ldots, x_n) \in \mathbb{R}^n: x_n > 0\}$ is defined by

$$g(\mathbf{v}, \mathbf{w}) = \mathbf{v} \cdot \mathbf{w}/x_n^2$$

for $\mathbf{v}, \mathbf{w} \in \mathbb{R}_p^n$ and $p = (x_1, \ldots, x_n) \in U$, where the dot product on the right hand side is the usual dot product on $\mathbb{R}_p^n$.

(a) Show that each of the following maps is an isometry of (U, g):

(i) $\psi(x_1, \ldots, x_n) = (\varphi(x_1, \ldots, x_{n-1}), x_n)$ where φ is any rigid motion of $\mathbb{R}^{n-1}$,
(ii) $\psi(p) = \lambda p$, where λ is any positive real number,
(iii) $\psi(x_1, \ldots, x_n) = (x_1, \ldots, -x_j, \ldots, x_n)$, where $j \in \{1, \ldots, n-1\}$,
(iv) $\psi(p) = p/\|p\|^2$.

(b) Use Exercise 24.8 to show that if $F_1, \ldots, F_k$ are fixed point sets of isometries $\psi_1, \ldots, \psi_k$ of U of types (iii) and (iv) above and if $\alpha: I \to U$ is a geodesic of (U, g) such that $\alpha(t_0) \in F_j$ and $\dot{\alpha}(t_0) \in (F_j)_{\alpha(t_0)}$ for all $j \in \{1, \ldots, k\}$ then $\alpha(t) \in \bigcap_{j=1}^k F_j$ for all $t \in I$.

(c) Conclude that the maximal geodesics of the half-space U with its Poincaré metric are

(i) vertical (i.e., parallel to the x_n-axis) straight lines in U, and

(ii) semi-circles centered on and orthogonal to the $(n-1)$-plane $x_n = 0$. [*Hint*: First show that unit speed parametrizations of the plane curves

$$\begin{cases} x_1 = \cdots = x_{n-1} = 0 \\ x_n > 0 \end{cases}$$

and

$$\begin{cases} x_1 = \cdots = x_{n-2} = 0 \\ x_{n-1}^2 + x_n^2 = 1 \\ x_n > 0 \end{cases}$$

are geodesics of (U, g) and then consider the images of these geodesics under isometries of types (i) and (ii), part (a).]

24.10. Let S be the n-sphere $x_1^2 + \cdots + x_{n+1}^2 = r^2$ of radius $r > 0$ in $\mathbb{R}^{n+1}$ with its usual metric.

(a) Show that the Riemann tensor R of S is given by

$$R(\mathbf{x}, \mathbf{y}, \mathbf{z}) = \frac{1}{r^2}((\mathbf{y} \cdot \mathbf{z})\mathbf{x} - (\mathbf{x} \cdot \mathbf{z})\mathbf{y}).$$

(b) Show that the Riemannian sectional curvature (Exercise 23.13) of S is constant: $\sigma(P) = 1/r^2$ for each 2-dimensional subspace P of S_p, $p \in S$.

(c) Conclude that the metric on $\mathbb{R}^n$ obtained from the usual metric on S via stereographic projection onto the equatorial hyperplane has constant sectional curvature $K = 1/r^2$ and that this metric is given by the formula

$$g(\mathbf{v}, \mathbf{w}) = \frac{4}{(1 + K\|p\|^2)^2} \mathbf{v} \cdot \mathbf{w} \qquad (\mathbf{v}, \mathbf{w} \in \mathbb{R}_p^n,\ p \in \mathbb{R}^n)$$

where the dot product on the right hand side is the usual dot product on $\mathbb{R}_p^n$. (*Remark*. This formula with $K < 0$ defines a metric g of constant negative sectional curvature K on the n-disc $x_1^2 + \cdots + x_n^2 < 1/|K|$. When $K = -1$, this metric is called the *hyperbolic metric* in the unit n-disc $x_1^2 + \cdots + x_n^2 < 1$.)

Bibliography

This bibliography is limited to a very few selections that may be particularly useful to a student seeking supplementary reading. The reader looking for a more complete list of works in differential geometry will find an excellent guided tour of the literature in Volume V of Spivak's *Comprehensive Introduction* (see below).

Advanced Calculus

Fleming, W. 1977. *Functions of Several Variables.* New York–Heidelberg–Berlin: Springer-Verlag.

Differential Equations

Hurewicz, W. 1958. *Lectures on Ordinary Differential Equations.* Cambridge, Mass.: M.I.T. Press.

Linear Algebra

Hoffman, K., Kunze, R. 1961. *Linear Algebra.* Englewood Cliffs, N.J.: Prentice Hall.

Differential Geometry

do Carmo, M. 1976. *Differential Geometry of Curves and Surfaces.* Englewood Cliffs, N.J.: Prentice Hall.

Millman, R., Parker, G. 1977. *Elements of Differential Geometry.* Englewood Cliffs, N.J.: Prentice Hall.

O'Neill, B. 1966. *Elementary Differential Geometry.* New York: Academic.

Singer, I., Thorpe, J. 1976. *Lecture Notes on Elementary Topology and Geometry.* New York–Heidelberg–Berlin: Springer-Verlag.

Spivak, M. 1970, 1975. *A Comprehensive Introduction to Differential Geometry.* Vols. I–V. Boston: Publish or Perish.

(With the exception of Spivak's *Comprehensive Introduction*, each of these differential geometry texts deals primarily with 2-surfaces. do Carmo's book is, among the above, closest in spirit to this text; it covers many topics not covered here. O'Neill's book uses differential forms as primary tool; Millman and Parker rely heavily on local coordinate computations. The geometry part of the book by Singer and Thorpe deals primarily with intrinsic geometry using the calculus of differential forms on the unit sphere bundle as primary tool. Spivak's book is really a graduate level text, but it does contain many items that are accessible to the reader of the text in hand: see especially Volumes II and III.)

Notational Index

Subject Index

Undergraduate Texts in Mathematics

Apostol: Introduction to Analytic Number Theory.
1976. xii, 370 pages. 24 illus.

Childs: A Concrete Introduction to Higher Algebra.
1979. xiv, 338 pages. 8 illus.

Chung: Elementary Probability Theory with Stochastic Processes. Third Edition
1979. 336 pages.

Croom: Basic Concepts of Algebraic Topology.
1978. x, 177 pages. 46 illus.

Fleming: Functions of Several Variables. Second edition.
1977. xi, 411 pages. 96 illus.

Halmos: Finite-Dimensional Vector Spaces. Second edition.
1974. viii, 200 pages.

Halmos: Naive Set Theory.
1974. vii, 104 pages.

Kemeny/Snell: Finite Markov Chains.
1976. ix, 210 pages.

Lax/Burstein/Lax: Calculus with Applications and Computing, Volume 1.
1976. xi, 513 pages. 170 illus.

LeCuyer: College Mathematics with A Programming Language.
1978. xii, 420 pages. 126 illus. 64 diagrams.

Malitz: Introduction to Mathematical Logic.
Set Theory - Computable Functions - Model Theory.
1979. 255 pages. 2 illus.

Prenowitz/Jantosciak: The Theory of Join Spaces.
A Contemporary Approach to Convex Sets and Linear Geometry.
1979. Approx. 350 pages. 404 illus.

Priestley: Calculus: An Historical Approach.
1979. 400 pages. 300 illus.

Protter/Morrey: A First Course in Real Analysis.
1977. xii, 507 pages. 135 illus.

Sigler: Algebra.
1976. xi, 419 pages. 32 illus.

Singer/Thorpe: Lecture Notes on Elementary Topology and Geometry.
1976. viii, 232 pages. 109 illus.

Smith: Linear Algebra
1978. vii, 280 pages. 21 illus.

Thorpe: Elementary Topics in Differential Geometry.
1979. 256 pages. Approx. 115 illus.

Whyburn/Duda: Dynamic Topology.
1979. Approx. 175 pages. Approx. 20 illus.

Wilson: Much Ado About Calculus.
A Modern Treatment with Applications Prepared for Use with the Computer.
1979. Approx. 500 pages. Approx. 145 illus.

Undergraduate Texts in Mathematics